Symmetry, Shape, and Space

Note to Instructors

For desk copies and information about supplementary materials to accompany this text, details can be found at the following addresses:

Materials include a complete guide for instructors, downloadable data files, activity solutions, sample syllabi and exams, and more.

Symmetry, Shape, and Space

An Introduction to Mathematics Through Geometry

L. Christine Kinsey
Canisius College

Teresa E. Moore
Ithaca College

L. Christine Kinsey
Department of Mathematics
Canisius College
Buffalo, NY 14208, USA
kinsey@canisius.edu

Teresa E. Moore
Department of Mathematics and
  Computer Science
Ithaca College
Ithaca, NY 14850, USA
moore@ithaca.edu

Key College Publishing was founded in 1999 as a division of Key Curriculum Press in cooperation with Springer-Verlag New York, Inc. It publishes innovative curriculum materials for undergraduate courses in mathematics, statistics, and math-

L. Christine Kinsey, Teresa E. Moore.
    p. cm.
  Includes bibliographical references and index.
  ISBN 1-930190-09-3
  1. Geometry.   I. Moore, Teresa E.   II. Title.   III. Series.
QA453. K46   2001
516—dc21   99-16038

Executive Editor: Richard J. Bonacci
Production Editor: MaryAnn Brickner
Manufacturing Manager: Joe Quatela
Text and Cover Design: Joe Piliero
Composition: Progressive Information Technologies, Emigsville, PA

Printed in the United States of America.   (HP)

9 8 7 6 5 4 3 2

ISBN 1-930190-09-3 Key College Publishing

*This book is dedicated to
Alex and Preston*

Preface

"For it is true, generally speaking, that mathematics is not a popular subject, even though its importance may be generally conceded. The reason for this is to be found in the common superstition that mathematics is but a continuation, a further development, of the fine art of arithmetic, of juggling with numbers."

—D. Hilbert
Geometry and the Imagination, 1932

This may, perhaps, be a stereotype, but all hikers can be divided into two categories. The first type is very goal oriented; "Point *A* to Point *B*" people who are always quite clear exactly where they are headed and most probably have a map and a compass and a stopwatch in their backpacks. Of course, hikers of this type may well admire the view at their (previously scheduled) rest breaks or at designated scenic overlooks. The second type of hiker has a much more "stop and smell the roses" attitude. These hikers may never reach their destinations, but are far more likely to wander down unmarked paths and get lost in the poison ivy. There are, obviously, inherent advantages and disadvantages to both approaches. The two types also tend to quarrel with each other.

The standard high school geometry course tends to be a forced march through geometry, with a clear goal and a lot of mandatory check-in waypoints. Our focus in this book is much more on the journey than on the destination and is full of detours and scenic overlooks. We are trying less to impart a body of facts than to inculcate a way of thinking and a recognition of the many ways that mathematics can help interpret manmade and natural patterns. Take your time and enjoy the view.

This book started with a two-foot tall pile of photocopied articles from journals such as the *American Mathematical Monthly, Scientific American* (especially Martin Gardner's columns), and *The Mathematical Intelligencer,* and various books, reacting on a passionate fondness for geometry and the encouragement of our institutions to develop new introductory level courses for nonmathematics majors. We had fun writing this and hope that the reader will also enjoy this leisurely tour through some of the applications of geometry.

WHY WE WROTE THIS BOOK

There is a need for new offerings in college mathematics for students not majoring in the sciences, mathematics, or business. For most of these students, the traditional first, and for many the only, college mathematics course is a college algebra or precalculus course. Because the high school curriculum generally emphasizes algebra, the instructors of such courses are frequently in the position of trying to teach a technique-oriented course to a class that has seen most of it before. Some of these students are bored and others have accumulated a history of failure which is difficult to overcome. When people say "I can't do math," most mean they had trouble with algebra. If a traditional college algebra course is the only exposure students have to mathematics, we will continue to graduate students who fear and hate math or consider math to be formula manipulation with no bearing on life outside the classroom. Whether this reputation is fairly deserved is debatable, but such remains the typical student perception. To begin changing such negative attitudes in the general population, mathematics departments should offer a variety of mathematical experiences. It is important that students understand the broad applicability of mathematics and have a chance to succeed in it.

In writing this book, we had in mind three groups of students: liberal arts majors who are required by the Higher Powers to take a mathematics course, students of the visual arts who would like to strengthen their visualization skills and learn mathematical methods of recognizing and classifying geometric patterns, and education majors and teachers looking for enrichment content and an approach easily adaptable to the elementary or secondary classroom. We are strong believers in the theory that one learns mathematics by doing mathematics, and we hope that we have created an opportunity for these students and others to realize the joy of figuring things out for themselves.

One often hears that mathematics is the study and analysis of patterns, yet seldom are students presented with patterns that they can study and analyze for themselves. In geometry, with its rich variety of patterns, this definition takes on new force and vitality. The analysis of such patterns leads one naturally to the invention of appropriate notation and the application of analytic skills. Moreover, geometry is a field where many people have a great deal of intuition, though they have not been encouraged to develop it and may not even recognize it as valid. Geometry draws not only on analytic abilities, but also on the ability to visualize and on creativity, something many people do not associate with mathematics.

In this text, we have attempted to provide interesting and varied content, with a common theme of geometry and suitably graduated activities, examples, and models that the reader can investigate. This provides the experience of doing and discovering mathematics as mathematicians do. The inquiry-based mode of instruction reinforces the scientific method: experimentation leads to a hypothesis, which must then be verified by further experimentation and an explanation or informal proof. We ask the reader to express his or her conclusions and explanations clearly and logically in writing. When appropriate, there will be natural transitions from purely geometric properties to their algebraic and analytic counterparts. The dual nature of mathematics, geometric and algebraic, is thus integrated throughout.

We chose to focus on geometry for many reasons. The first is that both authors love the subject, but that is not a convincing reason to anyone else. The beauty and accessibility of geometric patterns make the material approachable to most people. Applications of geometry occur across a wide variety of disciplines: similarity and questions of scale are studied by biologists, polyhedra by chemists and crystallographers; mechanical linkages must be understood in robotics and by back-hoe operators, imaging and cross-sections by medical technologists when learning to read CAT scans, perspective by artists and architects.

Students of architecture, design, and the arts will find here the terminology and concepts they need to understand in order to be able to express geometric relationships clearly and correctly. Geometry is a universal language for the communication of ideas on shape and space. We have tried to provide the mathematical background to describe symmetries and structures in two- and three-dimensional space, while developing the ability to recognize, analyze, and generate geometric patterns.

Future teachers who have used preliminary drafts of this text have recognized that much of the material and many of the activities and exercises can be easily modified for use in their classrooms, both at the elementary and secondary level. We believe that it is vital that teachers be exposed to a wide variety of mathematical experiences, and think it even more important that their attitudes toward math be positive:

> Too often, elementary teachers take only one course in mathematics, approaching it with trepidation and leaving it with relief. Such experiences leave many elementary teachers totally unprepared to inspire children with confidence in their own mathematical abilities. Teachers themselves need experience in doing mathematics—in exploring, guessing, testing, estimating, arguing, and

proving—in order to develop confidence . . . . Insecurity breeds rigidity, the antithesis of mathematical power. . . . All students, and especially prospective teachers, should learn mathematics as a process of constructing and interpreting patterns, of discovering strategies for solving problems, and of exploring the beauty and applications of mathematics.
—*Everybody Counts: a Report to the Nation on the Future of Mathematics Education*, National Research Council, 1989.

Although we would never claim that this book is a sufficient background for a teacher, we do hope to combat some small part of the problem by changing attitudes and widening exposure to different ways of mathematical thought.

The professional organizations are unanimous in their recommendations for implementation of a number of teaching strategies such as cooperative learning, the active involvement of the student, the use of appropriate technology, writing across the curriculum, the use of physical models. We have tried to incorporate these recommendations as an integral part of this text. We encourage the student to work through this text with a group of similarly-minded people. Unlike many textbooks, there are not very many traditional drill problems, nor are there many formulae. In general, we have tried to structure the exercises and activities so that the reader will discover the pattern or formula, rather than giving it explicitly. Many of the exercises require a drawing or physical model, and many require a careful written response. In a text designed, in part, for use by future teachers, we find it especially important that they should be taught as we wish them to teach.

ORGANIZATION

We have assumed a nodding acquaintance with Euclidean geometry, such as would be received in a traditional high school geometry course. Building on that base, we have tried to include topics which the reader will find interesting. We have designed this book so that each chapter (after the initial one covering basic ideas) is essentially independent of the others, allowing a great deal of flexibility in designing a course. Each section has at the beginning a list of the supplies needed. References for further reading are included at the end of each section. While we encourage the use of computers in the classroom, we have not required the use of any specific software. Suggestions and sources are given at the end of each section. Since web addresses seem to change rather frequently, we only give a few. We hope to institute a web page with links to a selection of web pages on the topics of this text.

ACKNOWLEDGMENTS

We would like to express our deep gratitude to the many people who have helped us put this together, encouraged our efforts and even tested some of the material. Among these must especially be mentioned our friends and colleagues Benita Albert, Terry and Mary Bisson, Dorothy Buerk, John Careatti, RoseMarie Castner, Connie Elson, Richard Escobales, Steve Ferry, Steve Hilbert, Dani Novak, Lauren O'Connell, Tom Rishel, Margaret and Eric Robinson, Ken Scherkoske, Ray Tennant, and Tony Weston. Sally Di-Carlo, the interlibrary loan librarian at Canisius, found everything we ever asked her for, no matter how obscure the reference. Thanks are also due to Dean Jim McDermott, Ray Clough, and the Center for Teaching Excellence at Canisius College for their support. We are grateful for the encouragement given by Tom Banchoff, Keith Devlin, Joan Hutchinson, and especially Stan Wagon. Kevin Lee deserves especial mention for providing some tiling drawings on very short notice. Thanks also to the many students at Canisius and Ithaca Colleges who have been through the material for providing feedback, for making suggestions, and for their enthusiasm and patience. And last, many thanks to the wonderful people at Springer-New York and Key College Press, especially MaryAnn Brickner, Richard J. Bonacci, Barbara Chernow, David Kramer, Mike Simpson, and Jerry Lyons.

L. Christine Kinsey
Canisius College

Teresa E. Moore
Ithaca College

To the Reader

In this section, we will try to give some very general advice on how to use this book. First, approach the book with an open mind. We want you to learn to explore and enjoy math. At first glance, many of the topics may not even look related to the subject. Let your concept of mathematics expand, and discover pattern where you never expected it. This is not a traditional textbook, and only a few answers are in the back (those marked with a ▶). In fact, some of the exercises will not have answers or at least not absolute, unchanging answers. Don't let that scare you!

Second, do not be afraid to try. If you can work in a group of three or four people and you are not afraid to express yourself, bits of ideas from different people will often lead to a plan of attack and then a solution. Mathematics doesn't need to be a solitary pursuit. In testing these materials, it has often been the self-proclaimed "bad at math" students who have had the best ideas. They are not hung up on preconceived notions of what an answer has to be in order to be mathematical. However, once they start seeing a pattern, many of these same "bad at math" students are surprised to find themselves using algebra to formalize their ideas.

Be persistent, but know when to quit. Okay, that sounds like a contradiction. The point is that many students in traditional courses find the exercises fairly straightforward. Any problem that isn't solved in the first three minutes is declared either too hard or something that has not been covered yet. One of the main principles of this book is that you will be asked to find answers to questions that have not been covered yet. You will be doing the discovering so do not expect solutions to be quick. If you work on a problem for fifteen minutes or more and are getting nowhere, leave it alone for awhile. Try looking at it again in an hour or even the next day. Time and distance may show you what you could not see before. However, if you have worked really hard at a problem, you may need to change your whole approach. Some ideas simply do not work. See if you have overlooked part of the problem. A word of caution: If you do solve a problem using a new idea after several hours (or months for many mathematicians) with an old idea, write up your solution without delay. If you put things away for awhile, it may be the mistake on which you wasted all that time that you remember, not the brilliant stroke of inspiration that made everything clear.

In order to work through this book, you must not be only a reader, but an active participant in your education. We offer some very general advice and a structured approach to problem-solving from George Polya's influential book *How to Solve It*. Polya breaks down the solution of any problem, be it a tremendous new discovery or a simple exercise, into four steps:

Step 1: Understanding the problem: To solve a problem, you must first be clear on what is being asked. Make sure you understand the conditions given in the problem. Draw some pictures or build some models. Try to think of an example.

Step 2: Devising a plan: Sometimes a problem may remind you of an earlier problem or example and that may give you an idea of how to proceed. Some of the exercises in this text ask you to find a pattern or formula after working out a series of examples; if you first make a table or chart to organize your data, the solution may become clear. If you get stuck, either try a similar simpler problem, which may give you insight on what is happening in the more complicated situation, or just guess. In any case, keep trying.

Step 3: Carrying out the plan: Once you have an idea, carry out your plan. Do not get discouraged if the solution does not work out immediately or if the pattern is not clear yet. Reorganize your data, or try some more examples. Above all, be careful and check your computations: a misplaced bit of data can obscure the desired pattern. A tidy mind goes a long way in mathematics. If your plan does not work out, think about what has gone wrong. Many times you will learn more from a mistake than from a lucky guess at the correct answer. Learn from your mistakes, and try to understand why your first attempt does not work. Then try again; either refine your plan or try something else, making use of the insight gained by the false start.

Step 4: Looking back: Once you have found a solution, think about it. Try to check your answer. Make sure your solution is appropriate in terms of units or scale. Make sure that you really did answer the question asked. Think about whether the solution feels right, if it reflects the basic nature of the problem. We do not ask for formal proofs in this text, but mathematics, above all, must make sense. State your conclusions clearly and completely. Then, push your idea around a bit: Can it be applied to other situations? Can you expand on your idea?

Polya's steps for problem solving are simple and natural and broadly applicable. An interesting aspect of his outline is that two different mental skills are needed. Certain parts of Polya's plan require very focussed thinking, drawing on your analytical skills and attention to detail. Other steps require more free-ranging thought, imagination, and creativity. To be very good at mathematics, you must learn to cultivate both types of skills. You must learn to use both halves of your brain.

Problem solving is a skill that you can apply to any situation or career, and mathematics is a very good place to work on it. Our intent in this text is not to have you learn or memorize any particular facts or formulae, but to learn to think mathematically. With a bit of practice you will gain confidence in your abilities and intuition.

Contents

Preface vii
To the Reader xiii

1. The Basics 1
   1.1 Measurement 1
   1.2 Polygons 13

2. Grids 20
   2.1. Billiards 20
   2.2. Celtic Knots 30

3. Constructions 40
   3.1. Ruler and Compass Constructions 40
   3.2. The Pentagon and the Golden Ratio 51
   3.3. Theoretical Origami 61
   3.4. Knots and Stars 72
   3.5. Linkages 78

4. Tesselations 85
   4.1. Regular and Semiregular Tilings 85
   4.2. Irregular Tilings 94
   4.3. Penrose Tilings 115

5. Two-Dimensional Symmetry 127
   5.1. Kaleidoscopes 127
   5.2. Rosette Groups: Point Symmetry 138
   5.3. Frieze Patterns: Line Symmetry 147
   5.4. Wallpaper Patterns: Plane Symmetry 156
   5.5. Islamic Lattice Patterns 171

6. Other Dimensions, Other Worlds 180
   6.1. Flatlands 180
   6.2. The Fourth Dimension 194

7. **Polyhedra** 208
   7.1. Pyramids, Prisms, and Antiprisms 208
   7.2. The Platonic Solids 216
   7.3. The Archimedean Solids 224
   7.4. Polyhedral Transformations 230
   7.5. Models of Polyhedra 235
   7.6. Infinite Polyhedra 253

8. **Three-Dimensional Symmetry** 263
   8.1. Symmetries of Polyhedra 263
   8.2. Three-Dimensional Kaleidoscopes 270

9. **Spiral Growth** 279
   9.1. Spirals and Helices 279
   9.2. Fibonacci Numbers and Phyllotaxis 290

10. **Drawing Three Dimensions in Two** 300
    10.1. Perspective 300
    10.2. Optical Illusions 315

11. **Shape** 329
    11.1. Noneuclidean Geometry 329
    11.2. Map Projections 336
    11.3. Curvature of Curves 349
    11.4. Curvature of Surfaces 359
    11.5. Soap Bubbles 372

12. **Graph Theory** 383
    12.1. Graphs 383
    12.2. Trees 393
    12.3. Mazes 401

13. **Topology** 411
    13.1. Dimension 411
    13.2. Surfaces 421
    13.3. More About Surfaces 431
    13.4. Map Coloring Problems 439

Hints and Solutions to Selected Problems 449
Bibliography 481
Index 489
Permissions 493

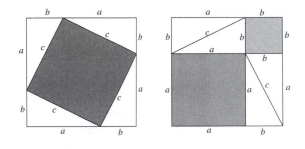

1. The Basics

◆ 1.1. MEASUREMENT

SUPPLIES

    tape measure
    ruler

These first two sections are intended to remind you of some basic tools and formulae which will be used throughout the text. The word geometry comes from the Greek words for earth and measurement. The first known application of geometry was surveying, most especially in Egypt where annual floods both enriched the soil and eradicated the demarcations between fields. Geometry provided the theoretical tools needed to accurately and efficiently divide up the land into individual holdings. Thus, we begin this book with some quick and easy exercises in measurement.

While the Egyptians developed a great deal of practical geometry, the Greeks developed its theoretical underpinnings, culminating in Euclid's *The Elements*, circa 300 B.C. Euclid's text contains most of the developments in mathematics known at the time, all contained in a superlatively clear and logical structure, wherein each result leads naturally to the next. *The Elements* is one of the most influential texts in the history of western civilization, perhaps only surpassed by the Bible. However, our purpose is not to recapitulate the standard results of Euclidean geometry. We are going to assume that you have already studied these, though we will provide significant prompting at times. We will make note of some of the particular results when we need to use them, such as the properties of isosceles triangles and the various ways of showing congruence.

It is often useful to be able to get a rough idea how big something is, even if you don't happen to have a ruler handy. The exercises that follow are intended to lead you to find readily portable substitutes for a ruler or yardstick.

▷ **Exercise 1.** Find a part of yourself an inch long. I use the first joint on the middle finger of my left hand which happens to be almost exactly one inch long. For some reason the corresponding joint on my right hand is a little longer.

▷ **Exercise 2.** Find a part of yourself a foot long. Use this and your "inch" measure of Exercise 1 to measure the length and width of your desk. Compare this with the actual length and width you find using your ruler. If you are too far off (more than 10%), explain why.

▷ **Exercise 3.** Find a part of yourself a yard long. Use this to measure the length and width of the blackboard. Compare this with the actual dimensions. If you are too far off (more than 10%), explain why.

▷ **Exercise 4.** Measure the length of your stride. Use this to measure the dimensions of your classroom. Compare this with the actual dimensions. If you are too far off (more than 10%), explain why.

The Pythagorean Theorem

One of the most useful tools in geometry is the Pythagorean theorem:

Pythagorean Theorem: If ABC is a right triangle with the right angle at vertex C, then the side AB opposite vertex C is called the hypotenuse and its length squared is the sum of the squares of the other two sides, BC and AC. I.e., $(AB)^2 = (BC)^2 + (AC)^2$, or if we label the sides a, b, and c as in the picture, then $c^2 = a^2 + b^2$.

Another way of stating the Pythagorean theorem in purely geometric terms is that the area of the square built with side c is equal to the combined areas of the squares built with sides a and b.

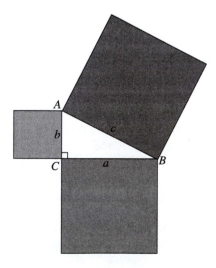

There are many different proofs of this theorem. Elisha Loomis has collected 367 of these in his book *The Pythagorean Proposition*, including one by President Garfield. My favorite of these is the following: Four copies of triangle $\triangle ABC$ can be arranged as in the picture below on the left to form part of a square with side $(a + b)$. Thus the big square has area equal to the areas of four triangles plus the area of a $c \times c$ square. But the same four triangles can be rearranged inside the same big square as on the right, so that the area of the big square with side $(a + b)$ is also equal to the areas of the four triangles plus the areas of an $a \times a$ square and a $b \times b$ square. Thus, $c^2 = a^2 + b^2$:

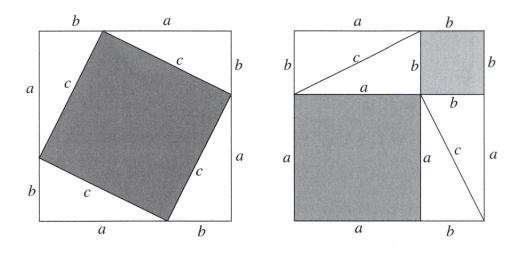

Another proof of the Pythagorean theorem involves cutting up the middle-sized square, which has side a. A copy of the triangle $\triangle ABC$ can be positioned inside the $a \times a$ square so the edge c goes through the point O at the center of the square as in the figure on the next page. The point O divides length c into two equal pieces, labelled x in the diagram, so that $c = 2x$. The side of the square is also divided into two pieces: a longer one marked y and a shorter one marked z. The diagram makes it clear that $y - z = b$, where b is the shortest side of the triangle. Draw another line perpendicular to the edge c as shown at the bottom left. These two lines divide the $a \times a$ square up into four identical pieces. These four pieces can be rearranged in the larger $c \times c$ square to leave a small square at the center. We must show that this small square is $b \times b$. From the lengths of the sides indicated, the small square has side equal to $y - z$, which we already know is equal to b. Thus the hole is a $b \times b$ square and so the largest $c \times c$ square has area equal to the sum of the areas of the middle-sized $a \times a$ square plus the $b \times b$ square. Therefore $c^2 = a^2 + b^2$:

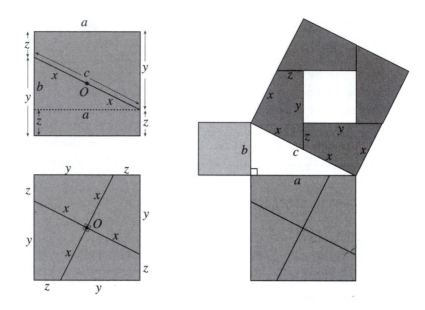

▷ **Exercise 5.** Redraw the figures for the proof above for the case in which *a* and *b* are the same length.

The converse of the Pythagorean theorem is also true:

Theorem: If a triangle has sides *a*, *b*, and *c*, and $c^2 = a^2 + b^2$, then the triangle is a right triangle with the right angle opposite the side *c*.

▷ **Exercise 6.** If a triangle has sides of lengths 3, 4, and 5, is it a right triangle, and, if so, where is the right angle?

▶ **Exercise 7.** If a right triangle has shorter sides of lengths 5 and 12, how long is the hypotenuse?

▷ **Exercise 8.** Triples of integers, such as those occurring in Exercise 6: 3-4-5 and Exercise 7: 5-12-?, are called Pythagorean triples, since they occur as the side lengths in right triangles and thus satisfy the Pythagorean theorem. Give one more example of an **integer** Pythagorean triple.

Since 1993 there has been a lot of publicity about Fermat's Last Theorem. This all started in 1637 when Pierre de Fermat, a French lawyer and very good amateur mathematician, was reading a book in which there was a proof that there are infinitely many Pythagorean triples. He scribbled in the book that he had a wonderful proof that this statement was false for higher powers, but there was not enough space in the margin for him to write it down. In other words, he was claiming that there were never nonzero integers *a*, *b*, and *c* satisfying the equation $a^n + b^n = c^n$ for any integer $n \geq 3$. For example, we showed above that a $c \times c$ square can be made out of an $a \times a$ square and a $b \times b$ square. Fermat's

Last Theorem says, for example, that a $c \times c \times c$ cube cannot be made from the pieces of a $a \times a \times a$ cube and $b \times b \times b$ cube for any integers a, b, and c. For many years, mathematicians tried to prove this theorem. It was finally proven in 1994 by Andrew Wiles. Wiles used extremely sophisticated techniques, so it is unlikely that Fermat's idea for a proof, whatever it was, was correct.

Perimeter

The *perimeter*, or *circumference*, of a geometric figure is the sum of the lengths of all of the sides, or the distance to walk all the way around the figure. For circles, one can use the formula

$$P = 2\pi r$$

to figure out the perimeter, or distance around the circle, where r is the radius of the circle (the distance from the center to the outside). This formula can be used to find the perimeter if you know the radius, or to find the radius if you know the perimeter.

▷ **Exercise 9.** (a) If a half-circle has radius 3, what is its perimeter? (Measure only the curved part of the semicircle.)
(b) If the perimeter of a circle is 8π, what is its radius?

▶ **Exercise 10.** Find the perimeter of each of these regions. Some of the figures in this exercise are composites of rectangles and portions of circles.

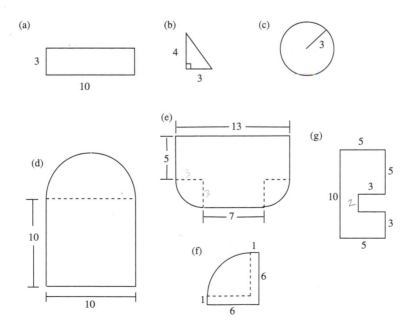

Area

The figure whose area is easiest to measure is the rectangle. For example, below is a 7 by 4 rectangle, which is 7 units wide and 4 units tall, and thus has area 7 times 4, or 28, square units. You can see in the picture that the rectangle contains 28 small squares.

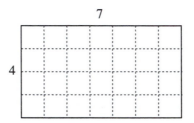

Thus the area of a rectangle is always the width times the height:

$$A(\text{rectangle}) = w \cdot h.$$

Next, we can figure out the area of a parallelogram labeled *ABCD* by dropping a perpendicular line from *D* to the point *E* on line *AB*.

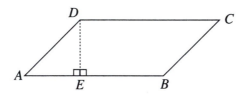

Notice that if the triangle *AED* is cut off and then glued back so that *AD* is glued to *BC*, a rectangle is formed that has the same area as the original parallelogram:

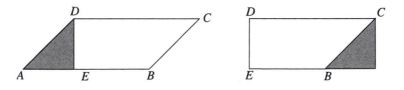

Thus, the area of a parallelogram is also width times height, but the height must be measured by dropping a perpendicular to the side that represents the width.

$$A(\text{parallelogram}) = w \cdot h.$$

▷ **Exercise 11.** You probably remember the formula for the area of a triangle from high school. State the formula and give an explanation of why this is the correct rule. You should model your explanation on the one above for the area of the parallelogram.

▷ **Exercise 12.** Figure out the area of the isosceles trapezoid shown below:

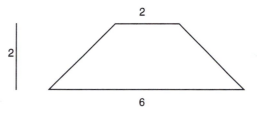

▷ **Exercise 13.** Figure out and state clearly a rule for the area of any trapezoid. Give an explanation of why your formula is correct. Be careful not to assume all trapezoids are as symmetrical as the one in Exercise 12.

The area formulae for other polygons are deduced from the ones above. The exception is the area of a circle.

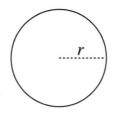

The radius r is measured from the center of the circle to the rim. The area is given by

$$A(\text{circle}) = \pi \cdot r^2.$$

Most other curved shapes require calculus in order to find their areas.

Now use these formulae to figure out the areas of the following figures. Some of the figures in Exercise 14 are composites of rectangles and portions of circles.

▶ **Exercise 14.** Find the area of each of these regions:

(a) 3 10

(b) 3 8

(c) 3 ⊢2⊣

(d) 5 12

(e) 5 10

(f) 4 5 3

(g) 3

(h) 10 10

(i) 13 5 7

(j) 1 6 1 6

(k) 5 5 10 3 3 5

(l) 6 2 4 2 8 4 2 5

Angles

Angles are used to measure steepness or inclination. A 0° angle is considered horizontal, while a 90°, or right angle, is perpendicular to the horizontal. To draw a right angle, an architect or builder uses ruler and compass, or, either a T-square (a sort of T-shaped ruler that has a precut right angle) or a triangle template. We shall assume that you had plenty of practice with angle measure in previous courses.

Some professions use other ways to describe angles or steepness. Builders talk of the pitch of a roof and will describe it, say, as a 4 in 12 pitch, meaning that for every 12 feet of horizontal distance, the roof goes up 4 feet.

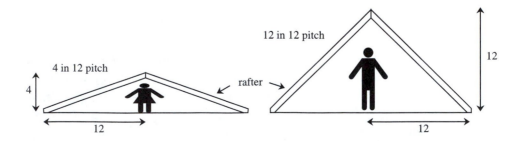

Different roof pitches have different advantages and disadvantages. A flat roof or a roof with low pitch (generally, a pitch less than 4 in 12 is considered low) minimizes the area of the roof, requiring less material for construction, but building codes require that leak-prone continuous membrane roofing be used on flat roofs and on roofs with pitch less than 3 in 12. Higher pitch roofs provide better runoff for rain and snow but require more building materials and form awkward interior spaces. Pitches greater than 12 in 12 are very rarely used.

▷ **Exercise 15.** What angle does a 12 in 12 pitch roof make with the horizontal?

▶ **Exercise 16.** If the building below has a roof with a 12 in 12 pitch and the building is 50 feet wide, how high is the peak of the roof from the attic floor? If the overhang is 1 foot, how long must the rafters be? What is the total area of the roof surface if the building is 75 feet long?

▷ **Exercise 17.** If a building has a roof with a 4 in 12 pitch and the building is 50 feet wide, how high is the peak of the roof from the attic floor? If the overhang is 1 foot, how long must the rafters be? What is the total area of the roof surface if the building is 75 feet long?

Stairs in buildings are commonly specified by riser and tread: the riser is the vertical height of the step, and the tread is the depth of the horizontal part of the step. Thus, an architect may specify a stair with a 7.5 inch riser and an 11 inch tread with a 1 inch nose, as pictured below:

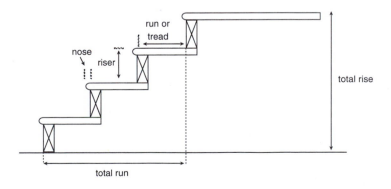

▷ **Exercise 18.** Design two different stairs from the first to the second story of a house if the distance from the floor to the ceiling is 7 feet and the interstitial space (the space between the ceiling and the floor of the story above) is 6 inches. Your stairs must satisfy the National Fire Codes for public buildings, which specify a maximum riser height of 7 inches and a minimum tread depth of 11 inches, and that treads must have uniform depth and risers uniform height. Use different risers and treads for each of your stairs.

Ramps are also specified by rise and run. The maximum steepness allowed for handicapped accessibility is a 1:12 slope, i.e., a handicapped ramp can go up at most 1 foot for every 12 feet of horizontal distance.

▷ **Exercise 19.** Check one handicapped ramp for compliance with the building codes.

In mathematics, slope is also described as the relation between rise and run, but is usually written as a fraction $\frac{rise}{run}$, so that instead of 1 in 12 or 1:12, a mathematician would describe the ramp as having slope $\frac{1}{12}$. A line 45° from the horizontal will have slope 1. Slopes more than 1 give steeper lines, while slopes less than 1 are more shallow.

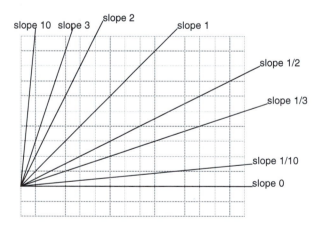

While finding the angle between two lines is a familiar task, sometimes one needs to find the angle between two curves. The angle between two curves is defined to be the angle between their *tangent lines* at the point where they intersect. The tangent line is the line that best approximates the curve at that point. The angle θ is formed between the curves shown:

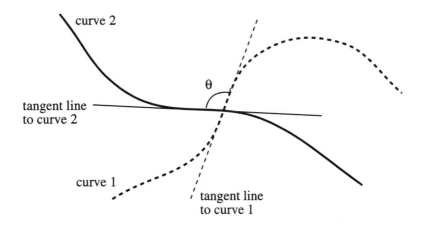

In general, one needs calculus in order to find a tangent line. The one easy case is a circle: it happens that the tangent line to a circle at a point is perpendicular to the radial line from the center to that point.

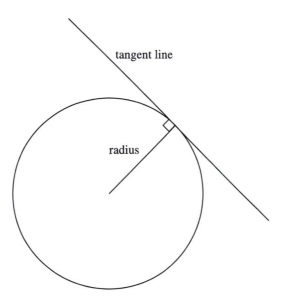

When one talks about the angle formed between a line and a circle, one refers to the angle θ formed by the line and this tangent line:

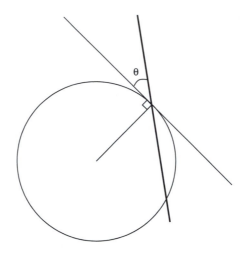

SUGGESTED READING

Elisha S. Loomis, *The Pythagorean Proposition*, NCTM, Washington, DC, 1968.

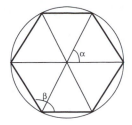

1. The Basics

◆ **1.2. POLYGONS**

SUPPLIES
> compass
> ruler

Note that most of the exercises are marked with a ▷, but a few are marked with a ★. The ★ exercises involve important formulae that will be needed later on in the course, so you should first make sure that you have gotten them right, and then either memorize the formula or record it somewhere where it can be easily found.

The word "polygon" is derived from the Greek for "many angles." The simplest polygon is the triangle, which has three straight sides. Different types of triangles have names. If one considers the angles, every triangle can be classified as one of three types: acute, obtuse, or right. An acute angle is one of measure less than 90° and an obtuse angle is one of more than 90°. A *right triangle* has one angle of 90°, and the other two must then be acute. An *acute triangle* is one in which all three of the angles are acute. An *obtuse triangle* has one obtuse angle and two acute angles.

One may also concentrate on the sides of the triangle and then classify it as scalene, isosceles, or equilateral. A *scalene triangle* is one with three unequal sides. An *isosceles triangle* has two equal sides. An *equilateral triangle* has three equal sides, and thus three equal angles. While no triangle can be both acute and obtuse, or scalene and isosceles, a triangle can easily be, for example, both acute and scalene.

Some relevant theorems about triangles are given below, with Euclid's numbering and wording. The angle which a side subtends is the angle opposite that side.

Proposition 4: [SAS congruence] If two triangles have the two sides equal to two sides respectively, and have the angles contained by the equal straight lines equal, they will also have the base equal to the base. The remaining angles will be equal to the remaining angles respectively, namely those which the equal sides subtend.

Proposition 8: [SSS congruence] If the three sides of one triangle are equal respectively to the three sides of another triangle, then the triangles are congruent.

Proposition 6: If in a triangle two angles be equal to one another, the sides which subtend the equal angles will also be equal to one another.

Proposition 26: [ASA and AAS congruence] If two triangles have the two angles equal to two angles respectively, and one side equal to one side, namely, either the side adjoining the equal angles, or that subtending one of the equal angles, they will also have the remaining sides equal to the remaining sides and the remaining angle to the remaining angle.

Proposition 32: In any triangle, the three vertex angles are equal to two right angles.

Quadrilaterals (from the Latin for four sides) have, of course, four sides. Again, certain types have names. A *trapezoid* has a pair of parallel sides. A *parallelogram* has two pairs of parallel sides. A *rhombus* is a parallelogram that has all four sides equal in length. A *rectangle* has four right angles, and thus the sides are parallel and equal in pairs. A *square* has four right angles and four equal sides. One can thus consider a square as a special type of rectangle.

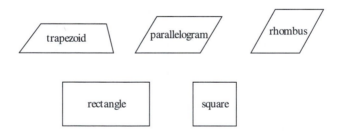

While it is true that any equilateral triangle must have three equal angles, this is not true for polygons with more than three sides. For example, a rhombus has four equal sides, but the angles are not all equal, while a rectangle has four equal angles, but the sides are not all equal. A *regular polygon* is one with all of its sides equal and with all of its angles equal; i.e., it must be both *equilateral* and *equiangular*. A regular quadrilateral must be a square.

Another distinction among types of polygons is *convexity*. A *convex polygon* is one in which the line segment joining any two points lies entirely inside the polygon. Here is a picture of a nonconvex polygon, and a line segment between two points in the polygon that does not lie inside the polygon:

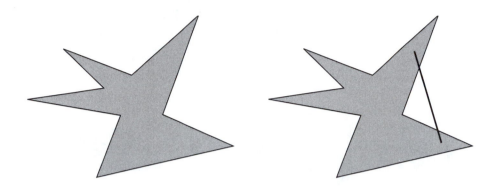

A tangram is a simple geometric puzzle, invented in China about two hundred years ago. In publicizing the puzzle in 1903, Mr. Sam Loyd also invented a history for the puzzle:

> According to the late Professor Challoner, whose posthumous papers have come into the possession of the writer, seven books of Tangrams, containing one thousand designs each, are known to have been compiled in China over four thousand years ago. These books are so rare that Professor Challoner says that during a forty years' residence in China he only succeeded in seeing perfect editions of the first and seventh volumes, with stray fragments of the second [quoted by Gardner, p. 29].

Loyd's hoax is still believed and perpetuated by many of the books on tangram puzzles written for children and their teachers.

▶ **Exercise 1.** Tangrams are formed by cutting a square up into pieces as shown below. Assuming that the smallest square is one inch on each side, find the lengths of the sides of each of the other pieces. Note that all of the angles are either 45°, 90°, or $135° = 45° + 90°$.

▷ **Exercise 2.** Find the area of each of the tangram pieces, assuming that the small square has sides of length 1.

▶ **Exercise 3.** Cut a square up as shown in Exercise 1 to make your own tangram puzzle. The pieces can be fit together to form exactly 12 other convex polygons as shown below. Show how each is done and sketch your solution. Each figure must use all seven tangram pieces with no overlap.

In any quadrilateral labeled *ABCD*, a line can be drawn connecting two opposite corners, for example *A* and *C*, and dividing the quadrilateral into two triangles, △*ABC* and △*ADC*.

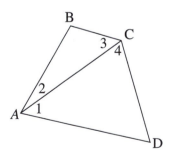

By Euclid's Proposition 32, the angles in a triangle add up to two right angles, or 180°. Note that ∢*A* = ∢1 + ∢2, and ∢*C* = ∢3 + ∢4. If we add up the vertex angles in quadrilateral *ABCD*, we get

$$\angle A + \angle B + \angle C + \angle D = (\angle 1 + \angle 2) + \angle B + (\angle 3 + \angle 4) + \angle D$$
$$= (\angle 1 + \angle D + \angle 4) + (\angle 2 + \angle 3 + \angle B)$$
$$= 180° + 180°$$
$$= 360°.$$

Proposition: The sum of the vertex angles in a quadrilateral is 360°.

▷ **Exercise 4.** Repeat the proof for the case of a nonconvex quadrilateral as shown below. Do the angles still sum to 360°?

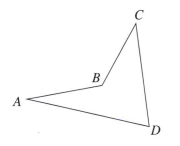

The names of other polygons are taken from the Greek number prefixes followed by -gon. For example, a five-sided figure is called a pentagon, since penta- is the prefix for five. A triangle could also be referred to as a trigon, but rarely is. A quadrilateral is occasionally called a tetragon.

$3 = $ tri	$9 = $ ennia or nona	$15 = $ pentakaideca
$4 = $ tetra	$10 = $ deca	$16 = $ hexakaideca
$5 = $ penta	$11 = $ hendeca	$17 = $ heptakaideca
$6 = $ hexa	$12 = $ dodeca	$18 = $ octakaideca
$7 = $ hepta	$13 = $ triskaideca	$19 = $ enniakaideca
$8 = $ octa	$14 = $ tetrakaideca	$20 = $ icosa

▶ **Exercise 5.** Figure out the sum of the vertex angles in an arbitrary pentagon by dividing it up into triangles as we did for the quadrilateral above.

▶ **Exercise 6.** Figure out the sum of the vertex angles in an arbitrary hexagon.

★ **Exercise 7.** Figure out the sum of the vertex angles in an arbitrary *n*-sided polygon.

★ **Exercise 8.** In Exercise 7 you figured out a formula for the sum of the angles of an *n*-sided polygon. A regular *n*-sided polygon will, of course, have *n* equal angles. Give the formula for one vertex angle of a regular *n*-sided polygon.

▶ **Exercise 9.** If a polygon is regular and one of the vertex angles is 165°, how many sides must it have?

In general, a regular polygon can be surrounded by a circle so that the vertices are equally spaced points around the circle. For example, a regular hexagon looks like this:

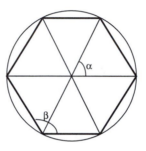

If we draw lines from the center to each of the vertices, the hexagon divides the circle into 6 pieces. The angle α is called the *interior angle* of the hexagon and will measure one-sixth of 360°, or 60°. By the formula of Exercise 8, the vertex angle, marked β in the diagram, measures 120°.

▷ **Exercise 10.** In the picture of the regular hexagon above, assume that the circle has radius 1. How long is each side of the hexagon?

▷ **Exercise 11.** What is the interior angle of a regular pentagon? What is the vertex angle?

▶ **Exercise 12.** What is the interior angle of a regular octagon? What is the vertex angle?

▷ **Exercise 13.** What is the interior angle of a regular dodecagon? What is the vertex angle?

A *tiling*, or *tesselation*, is a way of filling up the plane with repetitions of a basic tile meeting edge to edge. It is known that there are only three regular tilings: ones that use only one type of tile and that tile is a regular polygon. In making any tiling of the plane, we must make sure that the angles around any vertex add up to 360°. One such tiling by regular polygons is a pattern of square tiles, as shown on the next page. Each corner of a square contributes 90°, and four corners meet at each vertex to sum to 360°.

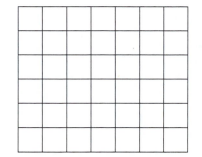

▷ **Exercise 14.** Find the other two regular tilings of the plane.

SUPPLIES

1. A set of plastic tangrams is marketed under the name *Tangoes* (their brochure says, "Tangoes is the modern version of the 4,000 year old Chinese tangram puzzle"), from Rex Games, 530 Howard St., #100, San Francisco, CA 94105-3007, and is available at many toy stores.
2. Sets of tangrams are readily available from suppliers of educational materials, such as Dale Seymour and Creative Publications.

SUGGESTED READINGS

Martin Gardner, *Time Travel and Other Mathematical Bewilderments*, W.H. Freeman, New York, 1988.

Sam Loyd, *The Eighth Book of Tan*, reprint of 1903 edition, Dover, New York, 1968.

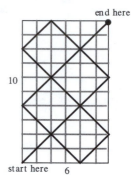

2. Grids

◆ 2.1. BILLIARDS

SUPPLIES

    graph paper (4 or 5 squares to an inch)
    ruler
    tracing paper

Let's play billiards on a rectangular table with no pockets to make things simpler. We'll also ignore any friction and english (the spin of the ball), and assume that the ball always travels in a straight line until it hits a corner, when it stops. We need to know the following:

Snell's Law: Assuming a frictionless environment and no spin on the ball, the ball traveling toward a wall at an angle will rebound at the same angle.

In other words, if a ball hits the bumper at a 30° angle, it will bounce off at a 30° angle.

▷ **Exercise 1.** Consider the billiards game pictured below. Figure out all of the angles marked.

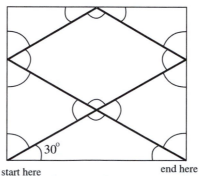

There are other ways to describe such a path besides angles: One can talk about the slope of a line. A line has slope $\frac{1}{2}$ if it goes up 1 unit for every 2 units it goes to the right. This is easiest to picture with graph paper:

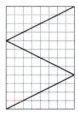

If a ball is traveling on such a path, it will bounce off at the same angle as it came in. Thus it will go up one unit for every 2 units it moves to the left:

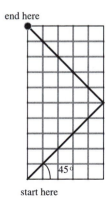

A traditional billiard table is twice as long as it is wide. To make life easier, let us first concentrate on problems where the ball starts at the lower left corner and is hit at a 45° angle, first on a table 10 feet long and 5 feet wide. Notice how using graph paper and a 45° angle facilitates figuring out the path.

end here

45°

start here

The ball bounces once, at a 45° angle, and lands in the upper left corner. For reasons that will become obvious later, count that as 3 hits: the start, the bounce, and the end. Try it again on a table that is 10 feet long and 6 feet wide:

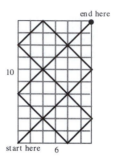

The ball bounces 6 times (8 hits), and lands in the upper right corner.

▶ **Exercise 2.** Use graph paper and track the path of a ball hit at a 45° angle on each of the tables below. Start in the lower left corner and follow each ball until it hits a corner, and mark the corner where it ends up.
a) 6 by 3 table
b) 9 by 3 table
c) 12 by 3 table
d) 6 by 6 table
e) 9 by 6 table
f) 10 by 8 table
g) 8 by 6 table
h) 7 by 3 table
i) 15 by 9 table

Next, compare the trajectories of the ball on each of the following tables:

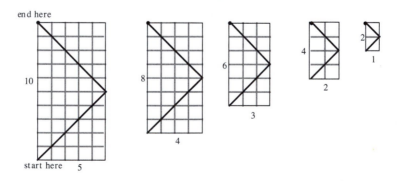

The size of the table does not seem to matter; what matters is the ratio of the length to the width. A 10 foot by 5 foot table has ratio 10 to 5, which is more easily understood as a fraction: $\frac{10}{5} = \frac{2}{1}$. Since

$$\frac{10}{5} = \frac{8}{4} = \frac{6}{3} = \frac{4}{2} = \frac{2}{1},$$

all the tables shown above have the same pattern.

▶ **Exercise 3.** For each table in Exercise 2, give another table on which the ball will have the same pattern.

In some of the following exercises, you are asked to come up with rules that dictate the behavior of the billiard ball. In stating such rules, you must be clear and complete. Your rules should be stated in terms of a table of arbitrary dimensions, such as for an n by m table where n and m satisfy some clearly stated condition. The rule should allow one to predict the behavior of the ball on any table whose dimensions satisfy the condition given.

▶ **Exercise 4.** Use the examples and exercises to figure out rules for when the ball will end up in
a) the upper right corner.
b) the upper left corner.
c) the lower right corner.

State your rules clearly in complete sentences. You may want to try some other tables to check your rules.

▷ **Exercise 5.** Note that in part (h) of Exercise 2, the ball went through every square of the graph paper. Figure out a rule for when this will occur. State your rule clearly in complete sentences. You may want to try some other tables to check your rule.

▶ **Exercise 6.** Figure out a rule for the number of hits the ball makes (count both the start and the end as hits as well as any bounces off the sides). State your rule clearly in complete sentences.

Examining all the different tables we have made, note that some paths exhibit certain types of symmetry.

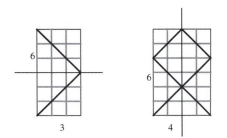

For example, the 6 by 3 table on the left in the previous picture has a horizontal mirror line as shown. The 6 by 4 table on the right has a vertical mirror line. The table on the left has *horizontal reflectional symmetry*, while the table on the right has *vertical reflectional symmetry*.

Another type of symmetry is shown below:

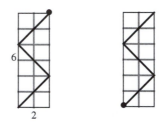

The path of the ball looks just the same if the 6 by 2 table is rotated by 180°. This is called *rotational symmetry*.

▶ **Exercise 7.** Identify the type of symmetry for each of the tables from Exercise 2.

▷ **Exercise 8.** Figure out a rule for when each type of symmetry occurs. State your rule clearly in complete sentences. You should notice that your rules seem quite similar to the rules that you found in Exercise 4. Explain in a paragraph why this is true.

▶ **Exercise 9.** Figure out the path of the ball on a table that is 5 by $2\frac{1}{2}$.

▷ **Exercise 10.** Figure out the path of the ball on a table that is $2\frac{1}{3}$ by 2.

▷ **Exercise 11.** Figure out the path of the ball on a table that is $1\frac{1}{3}$ by $1\frac{1}{2}$.

▷ **Exercise 12.** Figure out a general rule to deal with tables with fractional width and heights.

▶ **Exercise 13.** Must the ball always end up in a corner when shot from a corner at an angle of 45°? Try some peculiar rectangular tables. Explain your reasoning.

Angles Other than 45°

Now we return to trajectories with angles other than 45°. If the ball travels along a path with slope $\frac{1}{2}$ on a 10 by 8 table, it will follow the path shown:

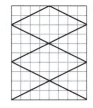

▶ **Exercise 14.** a) Use graph paper and track the path of a ball hit at a slope of $\frac{1}{3}$ on a 6 by 3 table.
b) Repeat for a 9 by 6 table.

▶ **Exercise 15.** a) Use graph paper and track the path of a ball hit at a slope of 2 (think of 2 as $\frac{2}{1}$, so the ball goes 2 units up for every 1 unit to the right) on a 6 by 3 table.
b) Repeat for a 10 by 6 table.

One way to think about different slopes is as a change of scale. Note the similarities in the following pictures:

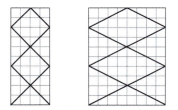

All we did was to change the scale along the width of the table by a factor of one half.

▷ **Exercise 16.** How would you rescale the tables in Exercises 14 and 15 to change the problem to a 45° trajectory?

Unfolding Trajectories

Consider the trajectory of the billiard ball on the 6 by 4 table as pictured below:

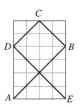

Copy the following pattern onto tracing paper, and fold the tracing paper, first along the horizontal dotted line and then in a zigzag along the vertical dotted lines, so that the final result is a 6 by 4 rectangle six layers thick. Hold the tracing paper up to a strong light, and you should be able to see the 6 × 4 table and the billiard trajectory pictured previously.

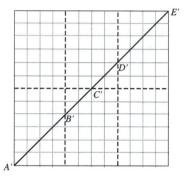

On the tracing paper you have 6 copies of the original table, forming a 12 by 12 square. The path of the 45° billiard ball is a straight line from the bottom left corner point A' to the top right E' on this square. Note that the line segment $A'B'$ is equal to the line segment AB on the original table, while $B'C'$ is the reflection of BC in a vertical line. Similarly, $C'D'$ is the reflection of CD first in the vertical line, then in the horizontal line, and $D'E'$ is the reflection of DE. When you fold up the tracing paper, each segment of the path of the ball is reflected by the fold. Since the 12 by 12 square is three 6 by 4 blocks wide and two 6 by 4 blocks tall, the ball must pass through two vertical dotted lines and one horizontal dotted line in order to get from corner to corner. These points, with the start and the end, account for the five hits that you know the ball must make. Furthermore, the total distance the ball travels, which is the same whether you view it on the original 6 by 4 table or in the unfolded 12 by 12 square, is easily seen by the Pythagorean theorem to be $12\sqrt{2}$.

▶ **Exercise 17.** Draw the unfolded trajectory for the following tables, assuming that the ball starts in the lower left-hand corner and is shot at an angle of 45°.
a) 6 by 3
b) 9 by 3
c) 4 by 8
d) 4 by 3
e) 10 by 4

▶ **Exercise 18.** Use the Pythagorean theorem to find the length of the trajectory for each table in Exercise 17.

▷ **Exercise 19.** Figure out a rule for the dimensions of the square unfolded table for a 45° billiard problem.

For angles other than 45°, i.e., slopes other than 1, the unfolded table need not be square. For example, for the billiard trajectory of a ball shot with slope $\frac{1}{2}$ on a 10 by 8 table, we get the following unfolded trajectory:

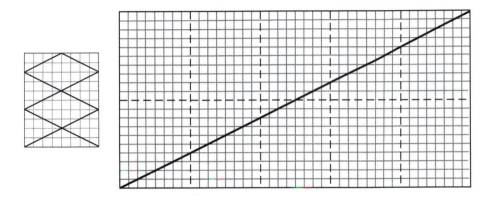

▷ **Exercise 20.** Unfold the trajectory of a ball traveling along a path with

a) slope $\frac{1}{3}$ on a 6 by 3 table.

b) slope 2 on a 6 by 4 table.

Application to Reality

You can use the idea of unfolding trajectories to decide how to hit the cue ball in order to impact another ball. Suppose that you want the white cue ball to impact the black ball in the next figure, but want to bounce off one of the cushions first. You can figure out where on the cushion to aim the cue ball by imagining the unfolded trajectory. On the left is a picture of how the two balls are situated, and one wishes to plot a shot that will make the cue ball hit the black ball after bouncing off the top cushion. The center illustration shows the reflection of the table and the black ball across the top cushion. The trajectory from the cue ball to the reflection of the black ball is shown. This pinpoints the spot on the top cushion where we need to aim the cue ball. On the right is the trajectory as executed. The cue ball bounces off the top cushion as planned and hits the black ball:

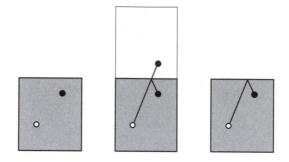

If in the same scenario one wished to hit the black ball with the cue ball, but after a bounce off the left cushion instead of the top cushion, we unfold the table to the left instead to plot the straight line from the real cue ball to the reflection of the black ball. The point where this straight line intersects the cushion is the point to aim for:

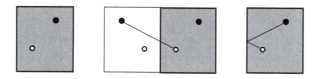

To do a right cushion bounce instead, we unfold the table to the right and plot the straight line from the real cue ball to the new reflection of the black ball:

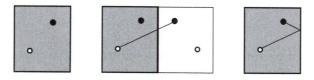

Similarly, a two-cushion shot can be analyzed by unfolding the trajectory. To make the cue ball bounce twice, first off the left cushion and then off the top cushion, we unfold the table first to the left and then to the top. Note how the image of the black ball is reflected in each side of the table. We aim from the original cue ball to the image of the black ball in the image of the table that has been unfolded twice. In folding the picture back up, the straight line forms the trajectory of the cue ball as it bounces off the cushions to hit the black ball:

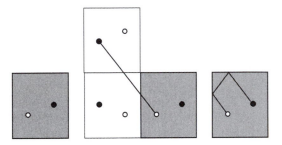

▷ **Exercise 21.** Plot the trajectory the cue ball must travel to hit the black ball, if you want to bounce first off the east wall then the west wall. Use tracing paper to trace the original table and situation of the two balls and to plan the trajectory.

▶ **Exercise 22.** Figure out all possible two-cushion shots for the following configuration. Use tracing paper to trace the original table and situation of the two balls and to plan each trajectory.

Note: The first part of this unit is adapted from H. Jacobs: *Mathematics: A Human Endeavor.*

SUGGESTED READINGS

Harold R. Jacobs, *Mathematics: A Human Endeavor*, W.H. Freeman, New York, 1994.
Serge Tabachnikov, *Billiards*, Société Mathématique de France, Marseille, 1995.

2. Grids

◆ **2.2. CELTIC KNOTS**

SUPPLIES
> graph paper (4 or 5 squares to an inch)
> non-photo blue pencil (optional)

Ornamental knotwork appears in Celtic manuscripts, carved stones, and metalwork. The most famous examples occur in illuminated manuscripts such as the *Lindisfarne Gospels*, the *Book of Durrow*, and the *Book of Kells*. Knotwork exhibits a beautiful geometric regularity while offering a great amount of flexibility for the artist. In this section, we have developed a system using dual grids that allows one to generate reasonably good drawings easily. Begin with a piece of graph paper, and color a section of the vertices so that the vertices are of alternate colors. In the absence of color, we have done this below alternating gray dots and black dots. We suggest using a non-photo blue pencil and a black pen to mark dots on graph paper. If you later photocopy the final drawing, these blue dots will not show. For the sake of consistency, begin all such grids with a gray (blue) dot in the upper left corner:

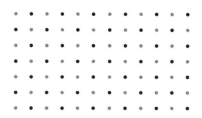

The celtic knots we will draw will be dictated by the grid of dots, generally going through (and so covering) the black dots, and avoiding the gray dots. In describing the dimensions of the knot, count squares or spaces, rather than dots. The rules we use to generate the knot follow:

Rule 1. Always begin the grid with a gray dot in the upper left-hand corner.

Rule 2. Never go through a gray dot, except at the corners of the diagram.

Rule 3. Go through all the black dots.

Rule 4. Alternate overpasses and underpasses at the crossings.

Rule 5. The cord or curve forming the knot should be of uniform width.

Rule 6. Corner points are V-shaped, rather than U-shaped.

Rule 7. *If you hit a bar, the crossings change as follows:*

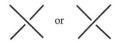

changes, depending on the orientation of the bar, to

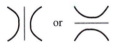

We construct a celtic knot following the steps outlined below:

Step 1: Decide on the dimensions of the drawing. We will begin with a simple 4 by 6 knot, so the appropriate grid, applying Rule 1, is:

Step 2: Next draw diagonals through all the black dots, applying Rule 3. Use a pencil and draw these in lightly, as some of these lines will be modified in Step 4.

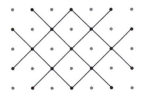

Step 3: It is traditional (though not invariable in practice) to square off the corners as required by Rule 6, which is why you are allowed to go through only the gray dots at the corners in Rule 2:

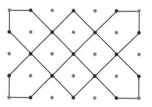

Step 4: Apply the crossing rule, Rule 4. We'll return to the question of why it is always possible to do this.

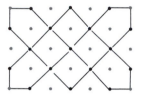

Step 5: Round off any angles, except those at the corners, and thicken the lines using Rule 5. The final design should consist of graceful curves and smooth transitions between straight and curved sections. All of the black dots should be covered by the design, while all of the gray dots, except those at the corners of the diagram, should be visible.

Here are two further examples, a 5 by 6 knot on the left, and a 3 by 7 knot on the right. Note that in all cases we begin the grid of dots with a gray dot in the upper left corner.

Note that both of these examples have loose ends, since in both cases there are black dots in some of the corners. Knots that have no loose ends are termed *closed knots*. In many classic examples of celtic knots, these loose ends are finished off as heads and tails of various monsters, but since we can't draw monsters, we will finish the ends off with curlicues:

▶ **Exercise 1.** Draw a 5 by 10 knot. Be sure to begin the grid of dots with a gray dot in the upper left corner.

▷ **Exercise 2.** Draw a 3 by 9 knot.

▷ **Exercise 3.** Draw a 6 by 9 knot.

▶ **Exercise 4.** Draw a 6 by 6 knot.

▷ **Exercise 5.** Draw a 6 by 8 knot.

▷ **Exercise 6.** Draw a 4 by 12 knot.

▶ **Exercise 7.** Draw a 8 by 12 knot.

▶ **Exercise 8.** Which of the knots in the examples and Exercises 1–7 have loose ends? State a rule (in complete sentences) for when a knot will have loose ends. Your rule will, of course, have something to do with the dimensions of the grid. Explain why your rule makes sense.

▷ **Exercise 9.** Since we begin the grid of dots with a gray dot in the upper left corner, there is always a traditional spade-shaped corner in the upper left corner. Thus the loose ends, if they exist, must be at two of the other three corners. State rules for which knots with loose ends have these ends in which positions.

▷ **Exercise 10.** Why is there always an equal number of over-passes and underpasses?

One factor in drawing celtic knots is how many strands, or pieces of string, are needed to form the knot. The 6 by 6 knot below has three strands, shown in varying shades of gray:

▶ **Exercise 11.** How many strands do the knots in the examples and Exercises 1–7 have? State a rule for how many strands a closed knot will have. You may want to do extra examples to test your theory.

The more famous examples of knotwork are rarely as simple as the plain knots we have generated so far. They are complicated by adding gray bars to the basic grid that connect some of the gray dots. These gray bars are considered as obstacles that the knot must avoid, just as the knot avoids the gray dots of the grid. As with the dots, when working with paper and pencil it is preferable to draw the bars with a non-photo blue pencil. In order to avoid these bars, we must apply Rule 7.

For example, the grid below, with one bar, gives the knot on the right. The original grid was 4 by 6 as in the first example of this section, so a plain knot on that grid will have one strand; this knot has two.

Next we ask what effect does adding a bar have on the number of strands? In the first picture below, adding the bar joins the black and dotted strands, while in the second picture, the bar splits the black strand.

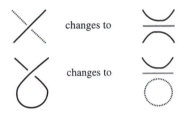

Adding a bar tends to split strands or to join different strands. A bar may either increase the number of strands by one or decrease by one, or leave the number of strands unchanged. Below is a plain 4 by 8 knot with two strands, then the same 4 by 8 knot with a bar that gives three strands, then with a bar that gives one strand.

The following grid gives the Josephine knot (named after Napoleon's wife), also known as the Carrick bend or true lover's knot. The original grid was 4 by 8, so a plain knot on that grid should have two strands; this knot also has two:

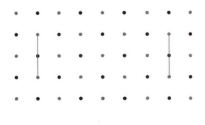

▶ **Exercise 12.** Draw a knot with the following grid.

▷ **Exercise 13.** Draw a knot with the following grid:

▷ **Exercise 14.** Draw a knot with the following grid. This knot appears in the *Lindisfarne Gospels*.

▶ **Exercise 15.** Draw a knot with the following grid. This knot appears in the *Book of Kells*.

▷ **Exercise 16.** Draw a knot with the following grid:

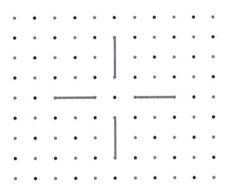

Why We Can Always Alternate Over- and Underpasses

To return to the unresolved question of why it is always possible to alternate over- and underpasses, let us consider any closed path (so that it ends at the point where it began) that crosses itself but never has a triple crossing, where three strands come together at a point.

The path with the crossings divides the plane into regions: one un-bounded region surrounding the scribble, and several others contained inside. The first thing we must show is that it is possible to color these regions gray and white, so that any two neighboring regions which share an edge are colored with different colors. Color the unbounded outer region gray and think of gray as water and white as land.

To see that such a coloring is always possible, note that each crossing can be replaced by either a channel connecting two bodies of water or a cause-way connecting two bodies of land. After replacing all the crossings, we have one or more closed paths or loops that do not cross themselves or each other. These essentially circular paths are either completely disjoint like separate islands, or nested inside each other, so that you could have an island with a lake, with an islet in the lake and a puddle on the islet. There are, of course, many different ways to do this, and two are shown below:

But notice that in each of the pictures above the land masses have the same coloring, even though different choices were made regarding whether to build causeways or dig water channels. Thus the coloring can be carried back to the original regions defined by the path with crossings. Both of the pictures above consist of disjoint (nonintersecting) arrangements of *simple closed curves*, curves that begin and end at the same point (*closed*) and that do not intersect themselves (*simple*). The ideal simple closed curve is a circle, and both pictures can be simplified and represented as arrangements of disjoint circles, as shown below:

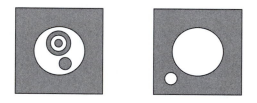

Thus, we have reduced the problem to show that we can color these circles properly. Let us start by considering a simpler problem: showing that we can color any three disjoint circles so that neighboring circles are different colors. For the purposes of this coloring, ○ ◎ is the same as ◎ ○. Three disjoint circles can be arranged in four ways: all nested one inside the next, all separate with no overlap, two nested and the third separate, or two inside the third as shown below. Each arrangement can be colored correctly.

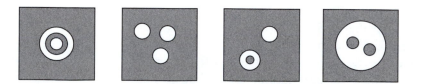

▶ **Exercise 17.** Find all the ways in which five circles that are either separate or nested can be arranged on a field of color. Color each arrangement so that neighboring circles are different colors.

An extension of Exercise 17 shows that all such arrangements of circles can be colored as specified, and this coloring can be carried back to the scribbled path with the crossings.

Once we have an appropriate coloring of the scribbled path, we can then build over- and underpasses by the following rule: travel around the path and as you approach a crossing, if there is water on the right, build an overpass; if there is land on the right, build an underpass.

Thus any such closed curve, which begins and ends at the same point and which has only two curves crossing at a time, can be given some system of alternate overpasses and underpasses.

▷ **Exercise 18.** First color the regions into which the scribble below divides the plane, with the unbounded region colored gray and so that neighboring regions that share an edge are different colors. Next, build over- and underpasses as explained above.

SUGGESTED READINGS

Colin Adams, *The Knot Book*, W.H. Freeman, New York, 1994.
Clifford W. Ashley, *The Ashley Book of Knots*, Doubleday, New York, 1944.
George Bain, *Celtic Art: The Methods of Construction*, Dover, New York, 1973.
Iain Bain, *Celtic Knotwork*, Sterling Publ., New York, 1992.
Peter R. Cromwell, "Celtic Knotwork: Mathematical Art," *Mathematical Intelligencer* 15(1), 1993.
Aidan Meehan, *Celtic Design: Knotwork*, Thames and Hudson, New York, 1991.

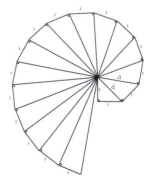

3. Constructions

◆ **3.1. RULER AND COMPASS CONSTRUCTIONS**

SUPPLIES
> compass
> straightedged or unmarked ruler (You can use a marked ruler if you
> promise not to look at the markings.)

The tools of any class involving geometry are the (unmarked) straightedge and the compass. The compass is used to draw circles and to measure line segments, while the straightedge is used to draw straight lines. The first postulate of Euclid's *The Elements* is:

Postulate 1: Through any two distinct points, it is possible to draw (exactly) one straight line.

That is done easily enough with the help of your straightedge. Then Euclid gives:

Postulate 2: A line segment can be extended at either end by an arbitrary amount.

Postulate 3: Given a length (the radius) and a point (the center), it is possible to draw (exactly) one circle.

With these basic theoretical tools called the *ruler and compass postulates*, and a knowledge of the basic theorems of Euclidean geometry, one can construct many figures, either with actual straightedge and compass or with a computer program such as *Geometer's Sketchpad* designed specifically for such constructions.

The postulates given above indicate the most elementary things one can do with a straightedge and compass. The natural question, at least to a mathematician, is what else can be done with these basic tools. Every ruler and compass construction, such as building a line perpendicular to a given line, bisecting an angle, or drawing a polygon, is a sequence of steps each of which involves one of the postulates listed above. The difference between the ruler and compass postulates above and the ruler and compass constructions below, is that the postulates give simple constructions that we are sure we can do without instruction. The constructions, on the other hand, are more complicated, but useful, things to do. Furthermore, it can be proven that each construction does indeed yield the desired result. The postulates are assumed true without proof. Therefore, mathematicians try to minimize the number of postulates, so that as little as possible is assumed without proof.

We first list the basic ruler and compass constructions, which you should remember from high school. Each construction has a list of steps for the construction.

Ruler and Compass Construction 1: Perpendicular Bisector.
Construct the perpendicular bisector of a line segment.

(1) Let AB be the given line segment.
(2) Set your compass to a radius somewhat larger than half the distance between A and B, and draw two circles with identical radii, centered at A and at B.
(3) These circles will meet at two points C and D, one above the line AB and one below the line.
(4) Now, using your straightedge, draw the line CD.
(5) This line will intersect AB at a point we will call E, which is the midpoint of AB, and is perpendicular to AB:

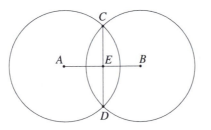

Why This Works. It is easy to show that triangles $\triangle ACE$ and $\triangle BCE$ are congruent, by first proving the congruence of triangles $\triangle ACD$ and $\triangle BCD$. Therefore, $\angle AEC = \angle BEC$, and since they form a straight line, $\angle AEC = \angle BEC = 90°$, and thus CE is perpendicular to AB.

▷ **Exercise 1.** Use your straightedge to draw a line segment and with straightedge and compass construct a perpendicular bisector to your segment.

R & C Constructions 2 and 3 also involve finding perpendicular lines: Each is a variation of R & C Construction 1. The key to each is locating two points on the given line that are equally spaced on either side of the point where we want to have the perpendicular line crossing.

Ruler and Compass Construction 2: Erect a Perpendicular. Given a line and a point on that line, construct a line perpendicular to the given line through the given point.

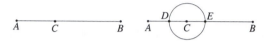

(1) Let AB be the given line segment, and C the point on AB where we wish to erect a perpendicular. The first thing to do is find two points on AB equally spaced on either side of C.
(2) Draw a circle centered at C and intersecting line AB at D and E.
(3) Construct the perpendicular bisector of DE as in R & C Construction 1.

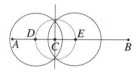

▷ **Exercise 2.** Draw a line segment AB and choose an arbitrary point C on your segment. Perform R & C Construction 2 with straightedge and compass.

Ruler and Compass Construction 3: Drop a Perpendicular. Given a line and a point not on the line, construct a perpendicular line from the point to the line.

(1) Let AB be the given line, and C the point that does not lie on AB.

▶ **Exercise 3.** Perform R & C Construction 3 with straightedge and compass. Give step-by-step instructions for the construction.

The next two constructions involve arbitrary angles instead of right angles.

Ruler and Compass Construction 4: Bisect an Angle. Given an angle, construct a line which bisects that angle.

▷ **Exercise 4.** Perform R & C Construction 4 with straightedge and compass. Give instructions for the construction.

Ruler and Compass Construction 5: Copy an Angle. Given an angle and a line segment, construct an angle on the given line segment congruent to the given angle.

(1) Let $\angle ABC$ be the given angle and let DE be the line segment.
(2) First, draw a circle with center D and radius equal to BC, and let F be the point where this circle intersects DE.
(3) Draw a circle with center D and radius AB.
(4) Draw a circle with center F and radius AC.
(5) These latter two circles intersect at a point G. (Actually, they will intersect at two points; either one can be chosen for G.)
(6) The angle $\angle GDF$ is congruent to $\angle ABC$.

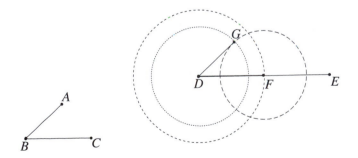

In R & C Construction 5, we are essentially copying the triangle $\triangle ABC$, not only the angle, but since congruent triangles have equal angles, this ensures that the angles will be equal. Recall that parallel lines form equal alternate interior angles: If AB is parallel to CD and both of these lines are cut by a transversal line EF, intersecting AB at G and CD at H, then $\angle AGH = \angle DHG$ (see the figure at the top of the next page):

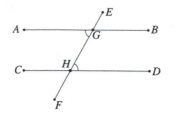

▶ **Exercise 5.** Given a line *AB* and a point *C* not on the line, use R & C Construction 5 to construct a line parallel to *AB* and passing through the point *C*.

•*C*

A •————————————• *B*

Constructible Lengths

One of the purposes of this unit is to explore what can be done with these tools. We will first discuss what lengths are constructible with straightedge and compass. If someone gave you an inch-long segment (since the ruler has no markings, you have no way to measure an inch), you can extend a line and, using the given line segment and your compass, mark off a line segment of any integral length.

▷ **Exercise 6.** The line segment below is 1 inch long. Using this to calibrate your compass, construct a line segment 3 inches long.

●————————●

R & C Construction 1 allows us to cut any length in half. The next ruler and compass construction involves cutting a line up into other fractions.

Ruler and Compass Construction 6: Partition a line segment. Given a line segment, construct points that divide the line segment into *n* equal pieces.

(1) Let *AB* be the given line segment.
(2) Draw another line *AC* that does not coincide with line *AB*.
(3) Set your compass on an arbitrary radius and draw a circle centered at *A*.
(4) The intersection of this circle and the line *AC* is called point P_1.
(5) Using the same radius, draw a circle with center P_1, and let P_2 be the intersection of this circle with *AC* as shown.
(6) Again, using the same radius, draw a circle with center P_2, and let P_3 be the intersection of this circle with *AC* as shown.

(7) Continue thus to find points $P_1, P_2, \ldots, P_n$ equally spaced along AC.

(8) Draw the line segment BP_n.

(9) Using Exercise 5, construct a line parallel to BP_n through each of the points $P_1, P_2, \ldots$. The line through P_i parallel to BP_n will intersect the line AB at a point we will label Q_i.

(10) The points $Q_1, Q_2, \ldots, Q_{n-1}$ divide AB into n equal pieces.

For example, let us cut the one-inch line segment AB below into one piece $\frac{1}{4}$ inch long and another $\frac{3}{4}$ inch long. We draw an arbitrary line AC, and using the compass we mark off four equally spaced points starting at A and label these P_1, P_2, P_3, and P_4. We next draw a line connecting P_4 and B. Lastly, we draw a line through P_1 parallel to BP_4, intersecting AB at a point we label Q_1. Then $AQ_1 = \frac{1}{4}$ and $Q_1 B = \frac{3}{4}$.

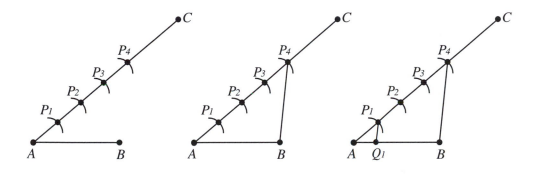

▷ **Exercise 7.** Use R & C Construction 6 to divide a line segment into 3 equal pieces.

Using R & C Construction 1, you can cut a one-inch line segment into two equal segments, creating a $\frac{1}{2}$ inch segment. With R & C Construction 6, you can divide the one-inch segment into n equal pieces, and then use these to find segments of any fractional length. For example, if you divided an inch-long segment into 10 equal pieces, you have markings for segments of length $\frac{1}{10}, \frac{2}{10}, \frac{3}{10}, \frac{4}{10}, \ldots$. Combining the construction of Exercise 6 and R & C Construction 6 and given an inch-long line segment to start from, you can construct line segments of any rational length (recall that the rational numbers are those that can be written as fractions, such as $0 = \frac{0}{1}, \frac{1}{2}, 47 = \frac{47}{1}, \frac{363}{182}$).

▷ **Exercise 8.** The line segment below is 1 inch long. Using this to calibrate your compass, use your straightedge and compass to construct a segment $\frac{4}{3}$ inch long.

What other lengths can be constructed? Assuming that one has an inch-long segment *AB* to work with, one can erect an inch-long perpendicular *AC* at *A*. By the Pythagorean theorem, the hypotenuse of the triangle formed is $BC = \sqrt{2}$, as shown below:

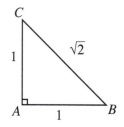

Next, construct another right triangle with one side of length $\sqrt{2}$ as just constructed and the other side of length 1 and the hypotenuse of this triangle will be length *c*, where:

$$
\begin{aligned}
c^2 &= 1^2 + (\sqrt{2})^2 \\
&= 1 + 2 \\
&= 3, \\
c &= \sqrt{3}.
\end{aligned}
$$

This construction can be continued indefinitely, forming a *Pythagorean spiral*. Note that all of the outside edges are 1 unit long and form right angles with the lines radiating from the center.

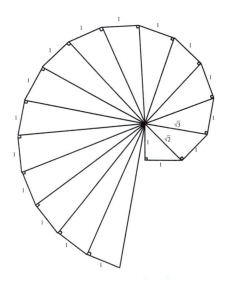

▶ **Exercise 9.** Find the lengths of each of the lines radiating from the center in the Pythagorean spiral.

Thus, with ruler and compass we can construct line segments of all possible lengths that can be expressed as the square root of a positive integer. We can also, by putting segments end on end, construct lengths of $1 + \sqrt{2}$, etc.

▷ **Exercise 10.** Use the inch-long segment below to calibrate your compass, and construct with straightedge and compass a line segment of length $1 + \sqrt{2}$.

If a right triangle had legs of length 1 and $1 + \sqrt{2}$, then the hypotenuse would have length

$$
\begin{aligned}
c^2 &= 1^2 + (1 + \sqrt{2})^2 \\
&= 1 + [1 + 2\sqrt{2} + (\sqrt{2})^2] \\
&= 1 + [1 + 2\sqrt{2} + 2] \\
&= 4 + 2\sqrt{2}, \\
c &= \sqrt{4 + 2\sqrt{2}}.
\end{aligned}
$$

So with straightedge and compass we can construct line segments whose lengths are any combination of integers, fractions, and square roots of positive integers. A famous problem of classical geometry was posed when, after a particularly nasty plague, an oracle instructed the Athenians to double the size of the cubical altar to Apollo at Delos. The original altar was 1 unit (I've forgotten exactly what kind of units the Greeks used) on each side, and they then built a new one in which each side had a length of 2 units. But the new altar was thus $2^3 = 8$ cubic units in volume, or 8 times as big as the old one, and of course the plague got worse. In order to precisely double the volume of the old altar, they would need to construct a cube with sides of length $\sqrt[3]{2}$. This cannot be done with straightedge and compass alone, as was proven in 1837 by Pierre Laurent Wantzel. Only lengths involving fractions and square roots are ruler and compass constructible.

Constructible Angles

The next question to ask is, which angles are constructible? Since by R & C Construction 1 (or 2 or 3) we can construct a 90° angle, we can bisect it by R & C Construction 4 to get an angle of 45°, and then bisect that to get 22.5°, 11.25°, 5.625°, ...

If all three sides of a triangle are equal (the triangle is then called equilateral), then the three angles must be equal. Since the sum of the angles must be 180°, each angle must be 60°.

▶ **Exercise 11.** Construct an equilateral triangle with straightedge and compass alone. Check your construction with the protractor to make sure that all three angles are equal.

Since we have now constructed a 60° angle, we can bisect it to get 30°, 15°, 7.5°, ... We can also use Construction 5 to copy angles side by side to get angles like 120° = 90° + 30°. However, we cannot construct an angle of 20°, for example, though we can approximate it quite closely, using, for example, $\frac{90°}{4} - \frac{90°}{32} = 22.5° - 2.8125° = 19.6875°$. Another famous problem for classical euclidean geometry, also solved by Wantzel, shows that it is impossible to trisect an arbitrary angle, i.e., one cannot, using unmarked ruler and compass, divide an arbitrary angle into three equal sectors. Of course, one can trisect some special angles, such as 90°, since we have already shown that we can construct a 30° angle.

The equilateral triangle has the interesting property that the angle bisector of any one of the angles is also the perpendicular bisector of the opposite side:

▷ **Exercise 12.** Bisect an equilateral triangle as above and use the Pythagorean theorem to find the lengths of all the edges (including the bisector) assuming that each side of the original triangle is 1 inch long. Find the measures of all the angles formed.

Constructible Polygons

Next we turn our attention to which polygons we can construct. You have already constructed an equilateral triangle.

▶ **Exercise 13.** With straightedge and compass, construct a square.

▷ **Exercise 14.** Construct a regular hexagon with straightedge and compass.

Once one has a regular polygon, it is (relatively) easy to construct another regular polygon with twice the number of sides by bisecting the interior angle of each side. More easily, if the first polygon was constructed within a circle with its vertices equally spaced around the circle, then after bisecting one of the interior angles and locating the point where the angle bisector intersects the circle, one can set the compass to the length between one of the neighboring vertices and the point where the angle bisector cut the circle to give the length of the side of the new polygon with twice as many sides. One can then use this compass setting to find the remaining vertices for the new polygon.

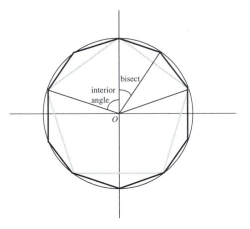

▶ **Exercise 15.** Construct a regular octagon with straightedge and compass.

We have constructed regular polygons of three, four, six, and eight sides in the exercises above. By bisection, we can construct polygons with 12, 16, 24, 32, etc. sides. The pentagon can be constructed with straightedge and compass, as will be discussed in the next section. Many regular polygons are, of course, possible, but some are not. For example, one cannot construct a regular heptagon (7 sides) with straightedge and compass alone. The question of precisely which polygons are constructible was solved in 1796 by Carl Friedrich Gauss (at the age of 19).

Gauss's Theorem: A regular n-sided polygon can be constructed with straightedge and compass alone if all the odd prime factors of n are distinct Fermat primes; i.e., if all the odd prime factors are different and if each is of the form $F_k = 2^{2^k} + 1$.

The known Fermat primes are:

$$F_0 = 2^{2^0} + 1 = 2^1 + 1 = 3$$
$$F_1 = 2^{2^1} + 1 = 2^2 + 1 = 4 + 1 = 5$$

$$F_2 = 2^{2^2} + 1 = 2^4 + 1 = 16 + 1 = 17$$
$$F_3 = 2^{2^3} + 1 = 2^8 + 1 = 257$$
$$F_4 = 2^{2^4} + 1 = 2^{16} + 1 = 65537.$$

Pierre Fermat thought that the formula for F_k always gave a prime number, but the next in the series, $F_5 = 2^{2^5} + 1 = 4294967297 = 641 \cdot 6700417$, is not prime. Gauss gave instructions for the construction of the regular 17-gon (and had it carved on his gravestone). The regular 257-gon was constructed by Richelot and Schwendenwein in 1832, while J. Hermes spent 10 years on the regular 65537-gon and left the manuscript in a large box at the University of Göttingen, where it remains on display.

By Gauss's theorem, we can construct regular n-sided polygons if the factors of n are the first five Fermat primes listed above and twos (two is, of course, the only even prime). Thus, we can construct polygons with 3, 4 = $2 \cdot 2$, 5, 6 = $3 \cdot 2$, 8 = $2 \cdot 2 \cdot 2$ sides, etc. The nonagon (9 sides) cannot be constructed with straightedge and compass (although 9 = $3 \cdot 3$ and 3 is a Fermat prime) since the factors are not distinct.

▷ **Exercise 16.** List all the regular polygons with up to 100 sides that can be constructed with straightedge and compass.

SOFTWARE

1. *Geometer's Sketchpad* is a dynamic interactive program for performing ruler and compass constructions. It allows the user to write macros for performing more complicated constructions, such as inversion in the circle. It is available from Key Curriculum Press.
2. *Cabri* has similar capabilities. It is available from Texas Instruments.

SUGGESTED READINGS

Benjamin Bold, *Famous Problems of Geometry and How to Solve Them*, Dover, New York, 1969.

Richard Courant and Herbert Robbins, *What Is Mathematics?*, Oxford University Press, London, 1941.

H.S.M. Coxeter, *Introduction to Geometry*, Wiley & Sons, New York, 1969.

Euclid, *The Elements*, translated by Sir Thomas Heath, Dover, New York, 1956.

Edward Wallace and Stephen West, *Roads to Geometry*, Prentice Hall, Upper Saddle River, 1992.

3. Constructions

◆ 3.2. THE PENTAGON AND THE GOLDEN RATIO

SUPPLIES
  compass
  straightedge

The regular pentagon can be constructed with straightedge and compass, but before we proceed to instructions for this, it is best to meditate for a bit on its structure. Using the formula $\frac{(n-2)180°}{n}$ with $n = 5$, we see that each vertex angle of a regular pentagon measures 108°.

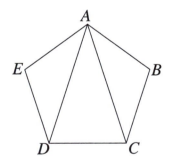

▷ **Exercise 1.** The pentagon above is divided into three isosceles triangles. Use this fact to figure out and label the measure of each angle formed in the pentagon above.

The isosceles triangle $\triangle ACD$ contained in the pentagon above is called a *golden triangle*: The base angles are twice the vertex angle of the triangle. The sides of a golden triangle bear a certain ratio, called the *golden ratio*, or ϕ. To see this, take a copy of $\triangle ACD$ by itself and bisect angle $\angle ADC$, with the bisector intersecting side AC at point F:

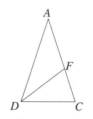

▷ **Exercise 2.** Use the results of Exercise 1 and the facts that *DF* bisects ∡ADC and that the angles in any triangle add up to 180° to figure out the measure of each of the angles in the diagram above.

By the results of Exercise 2, we see that triangles △*CDF* and △*ADF* are also isosceles, so *AF = DF = DC*. You should also have found that triangles △*ACD* and △*DCF* are similar (corresponding angles are equal even though corresponding sides may not be). Therefore, the sides are proportional:

$$\frac{AC}{DC} = \frac{DF}{FC}.$$

Since we can substitute *AF* for both *DC* and *DF* in this equation, we have:

$$\frac{AC}{AF} = \frac{AF}{FC}.$$

Side *AC* is divided at point *F*, *so the whole is to the larger piece as the larger piece is to the smaller.* This phrase describes what is called the *golden ratio*, denoted by ϕ. The golden ratio is also sometimes called the *divine proportion*, or, in the case where a line segment is divided into two pieces by the golden ratio, it is said to be divided in *mean and extreme ratio*.

Definition: Two lengths x and y with $x > y$ are related by the golden ratio if the whole is to the larger piece as the larger piece is to the smaller, i.e., $\frac{x + y}{x} = \frac{x}{y}$.

Thus, in the golden triangle △*ACD* above:

$$\frac{AC}{AF} = \frac{AF}{FC} = \phi.$$

If we assume that the base of the triangle *CD* has length 1, then *AF* is also of length 1, so $AC = \phi$ and $FC = \frac{1}{\phi}$. It follows using the quadratic formula that

$$AC = AF + FC,$$

$$\phi = 1 + \frac{1}{\phi},$$

$$\phi^2 = \phi + 1,$$

$$\phi^2 - \phi - 1 = 0,$$

$$\phi = \frac{1 \pm \sqrt{1^2 - 4(1)(-1)}}{2},$$

$$\phi = \frac{1 + \sqrt{5}}{2}.$$

Thus an alternate definition is the following:

Definition: Two lengths x and y with $x > y$ are related by the golden ratio if $\frac{x}{y} = \phi = \frac{1 + \sqrt{5}}{2}$.

The golden ratio ϕ has a number of interesting arithmetic properties, all derived from the basic formula $\phi^2 = \phi + 1$. For example,

$$\begin{aligned}
\phi^3 &= \phi \cdot \phi^2 \\
&= \phi(\phi + 1) \\
&= \phi^2 + \phi \\
&= (\phi + 1) + \phi \\
&= 2\phi + 1.
\end{aligned}$$

▶ **Exercise 3.** (a) Find an expression for ϕ^4 that involves only ϕ (no higher powers) and constants.
(b) Find an expression for ϕ^5 that involves only ϕ and constants.
(c) Find an expression for ϕ^6 that involves only ϕ and constants.
(d) Find an expression for ϕ^7 that involves only ϕ and constants.
(e) Can you find a pattern to the coefficients of the earlier parts of this exercise? Explain the pattern.

Similar calculations can be made for the quantity $\frac{1}{\phi}$. Since $\phi^2 = \phi + 1$, dividing by ϕ we get $\phi = 1 + \frac{1}{\phi}$, or $\frac{1}{\phi} = \phi - 1$. Therefore,

$$\begin{aligned}
\frac{1}{\phi^2} &= \frac{1}{\phi} \cdot \frac{1}{\phi} \\
&= (\phi - 1)(\phi - 1) \\
&= \phi^2 - 2\phi + 1 \\
&= (\phi + 1) - 2\phi + 1 \\
&= 2 - \phi.
\end{aligned}$$

▶ **Exercise 4.** (a) Find an expression for $\frac{1}{\phi^3}$ that involves only ϕ (no higher powers) and constants.

(b) Find an expression for $\frac{1}{\phi^4}$ that involves only ϕ and constants.
(c) Find an expression for $\frac{1}{\phi^5}$ that involves only ϕ and constants.
(d) Find an expression for $\frac{1}{\phi^6}$ that involves only ϕ and constants.
(e) Find an expression for $\frac{1}{\phi^7}$ that involves only ϕ and constants.
(f) Find a pattern to the coefficients of the earlier parts of this exercise. Explain the pattern.

To return to the golden triangle ADC discussed above, we have shown that $\frac{AC}{AF} = \frac{AF}{FC} = \phi$. Since $\triangle ADF$ and $\triangle DFC$ are both isosceles, $DC = DF = AF$, so $\frac{AC}{DC} = \phi$; i.e., the ratio of the side to the base of this golden triangle is ϕ. Similarly, $\phi = \frac{AF}{FC} = \frac{DC}{FC}$, so the golden triangle DFC has the same proportion of side to base.

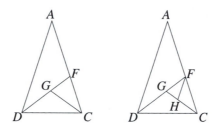

▷ **Exercise 5.** (a) For $\triangle CGF$ above, which two sides have ratio ϕ?
(b) For $\triangle FGH$ above, which two sides have ratio ϕ?

The key to constructing a regular pentagon is building a golden triangle, since any other central triangle will give either a short squat irregular pentagon or a thin irregular pentagon. Note that both of these pentagons have five equal sides but do not have five equal angles.

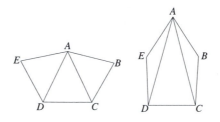

A golden triangle can be constructed if we can find two line segments whose ratio is ϕ, so that the whole is to the larger piece as the larger piece is to the smaller.

Ruler and Compass Construction 7: Divide a Line Segment by the Golden Ratio. Given a line segment, there is a point dividing it into two parts, so that the whole is to the larger piece as the larger piece is to the smaller.

(1) Let AB be a line segment. We wish to extend AB to a point C so that $\frac{AC}{AB} = \frac{AB}{BC} = \phi$.

(2) Construct a square $ABDE$ with AB as one side.

(3) Construct the midpoint F of AB.

(4) Draw a circle centered at F with radius FD.

(5) Extend the line AB to intersect this circle at C.

(6) The proportion of the line segments AC and AB is ϕ.

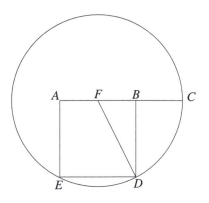

▶ **Exercise 6.** Use the Pythagorean theorem to show that if $AB = 1$ in the square $\square ABDE$, then $AC = \phi = \frac{1 + \sqrt{5}}{2}$. [Hint: First find FB, then FD, and use these to find AC.]

▶ **Exercise 7.** Use R & C Construction 7 to construct a golden triangle with base 1 unit long and sides each of length ϕ with straightedge and compass. [Hint: Perform R & C Construction 7 as above, then use your compass to construct an isosceles triangle with base AB and two longer sides both with length equal to AC.]

▷ **Exercise 8.** Construct a regular pentagon whose sides are each 1 unit long with straightedge and compass. [Hint: First construct a golden triangle as pictured in Exercise 1.]

If we build a rectangle using AC and AE from R & C Construction 7 for two of the sides, we have a figure called a *golden rectangle*. Since $AB = AE$, the proportion of the sides AC and AE is ϕ: i.e., $\frac{AC}{AE} = \phi$. If we assume that $AE = 1$, then $AC = \phi = \frac{1+\sqrt{5}}{2}$. Next consider the rectangle $BCGD$ that remains after we remove the 1×1 square $ABDE$. The longer side satisfies

$CG = AE = 1$, while for the shorter side $BC = \phi - 1 = \frac{1 + \sqrt{5}}{2} - 1 = \frac{1 + \sqrt{5} - 2}{2} = \frac{\sqrt{5} - 1}{2}$:

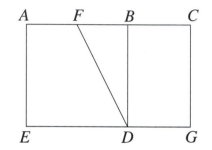

We can compute the ratio of the sides of rectangle $BCGD$ two different ways: using the numerical values, or algebraically. We shall do both so that you can compare them:

$$\begin{aligned}
\frac{CG}{BC} &= \frac{1}{(\sqrt{5} - 1)/2} \\
&= \frac{2}{\sqrt{5} - 1} \\
&= \left(\frac{2}{\sqrt{5} - 1}\right)\left(\frac{\sqrt{5} + 1}{\sqrt{5} + 1}\right) \\
&= \frac{2(\sqrt{5} + 1)}{5 - 1} \\
&= \frac{\sqrt{5} + 1}{2} = \phi.
\end{aligned}$$

We get the same result more quickly algebraically if we use the formula $\phi^2 = \phi + 1$ derived earlier. Dividing by ϕ, we get $\phi = 1 + \frac{1}{\phi}$. Thus,

$$\frac{CG}{BC} = \frac{1}{\phi - 1} = \frac{1}{\frac{1}{\phi}} = \phi.$$

▷ **Exercise 9.** (a) Consider the rectangle *HIGD* formed inside the figure above by building a square *BCIH* with side $BC = \phi - 1$. Recall that $CG = 1$. Find the ratio of side *HI* to side *IG*.
(b) Build another square *IGKJ* with side *IG* and consider the rectangle *JKDH*. Find the ratio of side *DH* to side *HJ*.

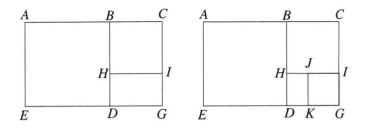

Thus, the golden rectangle has the property that removing a square always leaves a smaller rectangle that is also a golden rectangle. The process can be continued indefinitely. Similarly, as we saw in Exercise 5, removing an isosceles triangle from a golden triangle leaves a smaller golden triangle.

The golden ratio comes up everywhere in the pentagon. Consider the 5-pointed star, pentacle, or pentagram (often endowed with mystical significance).

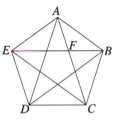

The triangles $\triangle CEF$ and $\triangle ABF$ are similar. Since $\angle AFB = \angle CFE$ and since $\angle EAB = 108°$ and $\angle DAC = \angle ECA = 36°$, it is easy to see that $\angle BAF = \angle ECF = 36°$. Therefore, the corresponding sides of these triangles must be in proportion, so $\frac{EC}{AB} = \frac{EF}{FB}$. The triangles $\triangle ECA$, $\triangle EBD$, and $\triangle DAC$ are golden triangles on equal bases, since $ABCDE$ is a regular pentagon, so $EC = EB$ and $AB = DC$. Using angle measures found in Exercise 1, one can show that $EFCD$ is a rhombus (since the opposite angles are equal), so that $DC = EF$. We can then substitute quantities:

$$\frac{EF}{FB} = \frac{EC}{AB} = \frac{EB}{AB} = \frac{EB}{DC} = \frac{EB}{EF}$$

Thus $\frac{EB}{EF} = \frac{EF}{FB}$, i.e., EB is cut at F by the golden ratio. In fact, throughout the figure above, each length is related to the next smaller length appearing in the pentagram by the golden ratio.

The golden ratio has been given enormous significance by many writers, perhaps due to a confusion with the ancient concept of the golden mean, which is used to indicate neither too much nor too little of something.

Confusing the golden mean with the golden ratio has led some authors to assign this specific numerical value to "neither too much nor too little." However, the term *divine proportion* dates to Fra Luca Pacioli's book *Divina Proportione* of 1509, which contains some illustrations credited to Leonardo da Vinci, and the term *golden ratio* does not appear until the nineteenth century. It has been claimed frequently that the pyramids and the Parthenon were designed to have certain parts with proportion to the golden mean, but such claims do not stand up to close scrutiny. Even when it is found that certain lengths are approximately proportional, the conclusion that the builders were aware of the golden ratio is suspect. Markowsky quotes Martin Gardner as saying, "If you set about measuring a complicated structure like the Pyramid, you will quickly have on hand a great abundance of lengths to play with. If you have sufficient patience to juggle them about in various ways, you are certain to come out with many figures which coincide with important historical dates or figures in the sciences."

For example, below is a photograph of the Parthenon surrounded by a golden rectangle, but you will notice that the lower edge of the rectangle runs, not along the floor of the building, nor along the base of the steps up to the building, but along the third step. If one is allowed to make adjustments such as this, it is fairly easy to place a golden rectangle frame around the building.

▶ **Exercise 10.** It has been claimed that the golden rectangle is the most visually pleasing of all rectangles. Which of the rectangles below do you find most appealing?

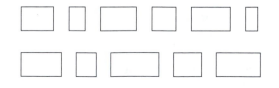

▷ **Exercise 11.** Show the rectangles of Exercise 10 to at least a dozen people and analyze their preferences.

Further claims are made that the golden ratio appears in both art and ideal natural beauty. It is, of course, expected that the golden ratio will appear in any object with five-fold symmetry, which is reasonably common in the natural world. But the extravagant claims in favor of the golden ratio are hard to uphold. For example, one such claim is that the proportions of the ideal human figure are such that the ratio of a person's height to the height to the navel should be the golden ratio.

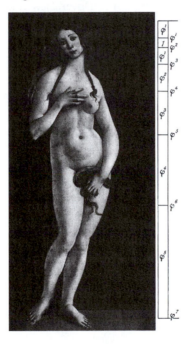

▷ **Exercise 12.** Measure your height and the height to your navel and figure out the ratio.

▷ **Exercise 13.** Measure the height and the height to the navel of at least a dozen people and figure out the ratios.

SOFTWARE

1. *Geometer's Sketchpad* is a dynamic interactive program for performing ruler and compass constructions. It allows the user to write macros for performing more complicated constructions, such as inversion in the circle. It is available from Key Curriculum Press.
2. *Cabri* has similar capabilities. It is available from Texas Instruments.

SUGGESTED READINGS

Theodore A. Cook, *The Curves of Life*, Dover, New York, 1978.

Richard Courant and Herbert Robbins, *What Is Mathematics?*, Oxford University Press, London, 1941.

H.S.M. Coxeter, *Introduction to Geometry*, Wiley & Sons, New York, 1969.

Euclid, *The Elements*, translated by Sir Thomas Heath, Dover, New York, 1956.

H.E. Huntley, *The Divine Proportion*, Dover, New York, 1970.

George Markowsky, "Misconceptions about the Golden Ratio," *College Mathematics Journal* 23(1), 1992.

Dan Pedoe, *Geometry and the Visual Arts*, Dover, New York, 1983 (a republication of *Geometry and the Liberal Arts,* St. Martin's Press, 1976).

Michael Schneider, *A Beginner's Guide to Constructing the Universe*, HarperCollins, New York, 1994.

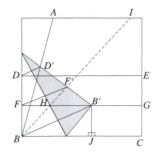

3. Constructions

◆ **3.3. THEORETICAL ORIGAMI**

SUPPLIES
> tracing paper
> origami paper

In Sections 3.1 and 3.2, we discussed which figures can be constructed with only a straightedge and a compass. With the addition of extra tools one can construct other figures. For example, if the ruler has markings, one can trisect any angle. Instead of pursuing the use of such supplementary tools, let us consider a completely different construction idea. This section is adapted from Robert Geretschläger's *Euclidean Constructions and the Geometry of Origami* and Thomas Hull's *A Note on "Impossible" Paper Folding*.

For this approach one needs only fairly thin paper, a pencil (to mark and label points), and nimble fingers. The basic ideas of theoretical origami are simple and completely natural. We try to parallel the development of what we could construct with ruler and compass. Of course, it is difficult to make the folds as precise as they need to be, but it is also difficult in practice to be precise with ruler and compass.

For these first exercises use tracing paper, since it is easier to see through to align the points or lines as needed. Be as precise in your folds as possible.

Origami Postulate 1: Given any two distinct points on a piece of paper, one can fold exactly one line passing through them.

▶ **Exercise 1.** Mark two points at random (but not too close together) on a piece of tracing paper and fold, forming a straight line passing through both points.

Origami Postulate 2: A folded line segment can be extended at either end by an arbitrary amount (assuming the paper is big enough).

Origami Postulate 3: Given any two distinct points on a piece of paper, one can fold the paper forming a single crease so that one point lies exactly on top of the other.

▷ **Exercise 2.** Mark two points at random (but not too close together) on a piece of paper and fold so that one point lies over the other. The crease thus formed is a line. Discuss the geometric relationship between this line and the line connecting the two points.

Origami Postulate 4: Given any two lines on a piece of paper, one can fold the paper forming a single crease so that one line lies exactly on top of the other.

▶ **Exercise 3.** Draw (or fold) two intersecting lines at random on a piece of paper and fold so that one line lies over the other. The crease thus formed is a line. Discuss the geometric relationship between this line and the two original lines.

▷ **Exercise 4.** Draw (or fold) two parallel lines on a piece of paper and fold so that one line lies over the other. The crease thus formed is a line. Discuss the geometric relationship between this line and the two original lines.

Origami Postulate 5: Given two points A and B and a line on a piece of paper with B closer to the line than it is to A, one can fold the paper forming a single crease so that point B lies on the crease while point A lies on the line.

▶ **Exercise 5.** Mark two points *A* and *B* and draw (or fold) a line on a piece of paper. Fold so that point *A* lies on the line while point *B* lies on the crease formed. Do this separately for the cases illustrated below:

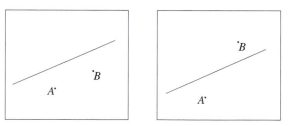

Using the origami postulates, let us try to see which of the ruler and compass constructions can be done.

Construction 1: Perpendicular Bisector. Construct the perpendicular bisector of a line segment.

(1) Let *AB* be the given line segment.
(2) Fold the tracing paper so that point *A* lies on top of point *B* and the line segment *AB* is doubled on itself.
(3) The line *CD* formed by the crease is the perpendicular bisector of *AB*. Let *E* be the point where *AB* intersects *CD*.

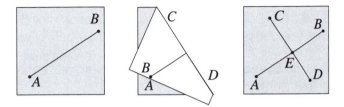

To see that this truly works, note that $AE = BE$, since *AE* lies directly over *BE*, so *E* is the midpoint of *AB*. Also, $\angle AEC = \angle BEC$, since these angles lie directly over each other, and since $\angle AEC + \angle BEC = 180°$, $\angle AEC = \angle BEC = 90°$, so *CD* is perpendicular to *AB*.

▷ **Exercise 6.** Draw a line segment *AB* on a piece of tracing paper and fold the perpendicular bisector.

Construction 2: Erect a Perpendicular. Given a line and a point on the line, construct a perpendicular line through the point to the line.

(1) Let *AB* be the given line, and *C* the point that lies on *AB*.

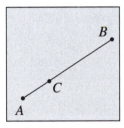

▷ **Exercise 7.** Complete the instructions for Construction 2. Perform the construction on tracing paper, labeling your lines.

Construction 3: Drop a Perpendicular. Given a line and a point not on that line, construct a line perpendicular to the given line through the given point.

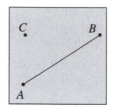

▷ **Exercise 8.** Give careful step-by-step instructions for Construction 3. Perform the construction on tracing paper, labeling your lines.

Construction 4: Bisect an Angle. Given an angle, construct a line that bisects that angle.

▶ **Exercise 9.** Give instructions for Construction 4. Perform the construction on tracing paper, labeling your lines.

Construction 5: Copy an Angle. Given an angle and a line segment, construct an angle on the given line segment congruent to the given angle.

▷ **Exercise 10.** Give instructions for Construction 5. Perform the construction on tracing paper, labeling your lines. Let ⦡*BAC* be the given angle and *DE* the given line segment. The object is to fold *DF* so that ⦡*FDE* = ⦡*BAC*. Note that it does not matter whether *FD* is as pictured or whether it lies on the other side of *DE*.

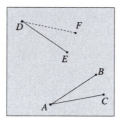

The first four constructions are quite easy in origami. Construction 5 may take a little experimenting to figure out but is not too difficult. R & C Construction 6 involves only laying out equally spaced points and repeated uses of Construction 5 to create parallel lines.

▷ **Exercise 11.** Fold a line on a piece of tracing paper and mark two points *A* and *B* on the line (about an inch apart or so). Using Origami Postulate 5, fold and mark three other points on the line, labeled *C*, *D*, and *E*, so that *AB* = *BC* = *CD* = *DE*.

By the preceding comments and Exercise 11, we have the origami analog of R & C Construction 6.

Construction 6: Partition a Line Segment. Given a line segment, construct points that divide the line segment into *n* equal pieces.

We have now shown that anything that can be constructed using an unmarked ruler and compass, can also (theoretically) be constructed using origami folds. We therefore should be able to use origami to construct any regular polygon that is constructible with ruler and compass.

▷ **Exercise 12.** Tear an irregular piece of tracing paper and fold a square.

Once you can do Exercise 12, then you can make squares anywhere, so use standard origami paper for the rest of the exercises.

▶ **Exercise 13.** Fold an equilateral triangle using square origami paper. Hint: Consider the diagram below:

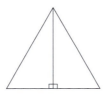

▷ **Exercise 14.** Fold a regular hexagon using square origami paper.

▷ **Exercise 15.** Fold a regular octagon using square origami paper. Hint: Consider the diagram below. What interior angles are formed?

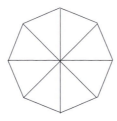

The key to folding a pentagon is, of course, finding two lengths whose ratio is the golden ratio or ϕ. The following construction is from Sundara Row's *Geometric Exercises in Paper Folding*.

Construction 7: Divide a Line Segment by the Golden Ratio.
Given a line segment, there is a point dividing it into two parts so that
the whole is to the larger piece as the larger piece is to the smaller:

(1) Let *ABCD* be a square piece of origami paper.
(2) Fold the paper in half, laying *AB* on top of *DC* and marking point *E*, the
    midpoint of *AD*.
(3) Fold through *E* and *B*.
(4) Fold so that line *AE* lies on top of *EB* and mark point *F* where this fold
    intersects *AB* and point *G* on *EB* so that *EG* = *EA*.
(5) Fold line *BA* on top of line *BE* and mark point *H* on *AB* so that
    BH = *BG*.
(6) Then *H* divides *AB* by the golden ratio.

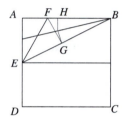

▶ **Exercise 16.** Assume that *AB* = 1.
  (a) Explain why $EA = EG = \frac{1}{2}$.
  (b) Show that $BE = \frac{\sqrt{5}}{2}$.
  (c) Use (a) and (b) to find the length of BH.
  (d) Show that $\frac{AB}{BH} = \phi = \frac{1 + \sqrt{5}}{2}$.

To construct a pentagon, first mark the point *H* on side *AB* as in Construction 7. Let *I* be the midpoint of *AH*, and mark point *J* on *AB* so that *AI* =
BJ. Fold so that point *J* lies on line *AD* while point *I* lies on the crease and
mark point *M* so that *IJ* = *IM*. Fold so that point *I* lies on line *BC* while
point *J* lies on the crease and mark point *K* so that *JK* = *JI*. Fold so that *JK* is
doubled on itself, so that *J* lies over *K* and *M* lies on the crease. Mark point *L*
where *I* lands. Note that *L* will not lie on edge *CD*:

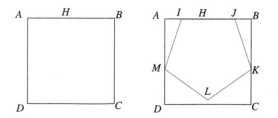

▷ **Exercise 17.** Construct a pentagon using the directions above.

Thus, we have successfully done with origami techniques everything that we could do with ruler and compass. Some things, such as folding perpendiculars and bisectors, are done more easily than with ruler and compass, and some things, such as the pentagon above, seem harder. In addition, there are some nice things that are very easy and natural with origami. One such result is called *Haga's theorem*, by Kazuo Haga. Take a piece of square origami paper with corners $ABCD$, and find the midpoint M of side AB. Fold corner C up to point M, forming the fold EG. In doing this, the point H on the bottom edge CD meets the point F on the left edge AD as shown. Three triangles are formed, $\triangle EBM$, $\triangle MAF$, and $\triangle GDH$. The following exercises will show that each of these triangles has the proportions of a 3-4-5 right triangle.

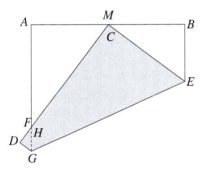

▷ **Exercise 18.** Fold a piece of origami paper as described in the previous paragraph.

▷ **Exercise 19.** Show that the triangles $\triangle EBM$, $\triangle MAF$, and $\triangle GDH$ are all similar.

We can use the exercise above and the ratios of the sides of these similar triangles to find all the other lengths involved. Assume that $ABCD$ is 1 unit on each side. Then $BM = AM = \frac{1}{2}$.

▶ **Exercise 20.** Let $BE = x$. Note that $BE + CE = 1$. Use the Pythagorean theorem to find values for $x = BE$ and $CE = 1 - x$.

▷ **Exercise 21.** Note that $AM = \frac{1}{2}$. Use Exercises 19 and 20 to find AF and FM.

▷ **Exercise 22.** Note that $DH + HC = 1$, $HG = FG$, and $AF + FG + GD = 1$. Use similar triangles to find DH, DG, and HG.

Yet another application of the origami constructions allows one to fold a rectangular piece of paper of arbitrary dimensions into thirds. This was

Problem of the Week #791 at Macalester College, posted by Stan Wagon in 1995, though the problem goes back to 1872, when Augustus De Morgan wrote: "It is easy to get half the paper on which you write for margin; but very troublesome to get a third. Show us how, easily and certainly, to fold the paper into three, and you will be a real benefactor to society."

▷ **Exercise 23.** The solution below is due to David Castro, then a student at Macalester College. Your job is to justify each step with one of the origami postulates. Let the rectangular piece of paper be labeled *ABCD*. Fold between corners *A* and *C*. This is allowed by Origami Postulate 1.
  (a) Unfold, and then refold the paper in half horizontally, and unfold. Let *M* be the midpoint of *BC* determined by this fold.
  (b) Unfold and fold through *M* and corner *D*.
  (c) Unfold and label the intersection of the first fold and the fold of (b) as point *E*. Fold through *E* so that point *C* lies on line *BM*. This folds the paper in thirds.

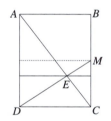

▷ **Exercise 24.** Fold an 8.5-inch by 11-inch piece of paper in thirds as shown above.

▷ **Exercise 25.** Show why Castro's solution works. Draw a line *JK* through *E* and perpendicular to *AB* and *CD*, Note that $AD = 2MC$.
  (a) Show that $\triangle ADE$ is similar to $\triangle CME$.
  (b) Show that $AE = 2CE$.
  (c) Show that $\triangle AJE$ is similar to $\triangle CKE$.
  (d) Show that $JE = 2KE$, and thus $KE = \frac{1}{3}KJ$.

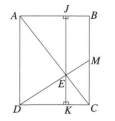

The Difference Between Ruler and Compass and Origami

There is one operation that is allowable in origami and that has no counterpart for ruler and compass.

Origami Postulate 6: Given two points A and B and two lines ℓ_1 and ℓ_2 on a piece of paper, one can fold the paper forming a single crease so that point A lies on top of line ℓ_1 while point B lies on top of line ℓ_2.

▷ **Exercise 26.** Lay out two points and two lines on a piece of tracing paper as shown in each of the four diagrams below. For each, find one fold so that point A lies on top of the line ℓ_1 while point B lies on top of line ℓ_2.

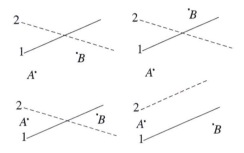

Origami Postulate 6 distinguishes the ruler and compass universe from the origami universe. In fact, with the use of this postulate we can trisect an arbitrary angle, which is impossible with an unmarked ruler and compass. This construction is due to Hisashi Abe.

Origami Construction 8: Trisecting an Angle. Given an angle, construct a line that trisects that angle.

(1) Let ABC be the given angle. We assume that the line BC forms the bottom edge of a piece of origami paper. See the figure on the following page.
(2) Fold the origami paper so that BC lies on top of the upper edge of the paper, forming crease DE parallel to BC and cutting the paper in half.
(3) Fold so that line BC lies on top of line DE, forming crease FG.
(4) Fold so that point D lies on line AB and point B lies on line FG.
(5) Without unfolding the crease formed in Step 4, extend the line formed by FG to form line HI.
(6) Unfold the crease formed in Step 4, and extend line HI. It will intersect point B.
(7) The angle $\angle ABI = \frac{1}{3}\angle ABC$:

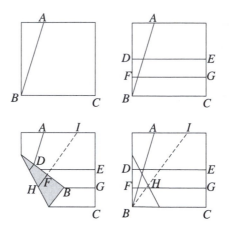

Proof: Let B', D', and F' be the points on top of which B, D, and F land in Step 4. Drop a perpendicular from point B' to line BC, intersecting BC at point J. Since $DF = BF$, we have $DF = FB = D'F' = B'F'$. Also, $B'J = BF$. Since FG is perpendicular to BD, and $F'H$ lands on top of FH, BF' is perpendicular to $B'D'$. Of course, BF' is equal to itself, so $\triangle BD'F'$ is congruent to $\triangle BB'F'$ by SAS. Thus $\angle ABF' = \angle B'BF'$. Since $\angle BF'B'$ and $\angle BJB'$ are both right angles, $BB' = BB'$ and $B'J = B'F'$, the triangles $\triangle BB'F'$ and $\triangle BB'J$ are congruent by Hypotenuse-Side congruence. Thus, $\angle JBB' = \angle F'BB'$, so angle $\angle ABC$ has been trisected.

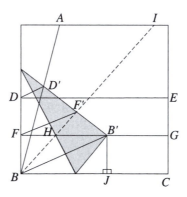

▷ **Exercise 27.** Trisect an angle as above using origami paper.

It is possible to trisect any angle with a *marked* ruler and a compass, as was shown by Archimedes, but not with the classical Euclidean unmarked ruler. There are many other constructions that can be done with origami, such as constructing a regular heptagon. By Gauss's theorem a heptagon cannot be constructed with ruler and compass.

SOFTWARE

1. Casady and Greene have a perfectly splendid CD-ROM called *Origami: The Secret Life of Paper*, created by Cloudrunner, Inc., which includes information on mathematical origami. It is sold by Key Curriculum Press.

SUGGESTED READINGS

David Auckly and John Cleveland, "Totally Real Origami and Impossible Paper Folding," *American Mathematical Monthly*, March 1995.

David Gale, "Egyptian Rope, Japanese Paper, and High School Math," *Math Horizons VI*, September 1998.

Robert Geretschläger, "Euclidean Constructions and the Geometry of Origami," *Mathematics Magazine* 68(5), 1995.

Thomas Hull, "A Note on 'Impossible' Paper Folding," *American Mathematical Monthly*, March 1996.

Humiaki Huzita, "Drawing the Regular Heptagon and the Regular Nonagon by Origami," *Symmetry: Culture and Science* 5, 1994.

Kunihiko Kasahara, *Origami Omnibus*, Japan Publ., Tokyo, 1988.

Kunihiko Kasahara and Toshie Takahama, *Origami for the Connoisseur*, Japan Publ., Tokyo, 1987.

Dénes Nagy, "Symmetro-Graphy: Bibliography: Origami, paper-folding, and related topics in mathematics and science education," *Symmetry: Culture and Science* 5, 1994.

Dénes Nagy, "Symmet-Origami (Symmetry and Origami) in Art, Science, and Technology," *Symmetry: Culture and Science* 5, 1994.

Sundara Row, *Geometric Exercises in Paper Folding*, Open Court Publishing Co., La Salle, 1958.

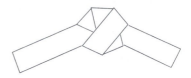

3. Constructions

◆ **3.4. KNOTS AND STARS**

SUPPLIES
> ruler
> colored pencils
> dot paper
> cash register tape

Knotted Polygons

There are ways of constructing regular polygons other than with ruler and compass or origami. One easy way of making a pentagon is by knotting a strip of paper in a simple overhand knot. We show both a drawing of the knot as it would appear if done with a bit of string and a picture of how the knotted strip of paper will look:

If the long sides of the strip of paper are parallel and the knot is snug but not too tight, this will form a regular pentagon. Cash register tape has parallel edges and is handy to use.

▷ **Exercise 1.** Knot two strips of paper to form a regular hexagon as shown below:

▷ **Exercise 2.** Knot a strip of paper to form a regular heptagon. Below is a picture of the knotted paper:

▷ **Exercise 3.** Tie an octagonal knot from a single strip of paper. The following instructions are taken from the *Ashley Book of Knots*, knot number 2590. First tie an overhand knot exactly as shown below on the left:

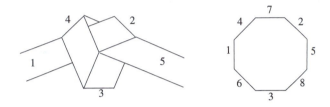

We wish to end up with an octagon as on the right. The folds of the knot you have made form three sides, labelled 2, 3, and 4 and the loose ends form two other sides labeled 1 and 5. Fold the loose end down at position 5 and thread the end through the fold at 3 to position 6. Next fold down at position 6 and go over the loose end at position 1 and through the fold at position 4 to position 7. Fold down at position 7 and go through the folds at positions 2 and 5 to position 8. Fold the loose end down at position 8 and go through the folds at 3 and 6 to end at position 1. After tightening, you will have an octagonal knot with both loose ends at position 1.

Star Polygons

A regular n-sided polygon can be drawn by locating n equally spaced points around a circle and connecting them, but other figures can be drawn using these n regularly spaced points. Below is a circle with 8 such points:

If we connect each point in order, we get a regular octagon. If we connect every other point, we get two squares. Thus, connecting every second point creates two *cycles* of length four.

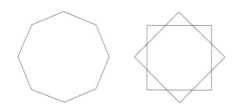

▶ **Exercise 4.** Place 8 dots equally spaced around a circle and figure out what happens if you connect every third dot. What happens if you connect every fourth dot? Every fifth dot? sixth? seventh? State your conclusions clearly in sentence form.

Note that sometimes the resulting figure connects every point in one path. It is then called a *regular star polygon*. Sometimes the figure consists of several separate cycles, such as the one above in which we connected every other point of 8 equally spaced points or in the familiar Star of David. These are *compound star polygons*. Also note that some of these star polygons are identical with others, the only difference being that the points are connected in a different order. The notation we shall use is $\{^n_k\}$ for the star polygon in which we connect every kth point of n equally spaced points around a circle.

▷ **Exercise 5.** Place 9 dots equally spaced around a circle, and draw $\{^9_1\}$, $\{^9_2\}$, $\{^9_3\}$, etc.

▷ **Exercise 6.** Which of the $\{^8_k\}$ star polygons are identical? Which $\{^9_k\}$ star figures are identical?

▷ **Exercise 7.** State a rule for when two star polygons are identical.

▶ **Exercise 8.** Place 12 dots equally spaced around a circle and draw all the different $\{^{12}_k\}$ star polygons. Note which are regular and which are compound. You need not duplicate identical figures.

▷ **Exercise 9.** Place 15 dots equally spaced around a circle and draw all the different $\{^{15}_k\}$ star polygons and note which are regular and compound. You need not duplicate identical figures.

If one were to walk around the polygon $\{^8_1\}$, one would make one trip around the center point. However, traveling around $\{^8_2\}$ requires two revolutions around the center.

▶ **Exercise 10.** For each of the star polygons drawn in Exercises 4, 5, 8, and 9, note how many revolutions around the center are made.

▷ **Exercise 11.** It was noted above that $\{^8_2\}$ consists of two cycles of length 4. Which $\{^9_k\}$ star polygons are compound? How many cycles do these have and how long are the cycles? Which $\{^{12}_k\}$ star polygons are compound? How many cycles do these have and how long are the cycles? Repeat for $\{^{15}_k\}$.

▷ **Exercise 12.** State a rule for when a star polygon is regular. State a rule for when a star polygon is compound, how many cycles it will have, and how long the cycles will be. Express your rules in complete sentences.

▷ **Exercise 13.** State a rule for how many revolutions around the center are made in a star polygon. Express your rule in complete sentences.

▷ **Exercise 14.** How many cycles does the compound star polygon $\{^{24}_9\}$ have? How long are those cycles? Try to do this without drawing the figure, though you may want to draw it to check your answer. If your rules in the previous exercises do not predict the right answers, figure out how to correct them.

▷ **Exercise 15.** List all the regular star polygons of the form $\{^{21}_k\}$.

Consider the star polygon $\{^5_2\}$, also called the pentagram, below:

The figure could also be considered as having 10 sides if one looks only at the outer perimeter. As a star polygon it is formed by only five straight lines, and when one talks about the vertex angles, one is referring to the pointy angles farthest out, not the obtuse angles made.

Notice that the edges of the star polygon form a smaller regular pentagon in the center of the circle. From our previous study of polygons in Section 1.2, we know that the vertex angles of this pentagon each measure 108°. Therefore, the supplementary angles measure 72°. The star polygon can be regarded as a small pentagon with isosceles triangles stuck onto each edge, and the base angles of these isosceles triangles must be 72°. Since the sum of the angles in any triangle must be 180°, the vertex angle of the isosceles triangle, and thus the vertex angle of the star polygon, must measure 180° − (2 × 72°) = 180° − 144° = 36°.

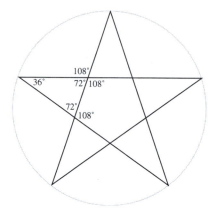

▶ **Exercise 16.** Figure out the vertex angle of the star polygon $\{^8_3\}$. A large and careful drawing will help.

▶ **Exercise 17.** Figure out the vertex angles of the star polygons $\{^9_2\}$ and $\{^9_4\}$.

▷ **Exercise 18.** From the examples and exercises above, fill in the following table. In the last column, write the difference between the vertex angle of $\{^n_k\}$ and $\{_k{}^n_+{}_1\}$.

Vertex Angles

$n\downarrow\ k\rightarrow$	1	2	3	4	Difference
5	$\measuredangle\{^5_1\} =$	$\measuredangle\{^5_2\} =$	•	•	
6	$\measuredangle\{^6_1\} =$	$\measuredangle\{^6_2\} =$	$\measuredangle\{^6_3\} =$	•	
8	$\measuredangle\{^8_1\} =$	$\measuredangle\{^8_2\} =$	$\measuredangle\{^8_3\} =$	$\measuredangle\{^8_4\} =$	
9	$\measuredangle\{^9_1\} =$	$\measuredangle\{^9_2\} =$	$\measuredangle\{^9_3\} =$	$\measuredangle\{^9_4\} =$	

▷ **Exercise 19.** Figure out a formula for the difference between the vertex angle of $\{^n_k\}$ and $\{_k{}^n_+{}_1\}$. [Hint: Most angle formulas for polygons involve either 180° or 360°.]

▷ **Exercise 20.** Figure out a formula for the vertex angle of the star polygon $\{^n_k\}$.

▷ **Exercise 21.** What do the knots of Exercises 1, 2, and 3 have to do with star polygons? Explain the connection carefully.

▶ **Exercise 22.** Can you make a nine-sided knot? Make one or explain why it cannot be done.

▷ **Exercise 23.** Make an eight-sided knot with two strips of tape.

SUGGESTED READINGS

Clifford W. Ashley, *The Ashley Book of Knots*, Doubleday, New York, 1944.

H.S.M. Coxeter, *Regular Polytopes*, Dover, New York, 1973.

H.M. Cundy and A.P. Rollett, *Mathematical Models*, Oxford University Press, New York, 1961.

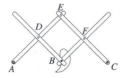

3. Constructions

◆ **3.5. LINKAGES**

SUPPLIES

Erector set or low budget equivalent, such as:
strips of cardboard, hole punch, and paper prong fasteners, or
popsicle sticks and pins

A mechanical linkage is an assembly of straight rods with pins at the joints. In this section we will play with some of the simpler ones, with special attention to their usage in drawing lines and curves. The conventions we use are:

⊗ a circle with an × is used to denote a fixed point: a point that is nailed to the table, but with enough free play that any bars through the point can rotate freely around the fixed point.

○ an open circle is used to denote a pivot that is not fixed to the table and can move freely.

● a black dot is a point whose movement is constrained in some way, usually by tracing a given path or curve.

⊙ a circle with a dot inside denotes a point where a pencil is inserted.

First, we investigate the *variable-based triangle linkage:* Build a linkage as pictured below, with $AB = BC = BD$ and E somewhere between B and D. This point E is pictured below at the midpoint of BD, but it can fall anywhere between those points. The point C must stay on the line AC as shown:

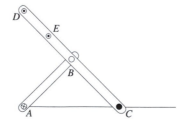

▶ **Exercise 1.** Build the linkage illustrated above and draw the line *AC*. Place a pencil at point *B*. As *C* moves back and forth along the line, describe the set of points that *B* traces out. Explain why this particular shape results.

The set of points traced out by the movement of a point is called a *locus of points*.

▷ **Exercise 2.** Move the pencil to the midpoint *E*. Describe the locus traced by the movement of *E*.

▶ **Exercise 3.** Move the pencil to point *D*. Describe the locus traced by the movement of *D*.

To understand the answer to Exercise 3, consider the following diagram, which is a geometrical picture of the linkage. Since $AB = BC = BD$, *B* can be considered the center of a circle passing through the points *A*, *C*, and *D*, and *CD* is a diameter of that circle.

▷ **Exercise 4.** (a) What is the relationship of $\angle BAC$ and $\angle BCA$?
(b) What is the relationship of $\angle BAC$, $\angle BCA$, and $\angle ABC$?
(c) What is the relationship of $\angle BAD$ and $\angle BDA$?
(d) What is the relationship of $\angle BAD$, $\angle BDA$, and $\angle ABD$?
(e) What is the relationship of $\angle ABD$ and $\angle ABC$?
(f) Use parts (a) − (e) to show that $\angle CAD = 90°$ as the picture implies.

Thus, in Exercise 4 you have shown that $DA \perp AC$. To see how this applies to our linkage, consider two positions of the linkage. As point *C* moves to position C' on the line, the point *B* moves to B' and *D* to D'. The point *A* doesn't go anywhere since it's nailed down.

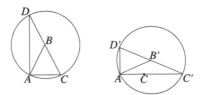

By Exercise 4, we know that $AD \perp AC$, and similarly $AD' \perp AC'$. Since AC and AC' are part of the same line, we conclude that AD and AD' are on the same line. Thus, the locus DD' traced out by the pen at D is a straight line perpendicular to AC.

▶ **Exercise 5.** Describe how an ironing board folds up.

A *pantograph* is a drawing device used to duplicate and enlarge a drawing, common before most illustrations were done on computers. It usually consists of 4 sticks of equal length, with a bolt at point A, a pointy thing at B used to trace over a given figure, a pencil holding device at point C, and movable pins at points D, E, and F, put together so that $BDEF$ is a parallelogram. Thus $AD = BD = EF$ and $DE = BF = FC$.

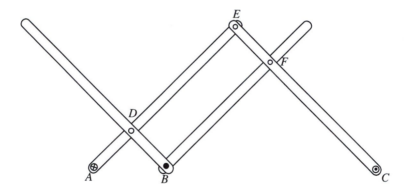

To begin analyzing the action of a pantograph, let us consider first the configuration below, where D is placed at the midpoint of AE and F at the midpoint of EC:

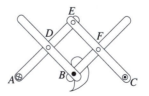

In this case, we have $AD = DE = EF = FC = BD = BF$.

▷ **Exercise 6.** (a) Show that $\triangle ADB$ is similar to $\triangle AEC$.
(b) Use (a) to show that $AC = 2AB$.
(c) Remember that point A is fixed. If point B moves 1 inch in a straight line to point B' and so C is forced to move to C', how far is it between C and C'?

▶ **Exercise 7.** Build or borrow a pantograph. Draw a scribble and trace it with the tracer at point *B*. What does the pencil at point *C* draw?

▷ **Exercise 8.** Switch the positions of the pencil and tracer, so that the pencil is at point *B* and the tracer at point *C*. Trace your scribble. What does the pencil at *B* draw?

The pantograph should give an enlargement or reduction in scale of the figure traced. The configuration of Exercise 7 has a scaling factor of $+2$, or $+\frac{1}{2}$ in the configuration of Exercise 8. Trace a circle with your pantograph, moving in a clockwise direction. If the image drawn by the pantograph is also drawn in the clockwise direction, the scaling factor is positive. A negative scaling factor indicates a reflection combined with an enlargement or reduction. With a negative scaling factor, the pencil will move counterclockwise as the tracer moves clockwise.

Next, reassemble your pantograph as pictured below with *A* fixed and the pencil at *C*, so that $DE = 3AD$ and also $AD = DB = EF$ and $DE = BF = CF$:

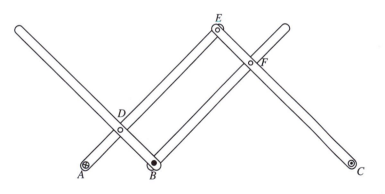

▷ **Exercise 9.** (a) Show that $\triangle ADB$ is similar to $\triangle AEC$.
(b) Find the scaling factor for this configuration of the pantograph.

▷ **Exercise 10.** Switch the positions of the pencil and tracer, so that the pencil is at point *B* and the tracer at point *C*. Trace your scribble. What does the pencil at *B* draw? What is the scaling factor?

▶ **Exercise 11.** Switch the positions of the bolt, pencil, and tracer, so that the point *B* is fixed, the tracer is at point *A*, and the pencil is at point *C*. Trace your scribble. What is the scaling factor?

▷ **Exercise 12.** Switch the positions of the bolt, pencil, and tracer, so that the point *B* is fixed, the tracer is at point *C*, and the pencil is at point *A*. Trace your scribble. What is the scaling factor?

▶ **Exercise 13.** Below is another parallelogram linkage. Point A is fixed, point B is used to trace a figure, and there is a pencil at point C. Using similar triangles, show that this linkage also acts like the pantograph and find the scaling factor:

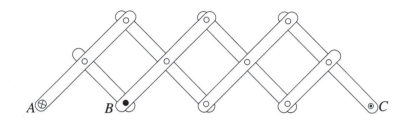

▷ **Exercise 14.** Figure out how to to move the fixed point, tracing point, and pencil around for the parallelogram linkage of Exercise 13 so that it acts as a pantograph with scaling factor:
(a) $+\frac{1}{4}$
(b) $+3$
(c) $+2$
(d) -2
(e) -3
(f) $\frac{4}{3}$
(g) $\frac{3}{4}$
(h) $\frac{2}{3}$
(i) $-\frac{1}{2}$

In Exercises 15, 16, and 17, consider the parallelogram linkages shown. For each diagram, the point A is fixed and the proportions of the sides changes.

▷ **Exercise 15.** Below is another parallelogram linkage. Point A is fixed, point B is used to trace a figure, and there is a pencil at point C. Using similar triangles, show that this linkage also acts like the pantograph and find the scaling factor:

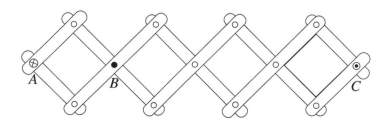

▷ **Exercise 16.** Below is another parallelogram linkage. Point A is fixed, point B is used to trace a figure, and there is a pencil at point C. As drawn, $BE = 3DB$. Using similar triangles, show that this linkage also acts like the pantograph and find the scaling factor:

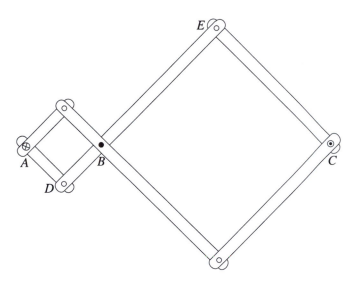

▷ **Exercise 17.** Below is another parallelogram linkage. Point A is fixed, point B is used to trace a figure, and there is a pencil at point C. As drawn, $BE = 2DB$. Using similar triangles, show that this linkage also acts like the pantograph and find the scaling factor:

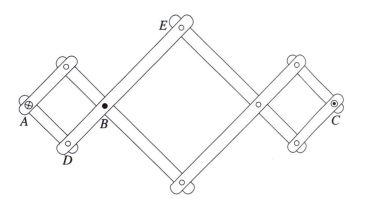

▷ **Exercise 18.** Design three different parallelogram linkages that act as pantographs with a scaling factor of 3.

Below is another linkage, called *Peaucellier's linkage*. While explaining why it works is beyond the scope of this text, see whether you can figure out what it does. Note that there are two fixed points at points X and Y in the line drawing. The linkage must be assembled so that $XY = YD$, $XA = XC$, and $AB = BC = CD = DA$:

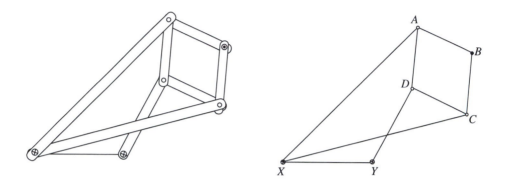

▷ **Exercise 19.** Build a Peaucellier linkage and place a pen at point B. Describe the locus of points traced by the movement of B.

SOFTWARE

1. *Geometer's Sketchpad* is a dynamic interactive program for performing ruler and compass constructions. It can also be used to find the locus of points traced out by a point in a given configuration. It is available from Key Curriculum Press.
2. *Cabri* has similar capabilities. It is available from Texas Instruments.

SUGGESTED READINGS

I.I. Artobolevski, *Mechanisms for the Generation of Plane Curves*, Macmillan, New York, 1964.

Brian Bolt, *Mathematics Meets Technology*, Cambridge University Press, New York, 1991.

H.M. Cundy and A.P. Rollett, *Mathematical Models*, Oxford University Press, New York, 1961.

A.B. Kempe, *How to Draw a Straight Line*, National Council of Teachers of Mathematics, Washington, D.C., 1977 (reprint of the Pentagon edition, which was a reprint of the 1877 edition of Macmillan, London).

Robert C. Yates, *Geometrical Tools*, Educational Publishers, St. Louis, 1949.

4. Tesselations

◆ 4.1. REGULAR AND SEMIREGULAR TILINGS

SUPPLIES
> tracing paper
> cardboard polygons
> scissors

A tesselation or tiling is an arrangement of polygons fitting together to cover the plane without overlap. The simplest example is the tesselation of squares. This is called a *regular tiling or tesselation*, one that consists of repeated copies of a single regular polygon, meeting edge to edge so that every vertex has the same configuration. Four squares must meet at each vertex, and since each square contributes a 90° angle, they join to form a 360° angle, thus lying flat on the plane. The figure below on the right does not count as a regular tiling, since the junction of two of the squares forces a vertex at the midpoint of the edge of the adjoining square:

▶ **Exercise 1.** Which other regular polygons can tile the plane?

A *semiregular or Archimedean tesselation* is one in which more than one type of regular polygon is used and each vertex has the same configuration. There are exactly three regular tilings (see Exercise 1), and eight semiregular ones. An example is the tiling where two regular octagons and a square, all with the same edge length, meet at each vertex. Each octagon has a 135° vertex angle and the square a 90° angle. Since 2 × 135° + 90° = 360°, they fit together to surround a vertex, and the pattern can be extended to form a tiling.

▶ **Exercise 2.** Fill in the following table carefully. You will need the answers later.

Vertex Angles of Regular Polygons

Polygon	Sides	Angle
Triangle	3	
Square	4	
Pentagon	5	
Hexagon	6	
Heptagon	7	
Octagon	8	
Nonagon	9	
Decagon	10	
Dodecagon	12	
Pentakaidecagon	15	
Octakaidecagon	18	
Icosagon	20	
Tetrakaicosagon	24	

So far, we know four tilings: the three regular tilings found in Exercise 1 and the octagon-square shown above. This latter semiregular tiling is denoted by 4.8.8, to indicate the vertex configuration (one square and two octagons). The regular tiling of squares is denoted 4.4.4.4. The rest of this section is devoted to finding all the other semiregular tilings. We begin by deriving the rules as found in 1785 by The Rev. Mr. Jones.

Rule 1: Every regular and semiregular tiling must have the angles of the polygons meeting at a vertex sum to exactly 360°.

▷ **Exercise 3.** What is the largest number of regular polygons that can fit around a vertex? [Hint: If you're going to have lots of polygons at a vertex, then the angles had better be small. What is the smallest angle measure among the regular polygons?] Only one tiling has this many. What is it?

From Exercise 3, we have a limit on how many polygons to look for. Clearly, we must have at least three polygons meet at each vertex. Thus, we have another rule that all semiregular tilings must obey:

Rule 2: Every regular and semiregular tiling must have at least 3 polygons and no more than six meeting at each vertex.

Next, let us think about how many different types of polygons can occur at each vertex.

▶ **Exercise 4.** Can there be four different polygons at a vertex? The angles of each would have to be rather small. Explain your answer.

▷ **Exercise 5.** Therefore, if there cannot be four different polygons at a vertex, then when there are four or more polygons at a vertex, what can you say about them?

This gives us another rule:

Rule 3: No semiregular tiling can have four different types of polygons meeting at a vertex.

A little additional thought gives us further limits on types of vertex configurations that can occur. Note that if we try to arrange one equilateral triangle and two other different polygons at a vertex, to make pattern 3.*n*.*m* they would be arranged as below:

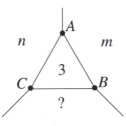

At vertex A, we have the vertex configuration 3.*n*.*m*. A triangle and an *m*-sided polygon meet at B, so the polygon marked "?" ought to have *n* sides. But at vertex C there are already a triangle and an *n*-sided polygon, so the polygon marked "?" should have *m* sides. Thus, we cannot have any vertex configuration of the form 3.*n*.*m* if $n \neq m$.

▷ **Exercise 6.** Show that one cannot have a vertex configuration of the form 5.*n*.*m* where $n \neq m$. Generalize the result to any vertex configuration of the form *k*.*n*.*m* where *k* is odd and $n \neq m$.

This gives us the following rule:

Rule 4: No semiregular tiling can have vertex configuration *k.n.m* where *k* is odd and *n* ≠ *m*.

Another configuration that cannot occur is 3.*k.n.m* unless *k* = *m*, as shown in the illustration below. At vertices *A* and *B*, we have the configuration 3.*k.n.m*. This forces the configuration 3.*k.n.k* at vertex *C*. If this is to be a semiregular tiling, we must have *k* = *m*:

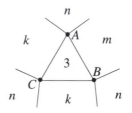

Rule 5: No semiregular tiling can have vertex configuration 3.*k.n.m* unless *k* = *m*.

Now we have set some parameters for our search. Next we will experiment to find out which polygons can fit around a vertex. Try fitting cardboard polygons together, or you can trace the polygons on the last page of this section, but check your answers by using the angles you calculated in Exercise 2. Write your answers to Exercises 7, 8, and 9 in the table below.

Vertex Configurations for Possible Tilings

Symbol	# polygons	Comment	Symbol	# polygons	Comment
	6	Exercise 3	3.7.42	3	Exercise 9
	5	Exercise 7		3	Exercise 9a
	5	Exercise 7		3	Exercise 9a
4.4.4.4	4	Exercise 1		3	Exercise 9b
	4	Exercise 8		3	Exercise 9b
	4	Exercise 8		3	Exercise 9c
	4	Exercise 8		3	Exercise 9c
6.6.6	3	Exercise 1		3	Exercise 9c
4.8.8	3	text			

▶ **Exercise 7.** There are two ways to fit five regular polygons around a vertex so that the angle sum is 360°. What are they? [Hint: Use lots of triangles.]

▷ **Exercise 8.** One tiling with four polygons at a vertex is 4.4.4.4, four squares meeting at each vertex. There are three other ways to fit four polygons around vertex. Find them. Remember what you figured out in Exercises 4 and 5.

▶ **Exercise 9.** There are, unfortunately, a lot of ways to fit three polygons around a vertex. One such pattern is the regular tiling consisting of three hexagons meeting at a vertex, each contributing 120°, to form 3 × 120° = 360°. We designate this pattern by 6.6.6. We also already know about 4.8.8. Another that I will give you, since I do not want to draw a 42-sided figure, is 3.7.42, consisting of an equilateral triangle (60°), a heptagon ($128\frac{4°}{7}$), and a 42-gon ($171\frac{3°}{7}$), so $60° + 128\frac{4°}{7} + 171\frac{3°}{7} = 360°$. Now there are seven more to figure out.

(a) There are two others that have duplicate polygons, besides 6.6.6 and 4.8.8. Find these.
(b) Two of the remaining patterns contain a square. Find these.
(c) The last three each contain one equilateral triangle. Find the patterns and list on the table.

In the exercises above we found 17 ways to fit regular polygons around a vertex. Unfortunately, not all of these extend to tilings of the plane. We have only found *possible* tilings: we know only that these vertex configurations add up to 360° at each vertex. Extending the configuration to cover the plane is another question.

For example, if we consider 3.7.42, the pattern starts like the picture below. The 42-gon is drawn with radial lines to help distinguish the sides, since a 42-sided polygon looks very much like a circle. We must have a triangle, a heptagon, and a 42-gon at each vertex. If we arrange this at vertex A and vertex B, then we are forced to have the triangle and *two* heptagons at vertex C. There is no room to squeeze in a 42-gon at vertex C; therefore, this pattern cannot be extended to a tiling of the plane.

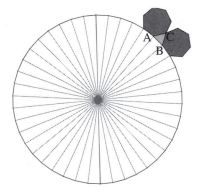

This is a geometric explanation of Rule 4, which already told us that the configuration 3.7.42 was not possible, since it is of the form $3.n.m$ with $n \neq m$. We have 16 other vertex configurations to check. All of the regular tilings work out, so we'll do those first.

▷ **Exercise 10.** Draw a section (at least 2 inches square) of each of the three regular tilings found in Exercise 1.

That takes care of three. And we already know that 4.8.8 tiles the plane, and that 3.7.42 does not, so we have 12 left to check. Answer the next set of exercises by fitting together cardboard polygons, by tracing the patterns on the last page of the section, or by applying Rules 4 and 5.

▶ **Exercise 11.** There are 10 patterns with three polygons at a vertex. We know that 6.6.6 and 4.8.8 tile the plane and 3.7.42 does not. Of the seven remaining, only two extend to a tiling.
(a) Use Rule 4 to eliminate the ones that do not work.
(b) Draw sections (at least 2 inches square) of the two remaining tilings of this type.

When we turn to the examples with four and five polygons at a vertex, there is an additional difficulty. One of the four solutions to Exercise 8 should be 3.4.4.6. There are two different ways that a triangle, two squares, and a hexagon can be arranged around a vertex: we designate these by 3.4.4.6 and 3.4.6.4.

Note that the pattern 3.4.4.6 violates Rule 5, but 3.4.6.4 does not.

▷ **Exercise 12.** Draw a section of the tiling 3.4.6.4.

▶ **Exercise 13.** Only one of the other two patterns found in Exercise 8 and their rearrangements with four polygons at a vertex can be extended to cover the whole plane.
(a) Find the one which does not work, and show why. Be sure to check all possible arrangements.
(b) The other configuration from Exercise 8 has one rearrangement that does not tile, and one that does. Show why the first fails and draw a section (at least two inches square) of the tiling.

▷ **Exercise 14.** Both of the patterns of five polygons around a vertex found in Exercise 7 can be extended to tilings. One gives two different tilings, depending on the order. Draw sections of these three tilings.

You should have found and drawn the three regular tilings and the eight semiregular tilings!

Dual Tilings

For each tiling we can associate another *dual tiling*, which might not be regular or semiregular. For example, below is the tiling 4.8.8 and its dual tiling, which is a tiling by nonequilateral triangles:

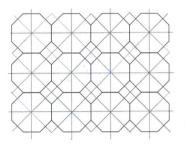

The process of forming the dual is as follows: place a vertex at the center of gravity of each polygon in the original tiling. Whenever two polygons share an edge in the original tiling, draw a dual edge connecting the new vertices at the centers of those polygons. In order to carry out this procedure, you will need to locate the centers of gravity for various regular polygons.

▷ **Exercise 15.** Find the center of gravity for a square and a regular hexagon. Figure out a general method for finding the center of gravity for any regular polygon with an even number of sides.

▷ **Exercise 16.** Find the center of gravity for an equilateral triangle and a regular pentagon. Figure out a general method for finding the center of gravity for any regular polygon with an odd number of sides.

▶ **Exercise 17.** Suppose somewhere in a tiling there are two regular polygons, one with n sides and one with k sides, that share an edge. Where will the line connecting the centers of the two polygons intersect the shared edge? What angle will this line make with the shared edge? Explain your answers.

▷ **Exercise 18.** In the dual tiling of 4.8.8, what are the angles of the triangles?

▷ **Exercise 19.** Draw and describe the dual tilings for each of the three regular tilings found in Exercise 1.

Here is the tiling 3.3.4.3.4 and its dual. Note that the dual contains only one type of tile: an irregular pentagon:

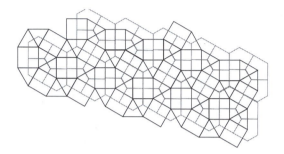

To find the angles of this irregular pentagon, note that it has two types of vertices: Two of the vertices are at the centers of squares of the original tiling and the other three are at the centers of equilateral triangles. Since the lines of the dual tiling divide the square into four equal sectors, these two angles must be $\frac{360°}{4} = 90°$. Since the dual tiling divides the triangles into three equal sectors, those angles must be $\frac{360°}{3} = 120°$. Thus, the irregular pentagon has angles 120°-120°-90°-120°-90°, which sum to 540°, as they should for a pentagon.

▶ **Exercise 20.** Draw the dual tiling for each of the tilings found in Exercise 11. Describe the types of polygons formed and find their vertex angles.

▷ **Exercise 21.** Draw the dual tiling for the tiling 3.4.6.4 of Exercise 12. Describe the types of polygons formed and find their vertex angles.

▶ **Exercise 22.** Draw the dual tiling for each of the tilings found in Exercise 13. Describe the types of polygons formed and find their vertex angles.

▷ **Exercise 23.** Draw the dual tiling for each of the tilings found in Exercise 14. Describe the types of polygons formed and find their vertex angles.

▷ **Exercise 24.** Explain why the dual tiling of a regular or semiregular tiling will always have only one type of polygon, though it may be irregular.

▷ **Exercise 25.** Of the examples and exercises above, which tilings have duals formed by triangles? squares? pentagons? Find a rule for the type of polygon in the dual.

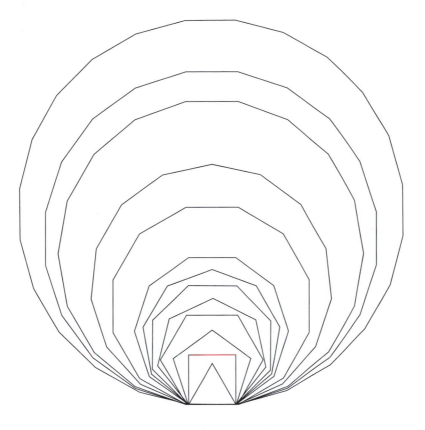

SOFTWARE

1. *KaleidoTile* by Jeffrey Weeks of the Geometry Center, University of Minnesota is
   freeware, for Macintosh only. ftp://geom.umn.edu/pub/software/KaleidoTile

SUGGESTED READINGS

Branko Grünbaum and G.C. Shephard, *Tilings and Patterns*, W.H. Freeman, New
   York, 1987.
Dale Seymour and Jill Britton, *Introduction to Tesselations*, Dale Seymour Publ., Palo
   Alto, 1989.

4. Tesselations

◆ 4.2. IRREGULAR TILINGS

SUPPLIES
    tracing paper
    cardboard
    scissors
    tape

In tiling, we need not restrict ourselves to regular polygons. It is easy to tile the plane with copies of any rectangle or parallelogram:

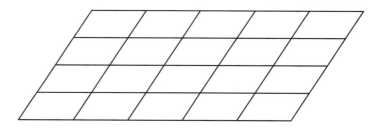

▶ **Exercise 1.** Show that any triangle can tile the plane.

▷ **Exercise 2.** Show that any trapezoid can tile the plane.

In fact, any quadrilateral will tile the plane, if they are put together right: Label sides a, b, c, and d as shown on the next page, and the angles α, β, γ, and δ. Since the sum of the angles in any quadrilateral must be 360°, $\alpha + \beta + \gamma + \delta = 360°$. Arrange four copies of the quadrilateral so that the sides match up and so that α, β, γ, and δ surround the center vertex. Then repeat this basic building block to create a tesselation:

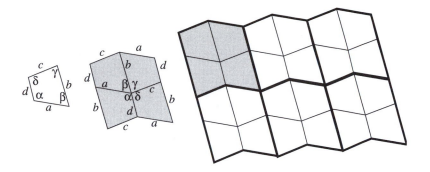

▷ **Exercise 3.** Draw the tiling generated by the nonconvex quadrilateral below. Your tiling should contain at least 12 repeats of the
quadrilateral.

A *reptile* (short for repeating tile) is a tile that can be arranged to form a
larger copy of itself. This larger copy must be an exact scaled replica of the
original. For example, a square is easily seen to be a reptile, since four
squares (or, for that matter, 9 squares, or 16, etc.) can be arranged to form a
larger square.

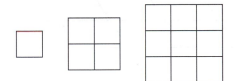

▷ **Exercise 4.** Below is a trapezoidal reptile. Find the angles of the
trapezoid.

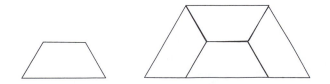

▶ **Exercise 5.** Below is a shaded rectangular reptile such that two repeats of the original rectangle gives a larger similar rectangle. Find the dimensions of the shaded rectangle if the longer side has length 1.

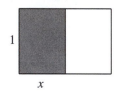

▶ **Exercise 6.** Below is a 30°-60°-90° triangle. Show how to put three copies of this triangle together to make a larger similar triangle.

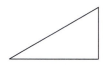

▷ **Exercise 7.** There is only one triangular reptile such that two repeats of the original triangle form a larger similar one. Which triangle is this? Draw a picture illustrating this reptile.

▷ **Exercise 8.** Show that each of the following is a reptile:

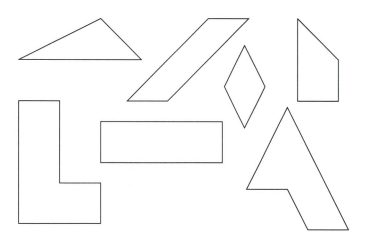

There are lots of irregular tilings, or tesselations. For example, one can tile the plane with any parallelogram, as above. Any reptile will tile the plane, since if, for example, four copies of the figure fit together to make a larger copy of itself, then four of the bigger copies will fit together to make an even larger one, etc. Although we showed that the regular pentagon does not tile the plane, it is easy to tile with the irregular pentagon below:

The Dutch artist M.C. Escher created many tiling pictures, initially inspired by the tiles at the Alhambra, in Spain. Many of these involved repetitions of pictures, rather than simply polygons, though the polygons are there in the background. In his notebooks, one sees the construction of the grid underlying many of these pictures. In this section, we will outline some processes for generating irregular tilings.

Recall that the regular tilings of the plane are by squares, equilateral triangles, and hexagons. The simplest irregular tilings are by rectangles or parallelograms. These grids can be modified to form other irregular tilings.

The first method we will use is a variation on a square, rectangular, parallelogram, or hexagonal grid. We want to make use of the parallel pairs of sides each of these has. Cut a basic unit (square, rectangle, parallelogram, or hexagon) out of cardboard, and use it to trace faint lines on a piece of paper, laying out your basic grid.

Now take the tile, and cut a section out of one side. Tape this piece onto the parallel side, as shown below. This move is called *parallel translation*. The cut-off piece moves in a straight line, without any rotation or reflection, to a position parallel to its original position.

Lay the new tile on your grid and trace it, to get a new irregular tiling. This process can be repeated for the other pair of parallel sides.

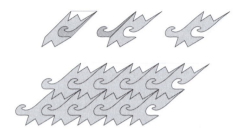

Escher used this technique in generating many of his pictures. He called a tiling a regular division of the plane. The picture below shows one of his tilings of simple stylized birds. In order to analyze this drawing as a tiling, we first note that the smallest repeating unit (ignoring the coloring) is a single bird. The birds are colored alternately light and dark to help the viewer figure out where one bird stops and another starts, but tracing will show you that the light and dark birds are indeed identical.

To find the grid that determines the tiles, connect some easily identifiable feature: We have chosen the birds' eyes. Connecting the eyes as shown on the next page on the left gives a grid of squares. We wish to find a single polygonal tile which generates the tiling shown, and describe how to modify it into a single copy of a bird. Thus the polygonal tile must contain enough parts to be assembled into a bird. In the picture on the left, there is a square tile containing the torso of a white bird. About half of the upper wing is cut off, but the square contains the upper portion of a dark wing to compensate. The square is missing the tip and one shoulder of the lower wing of the light bird and bits of the tail, but the square contains precisely those parts of a dark bird. The upper left corner of the square cuts the bird's head off, and in fact the grid slices the head up into four pieces. Duplicates of the missing three pieces can be found in the other three corners of the square. Thus the square contains exactly enough parts to make up a single bird. The bird is derived from the square by parallel translation on the top and bottom sides, and on the left and right sides. Note that a different choice for the grid lines, as shown below on the right, leads to a different tile. There are commonly several ways to generate a tiling such as this:

▷ **Exercise 9.** Create an irregular tiling from a parallelogram grid, using translation of both pairs of parallel sides.

▷ **Exercise 10.** Create an irregular tiling from a hexagonal grid, using translation of all three pairs of parallel sides.

Another way to modify a tiling when the original tile has a pair of parallel sides is called *glide reflection*: Take the original tile and cut something off of one side. Flip the cut piece over and tape it onto the parallel side:

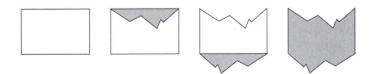

Then tile the plane with the new tile. You will have to turn over every other tile, so I have drawn the tile with two types of shading, one denoting copies of the basic tile in its original position, and lighter ones showing the tile flipped over:

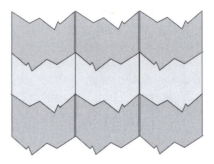

Here is an Escher-style tiling of unicorns that utilizes glide reflection. This was drawn by Jessica Adler using a program written by Kevin Lee for generating tilings. Drawing lines connecting the tips of the unicorns' horns, we get a grid of parallelograms. Each parallelogram contains enough parts to make a single unicorn. Using a glide reflection on the top and bottom edge and a translation on the left and right edges changes the parallelogram into a unicorn.

▷ **Exercise 11.** Create an irregular tiling from a rectangular grid, using glide reflection of **both** pairs of parallel sides.

Another way of creating an irregular tiling from one of the basic grids is *midpoint rotation*. For this one can use a basic grid formed by any quadrilateral or triangle, since we will not need parallel sides. Take one side of the tile and find its midpoint. Cut a pattern out of the tile that begins at one endpoint and ends at the midpoint of the same side.

Then take the piece that was cut out, rotate it by 180° about the midpoint, and tape it back along the edge on the other side of the midpoint.

Then tile the plane with the new tile.

This maneuver can be repeated on the other sides of the basic unit or can be combined with the translation move of the first example.

Below is a draft drawing from Escher's notebooks showing a tiling of birds generated from a grid of parallelograms by midpoint rotation. You can see the lines he drew to aid the construction and the small circles he placed at the rotation points.

▷ **Exercise 12.** Create an irregular tiling from an equilateral triangle grid, using midpoint rotation on each side.

▷ **Exercise 13.** Create an irregular tiling from a parallelogram grid, using a translation on one pair of parallel sides and a midpoint rotation on each of the other sides.

The fourth technique we will investigate, *side rotation*, works nicely only on the regular grids: squares, equilateral triangles, and regular hexagons. Cut a pattern out of the tile that begins at one end of one of the sides and ends at the other end of that side.

Then take the piece you cut out, rotate it about the endpoint, and tape it back along an adjacent edge of the tile.

Then tile the plane with the new tile.

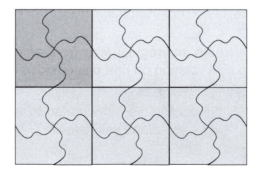

This maneuver can be repeated on the other sides of the basic unit or can be combined with translation, glide reflection, or midpoint rotation.

Next we exhibit another Escher-style tiling generated by side rotation, drawn using the program *TesselMania*. On the facing page is a drawing of turtles. There are some fairly obvious rotation points: where the heads of four turtles meet, and again where the back feet of four turtles meet. Draw a set of horizontal and vertical lines through the points where the four heads meet and another through the back feet points. The head lines meet the feet lines at another, less obvious, rotation center where the tails of two turtles and the front feet of two other turtles meet. All of these lines give a square grid. Each square contains the body of one turtle. For example, the square outlined below contains most of a dark turtle. The back foot and tail of this turtle lie outside the square, but a back foot and tail belonging to a light turtle compensate for that loss. The head and part of the front foot is cut off, but a light turtle's foot and head make up for that. A side rotation about the head point rotates the light turtle head and front feet around to where we want the dark turtle head and foot to be. A side rotation around the back foot point rotates the light turtle's foot and tail into the correct position for the dark turtle's rear.

▷ **Exercise 14.** Create an irregular tiling from an equilateral triangle grid, using side rotation on one pair of sides and midpoint rotation on the other side.

▷ **Exercise 15.** The following tile was created by side rotation on a rhombus. Can you tile the plane with this tile?

▷ **Exercise 16.** Analyze the tiling of frogs by Kevin Lee shown below. Describe the underlying grid and the polygonal tile, and how this tile was modified to form a single copy of the frog.

▶ **Exercise 17.** Analyze the tiling of dogs by Kevin Lee shown below. Describe the underlying grid and the polygonal tile, and how this tile was modified to form a single dog.

3 Fold dogs by Kevin Lee

▷ **Exercise 18.** Analyze the tiling shown of the crowd below. Describe the underlying grid and the polygonal tile, and how this tile was modified to form a single copy of a person.

▷ **Exercise 19.** Analyze the tiling shown in M.C. Escher's *Reptiles*. Describe the underlying grid and the polygonal tile, and how this tile was modified to form a single copy of the reptile.

Many of Escher's tiling pictures make use of more than one tile. For example, in the drawing on the next page there are both fish and birds. Thus, the basic repeating unit must contain one fish and one bird. The parallelogram grid chosen gives one of each after a translation is applied in each pair of parallel sides.

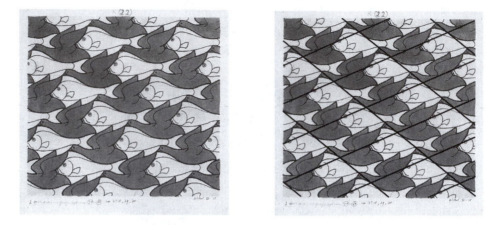

There are many shapes, in fact most shapes, that cannot tile the plane, for example, regular pentagons, the modified rhombus of Exercise 15, etc. Here is one rule for deciding whether a figure will tile the plane. However, note that there are lots of figures (such as the one you developed in Exercise 9) that do not fulfill the criterion but succeed in tiling.

Conway Criterion: A simple region (the boundary must form a loop without crossings and there can be no punctures) will tile the plane if the boundary can be divided into six arcs by six points labeled A, B, C, D, E, and F in order as one travels around the boundary, and
(1) The arc AB from A to B is the parallel translate of the arc ED.
(2) The arcs BC, CD, EF, and FA have rotational symmetry about their midpoints.

Here is an example of a tile fitting the Conway criterion. The points marked with a ○ are the midpoints (centers of rotation) for the arcs:

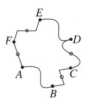

▶ **Exercise 20.** Trace nine copies of the tile above and fit them together to cover a roughly square region to show a piece of the tiling.

▷ **Exercise 21.** Show that the tile below fits the Conway criterion by marking the points *A, B, C, D, E,* and *F,* and the centers of rotation for the arcs *BC, CD, EF,* and *FA*:

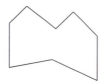

▷ **Exercise 22.** Trace nine copies of the tile above and fit them together to cover a roughly square region to show a piece of the tiling.

▷ **Exercise 23.** Show that the tile below fits the Conway criterion by marking the points *A, B, C, D, E,* and *F,* and the centers of rotation for the arcs *BC, CD, EF,* and *FA*:

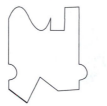

▶ **Exercise 24.** Trace nine copies of the tile above and fit them together to cover a roughly square region to show a piece of the tiling.

An extension of the Conway criterion allows some of the points *A, B, C, D, E,* and *F* to be the same, as long as there are at least three distinct points. Of course, if $A = B$, then the arc *AB* is a single point, and then we must have $D = E$. In this case there are no translated sides, and we have a quadrilateral *ACDF* with a midpoint rotation on each side, such as the one pictured below:

▷ **Exercise 25.** Trace nine copies of the tile above and fit them together to cover a roughly square region to show a piece of the tiling.

▷ **Exercise 26.** Describe what happens if a tile fits the Conway criterion with $A = B = C$ and $D = E$.

▷ **Exercise 27.** Describe what happens if a tile fits the Conway criterion with $B = C$ and $E = F$.

Parquet Deformations

In 1937 the Dutch artist M.C. Escher began to experiment with the metamorphosis of his tiling patterns. The simplest of such works are those like *Sky and Water II*:

In *Sky and Water II*, note that in the center of the drawing there is a tiling with two tiles: a bird tile and a fish tile. If you consider a basic tile consisting of one copy of the bird and one copy of the fish, you have a basic tile derived from a parallelogram by parallel translation. Moving away from the center, the motifs become less stylized and are spaced farther apart so that the space between birds is less and less fishlike, and the space between fish is less and less birdlike.

The idea of transformation of a basic tiling pattern is carried further in his later drawings. In an essay entitled *The Regular Division of the Plane* written in 1957, Escher describes one such transformation.

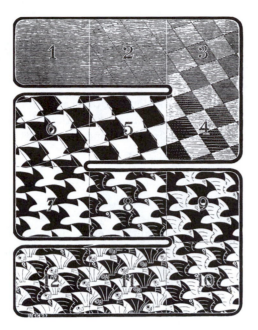

An abbreviated version of his comments on the numbered stages of this work follows:

[1] The beginning is grey.

[2] Two systems of straight parallel lines emerge from the indeterminate grey mists. They form the guidelines for the division of the plane. The distance between them and the angles at which they intersect reveal something of the character of the figures that will grow out of them later. The surface area of each of these, in particular, is henceforth totally determined by the area of each parallelogram and will remain constant during the succeeding transformations.

[3 and 4] Visual delimitation or bordering is not achieved by "lines" but by the effect of the contrast between planes of different shades. . . white and black suffice, and grow to maximum contrast by the end of section 4.

[5 and 6] The straight lines of the borders between black and white in section 4 change gradually in sections 5 and 6 [by sliding or parallel translation]. They become increasingly bent and broken; where "white" advances, "black" automatically recedes.

[7] In this section, the gradual growth of the figures has come to an end. They have achieved their definitive form and will retain it to the end of the strip. Although nothing seems to remain of the original parallelogram, its characteristics are nevertheless incorporated in the completed figure or "motif." Thus the area of every motif is still the

same as that of the earlier parallelogram, and the points where four figures meet are still in the same place in relation to one another.

[8] Uncertainty ends as soon as the black silhouettes are filled with a few detail lines. It leaves no room for doubt: there are black birds flying against a white background.

[9] Obviously this can be done the other way around just as well. When we move the detail lines from the black to the white motifs, the representation appears in reverse and white birds can be seen against a black background.

[10] A completely regular division of the plane in the sense I mean comes about only when the function of an "object" can be attributed to each of the congruent figures.

[11] The white and black marks that we read as birds can also be seen as something else. If we shift the eye and the mouth from right to left and turn the wing into a fin, the birds become fish.

[12] Finally, it is of course also possible to unite the two types of animals in a single division of the plane. In the solution put forward here, black birds fly to the right and white fish swim to the left, but they can be made to change places at discretion.

In sections 5 and 6 of the drawing, you can see the gradually deepening cuts and the parallel translations that form the transition from the tiling of parallelograms to the tiling by fish and birds. We are particularly interested in studying such transformations, instead of studying only a particular static tiling. Escher explains his interest in such transformations:

It showed clearly that a succession of gradually changing figures can result in the creation of a story in pictures. In a similar way the artists of the Middle Ages depicted the lives of the saints in a series of static tableaux, each showing an important moment and together telling their story, either in separate sections or in a continuous landscape, within a single frame. The observer was expected to view each stage in sequence. The series of static representations acquired a dynamic character by reason of the space of time needed to follow the whole story.

The woodcut *Metamorphosis I* was executed in 1937.

It begins on the left with a rather medieval town, whose buildings become more and more stylized and regular and merge into a tiling by hexagons resembling the traditional quilt pattern called Tumbling Blocks.

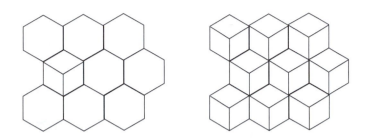

which are then altered gradually to form a pattern of interlocking figures that Escher had earlier developed as a tiling.

The idea of linking a sequence of tilings or *parquet deformation* resulted in one of Escher's most famous works: *Metamorphosis III* created in 1967 as a frieze for the Dutch Post Office.

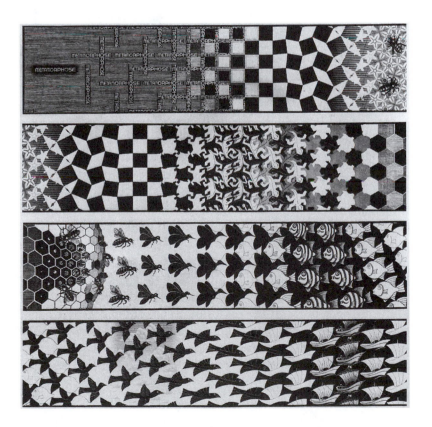

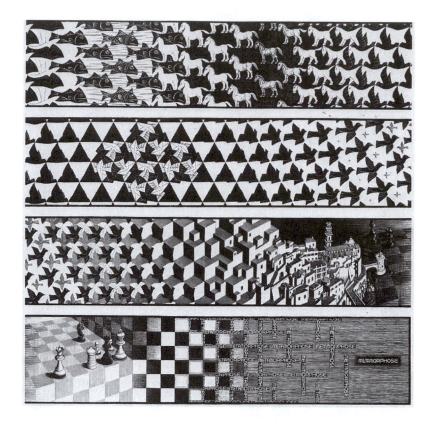

Again, there is a progression of tilings with transitional deformations linking them. The component tilings had already been worked out by Escher, some years before, others recently (most notably the central winged envelopes) with this project in mind. Reading from left to right, the sequence is the word "METAMORPHOSIS" to a tiling of squares to a tiling by squares and rhombi to 5-fold flowers back to squares and rhombi to squares to lizards to a hexagonal tiling to bees and fish to fish and birds to flying fish and boats to boats and fish to fish and horses to horses and birds to triangles to winged envelopes to triangles to birds to tumbling blocks to a village to a chess board to squares back to the work "METAMORPHOSIS."

The transitions between the tilings of *Metamorphosis III* are sometimes progressive deformations of the tilings, as in the transition from the squares to the lizards and the lizards to the hexagons, or sometimes purely imaginative, as the lovely transition from hexagons to bees and fish, where the tiling of hexagons becomes a honeycombed sphere that launches realistic bees, which become more stylized until the space between them is filled by the fish.

A later development of the idea of transforming tiling took place in the architecture studios of William Huff, at Carnegie Mellon and SUNY-Buffalo. He coined the phrase *parquet deformation* and dictated two rules [quoted from Hofstadter]:

(1) There shall be change only in one dimension, so that one can see a temporal progression in which one tesselation gradually becomes another;
(2) At each stage, the pattern must constitute a regular tesselation of the plane (i.e., there must be a unit cell that could combine with itself so as to cover an infinite plane exactly).

These rules are far more formal than the transitions used by Escher, but they lead to some beautiful work, some examples of which are owned by the Museum of Modern Art. The interested reader is encouraged to investigate the Hofstadter book from the suggested readings for this chapter.

SOFTWARE AND VIDEOS

1. *TesselMania!*, by Kevin Lee, of MECC Software, 1996. This program is very easy to use and allows the user to manipulate tiles as we have done in this section. It includes a tutorial and is available for Macintosh or PC Windows. It is available from Key Curriculum Press.
2. A video, *The Fantastic World of M.C. Escher*, by Michele Emmer is available from Atlas Video.

SUGGESTED READINGS

F.H. Bool, J.R. Kist, J.L. Locher, and F. Wierda, *M.C. Escher: His Life and Complete Graphic Work*, Harry N. Abrams, New York, 1992.

M.C. Escher, *The Regular Division of the Plane*, from F.H. Bool, J.R. Kist, J.L. Locher, and F. Wierda, *M.C. Escher: His Life and Complete Graphic Work*, translated by T. Langham and P. Peters, Harry N. Abrams, New York, 1992.

Martin Gardner, "On tessellating the plane with convex polygon tiles," *Scientific American*, July 1975.

Martin Gardner, *The Unexpected Hanging and Other Mathematical Diversions*, Simon & Schuster, New York, 1969.

Branko Grünbaum and G.C. Shephard, *Tilings and Patterns*, W.H. Freeman, New York, 1987.

Douglas Hofstadter, *Metamagical Themas: Questing for the Essence of Mind and Pattern*, Basic Books, New York, 1985.

John A.L. Osborn, "Amphography: The Art of Figurative Tiling," *Leonardo* 26(4), 1993.

Doris Schattschneider, "Tiling the Plane with Congruent Pentagons," *Mathematics Magazine* 51(1), 1978.

Doris Schattschneider, *Visions of Symmetry: Notebooks, Periodic Drawings, and Related Work of M.C. Escher*, W.H. Freeman, New York, 1990.

Doris Schattschneider, "Will It Tile? Try the Conway Criterion!," *Mathematics Magazine* 53(4), 1980.

Dale Seymour and Jill Britton, *Introduction to Tesselations*, Dale Seymour Publ., Palo Alto, 1989.

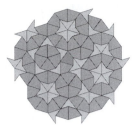

4. Tesselations

◆ **4.3. PENROSE TILINGS**

SUPPLIES
> tracing paper
> cardboard

The tilings with which we are most familiar tend to be *periodic*: that is, there is a fixed pattern that is repeated over and over again in a predictable way to cover the plane. A tiling is periodic if there is a finite section of the tiling that can be translated in two nonparallel directions to recreate the entire tiling. In other words, if you can make a rubber stamp out of some tiles and use it to cover the rest of the plane without rotating or reflecting the stamp and with no gaps or overlap, then you have a periodic tiling. Most of the tilings from the previous sections are periodic. The regular tilings of the plane by squares and by hexagons are periodic, and the simplest rubber stamp would be a single tile. The regular tiling by triangles is also periodic, but the rubber stamp would be made of two triangular tiles (a parallelogram) since you are not allowed to rotate the stamp. In general, if a tiling is periodic, your rubber stamp will need at least one of each type of tile in your tiling and more than one if there are rotations or glide reflections in your pattern. The tiling of the plane by an arbitrary quadrilateral in Section 4.2 would have four tiles in the stamp: (You could also use 8 or 12 or 16, etc., but we look for the smallest repeating region.)

▷ **Exercise 1.** Find the smallest repeating region for the tilings you made for Exercises 4.1.12, 4.2.11, and 4.2.14.

Note that these regions are not unique. For the octagon-square tiling, we could choose several different regions for the rubber stamp, as the figures below indicate:

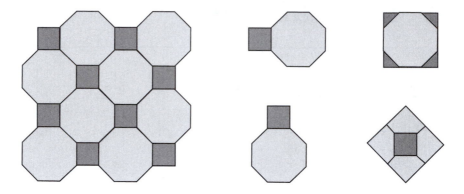

Note also that the Escher metamorphosis etchings or parquet deformations are not periodic, even though small sections seem to be. Your stamp is not allowed to change as you move along the tiling. However, these prints are only tilings of finite sections of the plane and so periodicity is not an issue.

The question arose whether all tilings are periodic. The answer is no, as the figure below on the right demonstrates. While the pattern is very symmetric, there is no one section that can be copied and translated to cover the whole plane because there is only one point that has rotational symmetry. However, it is possible to tile the plane in a periodic way using the shapes from the figure as we saw in Chapter 4.2 in the tiling of the plane by parallelograms. These tiles allow both a periodic and a nonperiodic tiling.

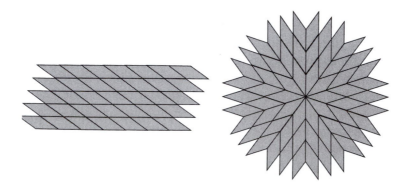

The next logical question is whether there is a set of tiles that can tile the plane only in a nonperiodic way. We adopt the language of Grünbaum and

Shephard and call such a set of tiles aperiodic. Upon hearing the question, most people suspect that there is not an aperiodic set of tiles. Most mathematicians agreed until 1964, when Robert Berger produced such a set. His original set contained over 20,000 different tiles. He continued to work on the problem and produced a set of 104 tiles that could cover the plane only in a nonperiodic way. Another mathematician, Raphael Robinson, found an aperiodic set containing only six tiles.

The most well-known set of aperiodic tiles are the Penrose tiles. The set was introduced by Roger Penrose in 1974 and contains only two tiles along with a set of rules for how the tiles must be put together. Penrose began his search for an aperiodic set of tiles by looking at pentagons. While it is true that the plane cannot be tiled by regular pentagons, Penrose studied the gaps left when one tried. He then took smaller pentagons and tried to fill in the holes. After several subdivisions, he found that the holes could have only a few shapes. These are pictured below, and he called the shapes diamonds, paper boats, and stars. Significant insight and refinement led to Penrose's first aperiodic set of six tiles and finally to the set of two tiles most commonly known as the *kite* and *dart*, names suggested by John Conway, another mathematician who has contributed a great deal to what is known about Penrose tilings.

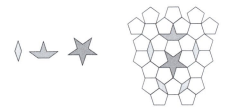

Anyone who has studied stars and pentagons should expect to find the golden ratio somewhere in a situation involving these two shapes, and indeed it makes several appearances here. In any infinite tiling of the plane by kites and darts, the number of kites used is the golden ratio times the number of darts. In any tiling of a finite section of the plane, the ratio of kites to darts will approximate the golden ratio. The approximation improves as the area tiled increases. The area of a kite is the golden ratio times the area of a dart. With these facts in mind, it is probably not surprising that the golden ratio is part of the actual measurements of the tiles.

The kite and dart are cut from a rhombus with side lengths equal to the golden ratio ϕ and main diagonal length equal to the golden ratio plus one, $\phi + 1$. Connect the vertices at the obtuse angles to the main diagonal at a distance of the golden ratio from an acute angle (and hence a distance of one from the other acute angle). The large piece is the kite; the small piece is the dart. Notice that the kite is made of two isosceles triangles with the golden ratio for the length of the equal sides. The length of the third side of these triangles is equal to one. These triangles are sometimes called golden triangles. Consequently, the kite is also made of two isosceles triangles with equal sides of length one and third side of length equal to the golden ratio.

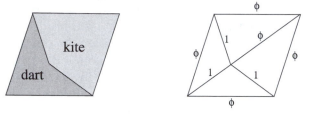

▶ **Exercise 2.** The measurements of two of the angles are given on the figure below. Find the measure of all other angles.

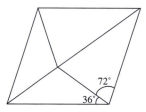

▷ **Exercise 3.** The Pythagorean pentagram (pictured below) is a familiar figure involving both the five-pointed star, or pentagram, and the pentagon. The length of any line segment in this figure is the golden ratio times the length of the next smaller line segment. Using this information and what you know about the lengths and angles of the Penrose tiles, find both a kite and a dart in the Pythagorean pentagram.

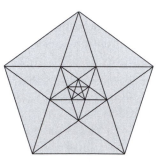

The rules for constructing a nonperiodic tiling with the Penrose tiles are simple. However, following them consistently is not necessarily easy. First, as with many tilings, only sides of the same length can be put together. This rule ensures that no vertex of one tile can occur in the middle of the side of another tile. The second rule requires a certain direction along the sides of the tiles. To enforce the direction rule, some people have put notches and bumps on the tiles, some use dots or holes, and some reshape the tiles to fit only in the correct way. Inspired by some of Escher's prints, Penrose reshaped the tiles as chickens, as below. John Conway puts arcs on the tiles and requires that the arcs of the same color must meet to form continuous curves. Thus, in constructing a nonperiodic tiling, dark arcs must join dark arcs, and light arcs must join to light. Conway went on to prove a number of results involving the way the arcs connect.

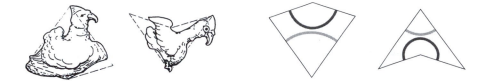

Dots are easy to make and easy to talk about, so we will simply put dots on our tiles. Cut lots of tiles out following the pattern below:

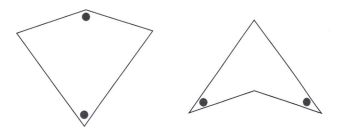

The next few exercises are designed to make you think about how the tiles fit together so we will not introduce the Penrose restriction on direction until a bit later. We always enforce the rule allowing only sides of the same length to be put together.

▶ **Exercise 4.** Tile the plane using only darts. Is your tiling periodic? How do the dots match?

▷ **Exercise 5.** Tile the plane using only kites. Is your tiling periodic? How do the dots match?

▶ **Exercise 6.** Can you tile the plane using only darts if you must put dots next to dots?

▷ **Exercise 7.** Can you tile the plane using only kites if you must put dots next to dots?

▷ **Exercise 8.** If you do not allow dots to be put next to dots, can you tile the plane periodically using some of each kind of tile in a way other than a rhombus tiling (where a rhombus is made from one kite and one dart)?

The Penrose rules allows two tiles to be put together only if the sides are the same length and the dots match. That is, dots are always next to other dots. In a lesson plan included with *Kites and Darts* (a foam rubber puzzle produced by Tesselations), Robert Fathauer suggests other combining rules.

▷ **Exercise 9.** Can you tile an infinitely large region of the plane using both kites and darts if dots must match between two of the same tile but cannot match between two different tiles? (That is, if you put two kites or two darts together, the dots must match. If you put a kite next to a dart, the dots must not match.) If not, what regions can you tile?

▷ **Exercise 10.** Can you tile an infinitely large region of the plane using both kites and darts if dots cannot match between two of the same tile but must match between two different tiles? If not, what regions can you tile?

▷ **Exercise 11.** What regions can you tile if dots are allowed to match only on kites? (That is, not between two darts or between a kite and a dart.) Only on darts?

▷ **Exercise 12.** What regions can you tile if dots are allowed not to match only on kites? (That is, they must match between two darts or between a kite and a dart.) Only not on darts?

Recall the angle measurements you found for the tiles in Exercise 2. We will label the vertices of the tiles by kite head (KH), kite tail (KT), kite wing (KW), dart head (DH), dart tail (DT), and dart wing (DW) as shown below. We know that angles totalling exactly 360 degrees fit around any vertex in the plane.

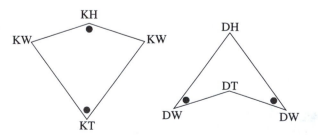

▶ **Exercise 13.** What is the maximum number of kite heads that can be put at any vertex in the plane? of kite tails?

▷ **Exercise 14.** What is the maximum number of dart heads that can be put at any vertex in the plane? of dart tails?

▷ **Exercise 15.** What is the maximum number of kite wings that can be put at any vertex in the plane? of dart wings? Be careful here. You need to think about more than just angle measures.

By now you should be fairly comfortable putting tiles together, so we will begin using the Penrose rules: dots and lengths must match for any two tiles to be put together. With these rules, there are only seven ways tiles can be put around a vertex. One way is to put all dart heads together. This figure is typically called *the star*. Another, called *the sun*, places five kite tails together at a vertex. Yet another, called *the king*, is to put two kite wings and three dart heads together. Notice that the two kites must touch each other in order for the other pieces to fit.

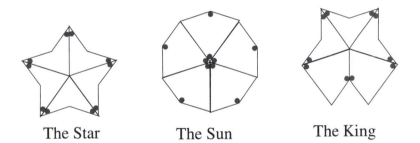

The Star   The Sun   The King

▶ **Exercise 16.** Find the other four ways that tiles can be put around a vertex following the Penrose rules.

The star and the sun have fivefold rotational symmetry. All of the vertex groupings have at least one line of reflectional symmetry. (The groupings with rotational symmetry will have five reflecting lines.) It is possible to maintain the rotational symmetry for star and sun, but there is only one way to do this for each vertex. That is, there are only two Penrose tilings that have rotational symmetry. It is also possible to maintain reflectional symmetry if you start with one of the other vertex groupings. Specifying reflectional symmetry does not yield a unique tiling for any vertex, but there are many more nonsymmetric tilings than there are symmetric ones.

As you work out from a vertex, sometimes there is only one way to place a tile at a certain point. Sometimes there are several choices. Sometimes it seems as if you have a choice, but farther out in your tiling you will find a place where no tile will fit and you will have to go back and change your original choice. If there is only one choice for fitting a tile around a vertex grouping, we say that tile is in the *empire* of the vertex.

If you begin a tiling with the sun grouping and do not insist on maintaining rotational symmetry, you can choose your next tiles in several ways. Therefore, the empire of the sun is empty. However, if you start with the star grouping, there is only one way to add the next tile. (Try it!) There is no room for another dart, and a kite can be placed in only one way. In fact, you are forced to fit ten kites around the star as the figure below shows before you can choose any other tile. After these ten kites, there are several ways to fit tiles. Hence, the star has ten tiles in its empire.

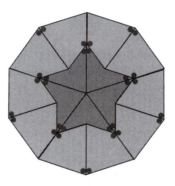

Empire of the Star

The vertex group called the King has 126 tiles in its empire. Fifty of these tiles are connected to the vertex by forced tiles. The rest are further out—after you have made some choices. Of the other four vertex groupings found in Exercise 16, one has no tiles in its empire, one has two tiles, one has 11 tiles, and the last has 22 tiles directly connected to the vertex grouping.

▷ **Exercise 17.** Find the connected empires of the four vertex groupings of Exercise 16.

Earlier, we discussed the possible symmetries of Penrose tilings. From that discussion, we have two tilings with rotational symmetry and tilings from the other vertex groupings that have reflectional symmetry. We commented that there are more nonsymmetric tilings than symmetric ones. In fact, there are infinitely many Penrose tilings that cover the plane. However, if you can look only at finite sections of a tiling, it is impossible to tell which tiling it is. John Conway proved that if you choose any region of diameter d from one Penrose tiling and choose any point P in another Penrose tiling, you will find an exact copy of your region in the other tiling less than a distance $2d$ from your point P. Actually, it is worse than that. There are infinitely many copies of your region in the other Penrose tiling and also in the tiling where you originally found the region. If you move a distance of no more than $2d$ from the edge of your original region in the right direction, you will encounter another exact copy of it. Two Penrose tilings can be distinguished only by considering the complete infinite plane.

If there are so many copies of each finite region floating around, how do we know that the Penrose tiling is not periodic? How do we know that we just have not found a big enough section to translate if these tilings are infinite? One proof involves the concept of *inflation*.

Given any Penrose tiling, you can cut the pieces and reattach the parts to form another Penrose tiling where the kites and darts are bigger than in the initial tiling.

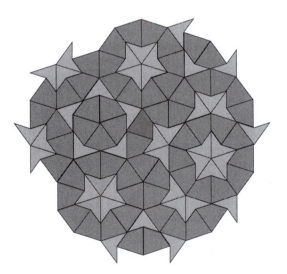

Cut all darts from head to tail. Then glue together any pieces of tiles that originally touched along the "short" side. (Remember that all sides of kites and darts had length equal to either one or the golden ratio. Glue along the "one" lengths.) DO NOT reglue the two halves of the darts you cut apart. While the new parts do not have lengths equal to one and the golden ratio, the ratio between the lengths is still the golden ratio. That is the magic of the ratio. Each reassembled piece will be either a kite or a dart, and they are positioned according to Penrose's rules. You have a new inflated tiling by bigger pieces.

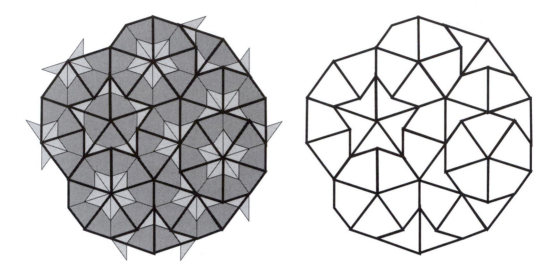

▷ **Exercise 18.** There is a reverse process called *deflation*. How would you deflate a tiling?

Suppose the original tiling was actually periodic. When you inflate the tiling, you do the same thing to all parts of the tiling. If the original tiling was periodic, the new tiling must be, too. If a tiling is periodic, there is a section of tile that can be translated a fixed distance over and over to cover the plane. Call this distance d. Pick any point P. Call one of the closest translations (distance d) of this point P'. By periodicity, if we slide the whole tiling from P to P', all the translated tiles will match up exactly with the original tiles. Even if we inflate the tiling, the distance between P and P' is still d, so if the original tiling was periodic with period translation distance d, the new tiling will be, too. If we slide the new tiling from P to P', the translated new tiles should exactly overlap the untranslated new tiles. We can continue to inflate the tiling. However, if we inflate enough times, the tiles will be so big that the both P and P' will be in the same tile. Here is the problem. If both P and P' are in the same tile, when we slide the whole tiling from P to P', the translated tiles cannot overlap the original ones. Therefore, no Penrose tiling of kites and darts can be periodic.

▷ **Exercise 19.** Create your own Penrose kite and dart tiling, carefully obeying the rule about which vertices can touch. Use tracing paper and the tiles above, or cut tiles out of cardboard.

Periodicity is a naturally occurring phenomenon in crystal structures and as such is interesting to chemists and physicists, among others. A crystal is formed when identical molecules align themselves to fit a lattice. It was long believed that these lattices must be derived from the tetrahedron, the cube, or the octahedron. Such crystals could have only two, three, four, and sixfold rotational symmetry.

Aperiodic tilings were considered interesting only to geometers and recreational mathematicians until they made an appearance in the natural world. In 1984, chemist Daniel Shechtman of the Israel Institute of Technology, in Haifa, was working on a project at the U.S. National Bureau of Standards. Using rapid cooling, Shechtman and colleagues produced an aluminum-manganese alloy that displayed fivefold rotational symmetry. This alloy could not be a true crystal, since fivefold rotational symmetry is not allowed, but it had more structure than a noncrystalline substance should. Soon, physicists Paul Steinhardt and Don Levine found that the spacings between the atoms were related to the spacings in a Penrose tiling. Scientists found it hard to believe that in such a rapidly cooled situation, rules as complicated as the Penrose rules could apply, but the similarities were striking.

Recent theories use the work of mathematician Petra Grummelt. She replaced Penrose's two tiles with a single decagon tile and rules for how it can overlap itself. (John Conway proved that every point in any Penrose tiling is either in the interior or on the boundary of a decagon.) According to physicists Steinhardt and Hyeong-Chai Jeong, the decagons become clusters of atoms in the quasi-crystal. The overlap translates as a sharing of atoms. The presence of overlapping clusters lowers the energy of the structure, and maximum overlap guarantees both a nonperiodic structure and the lowest energy. These rules are much more promising for structures created by rapid cooling, though work continues in the area.

Yet another unexpected application of Penrose tilings: "In April, Sir Roger Penrose, a British math professor who has worked with Stephen Hawking on such topics as relativity, black holes, and whether time has a beginning, filed a copyright-infringement lawsuit against the Kimberly-Clark Corp., which Penrose said copied a pattern he created (a pattern demonstrating that "a nonrepeating pattern could exist in nature") for its Kleenex quilted toilet paper. Penrose said he doesn't like litigation but 'When it comes to the population of Great Britain being invited by a multinational to wipe their bottoms on what appears to be the work of a Knight of the Realm, then a last stand must be taken.'" (from Chuck Shepherd's *News of the Weird*).

SUPPLIES

1. *Kites and Darts* tesselation puzzle, from Tesselations, Tempe, AZ, includes an excellent instruction booklet, including examples of lessons based on the puzzle, written by Robert Fathauer (June 1996).
2. Another Penrose tiling puzzle, based on the kite and dart tiling modified by edge and midpoint rotations to form fat chickens (the kites) and skinny chickens (the darts), is called *Perplexing Poultry*, from Pentaplex Limited, Royd House, Birds Royd Lane, Brighouse, West Yorkshire, HD6 1LQ.

SUGGESTED READINGS

COMAP, *For All Practical Purposes: Introduction to Contemporary Mathematics*, W.H. Freeman, New York, 1994.

Robert Fathauer, *Kites and Darts Instruction Booklet* (see above), Tesselations, Tempe, 1996.

Martin Gardner, "Mathematical Games: Extraordinary nonperiodic tiling that enriches the theory of tiles," *Scientific American* 236, January 1977.

Martin Gardner, *Penrose Tiles to Trapdoor Ciphers*, W.H. Freeman, New York, 1989.

Branko Grünbaum and G.C. Shephard, *Tilings and Patterns*, W.H. Freeman, New York, 1987.

Jay Kappraff, *Connections: The Geometric Bridge Between Art and Science*, McGraw-Hill, New York, 1991.

George E. Martin, *Transformation Geometry: An Introduction to Symmetry,* Springer-Verlag, New York, 1982.

Michael Naylor, "Nonperiodic Tilings: The Irrational Numbers of the Tiling World," *The Mathematics Teacher* 92(1), January 1999.

Roger Penrose, "Escher and the Visual Representation of Mathematical Ideas," in *M.C. Escher: Art and Science,* ed. H.S.M. Coxeter et al., North Holland, Amsterdam, 1986.

Roger Penrose, "Pentaplexity: A Class of Non-Periodic Tilings of the Plane," *Mathematical Intelligencer* 1, 1974.

Ivars Peterson, "Clusters and Decagons: New rules for constructing a quasicrystal," *Science News* 150, October 12, 1996.

Charles Radin, "Symmetry and Tilings," *Notices of the American Mathematical Society* 42(1), 1995.

Chuck Shepherd, *News of the Weird*, The Tennessean, Nashville, July 6, 1997.

5. Two-Dimensional Symmetry

◆ **5.1. KALEIDOSCOPES**

SUPPLIES
4 mirrors
tape
protractor
cardboard tube

Using One Mirror

Many natural and man-made objects exhibit *bilateral symmetry*: Each half is the mirror image of the other. Place your mirror on the dotted line below:

▷ **Exercise 1.** Place your mirror on the dotted line, so that it is perpendicular to this piece of paper, and consider the line from the dot to its image. What angle does this line form with the dotted mirror line?

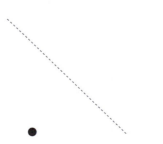

▷ **Exercise 2.** Place your mirror so that it is at one end of the ruler drawn below and perpendicular to it. How far is the perceived distance from the dot to its reflection? How far is the actual distance from the dot to its image? Try placing the mirror at different distances from the dot. How does the perceived distance from the dot to its image compare to the actual distance?

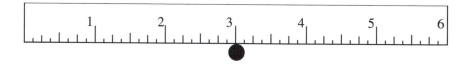

▷ **Exercise 3.** Place your mirror on the dotted line and consider the solid line and its image. What can you say about the angle formed between the solid line and the mirror line compared to the angle between the reflection of the solid line and the mirror line?

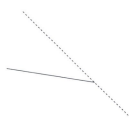

People have approximate bilateral symmetry, though the extent of this is often overestimated. In this, as in many other regards, bugs are superior. It is, however, somewhat amusing to take a full frontal photograph of someone you are not too fond of, scan it into a computer, and use a program like *Adobe Photoshop* to manipulate the image and put together two right halves and two left halves:

▷ **Exercise 4.** Look into a mirror and wink with your right eye. Notice that the person in the mirror is winking his or her left eye. Why does a mirror reverse right and left but not up and down? Explain your theory in a paragraph.

If one half of a figure is the mirror image of the other, we say that the figure has reflectional or mirror symmetry, and the line marking the division is called the *line of reflection,* the *mirror line,* or the *line of symmetry*. Many figures have more than one mirror line, of course.

▷ **Exercise 5.** Draw the line of reflection:

▶ **Exercise 6.** Draw the lines of reflection, if any, for each of the following figures:

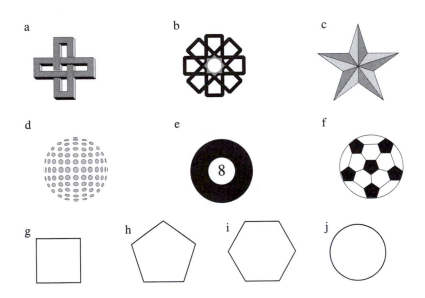

As an example, consider the equilateral triangle below. It has three lines of reflection, each passing through a vertex and the midpoint of the opposite side.

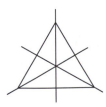

▷ **Exercise 7.** How many lines of reflection does an *n*-sided regular polygon have? Describe them (by describing where each line slices the polygon).

Using Two Mirrors

▶ **Exercise 8.** Place two mirrors so that they are parallel and facing each other, perpendicular to the ruler drawn below and positioned at opposite ends of the ruler. Look over the first mirror into the image in the second. How far is the perceived distance between reflections of the dot?

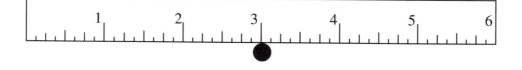

This works best if the mirrors are wall-sized (or if you're the size of a moderate bug) and you stand between them. Barring this eventuality, keep the mirrors about 4″ to 6″ apart and parallel to your body, and bend over so that your eyes are barely above the mirror that has its back to you and look straight into the other one. A candle works even better than a dot.

▷ **Exercise 9.** Place your mirrors so that they are parallel and facing each other on either side of the image below. Describe the pattern of reflected images that you see.

Now tape your mirrors together like a hinge. Hold the mirrors perpendicular to the paper. Spread the mirrors to form various angles. The best view is usually from straight ahead, but you should also try viewing from the sides and note any distortion of the images that occurs.

▷ **Exercise 10.** Place your hinged mirrors so that the hinge is at the dot and draw what you see. What happens when you change the angle?

Below is a triangle with three dotted lines of reflection. There are three different ways of placing your mirrors so that you see an intact equilateral triangle in the hinged mirror. These are marked by the heavy lines in the figures. Note that we are not counting the 180° reflection across the center, where a mirror could be placed along the vertical line cutting the triangle in two, since you do not need a hinged mirror for that.

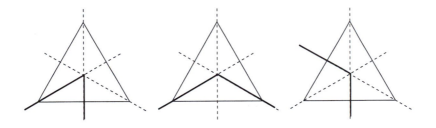

▶ **Exercise 11.** Below is a square with dotted lines of reflection. Try different ways of placing your hinged mirrors along these lines. Find three different ways of placing your mirrors so that the view is of an intact square. Mark these on the diagram using three different colors. What angles are formed in each case?

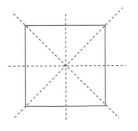

▷ **Exercise 12.** Draw three different wedge-shaped sectors that when viewed with a hinged mirror will form a regular pentagon. What angles are formed in each case?

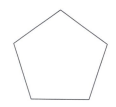

▷ **Exercise 13.** Draw three different wedge-shaped sectors that when viewed with a hinged mirror will form a regular hexagon. What angles are formed in each case?

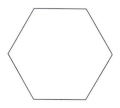

When you look into a hinged mirror placed on a plain sheet of paper, or at the circle of Exercise 10, you have the illusion of an unobstructed view. In other words, you see a pattern filling 360°. Of course, what you are really seeing is the triangular wedge formed by the mirrors reflected in each mirror. Let us see what happens in the case where copies of the triangular wedge do not form a circle or 360°. For example, let us use a 55° wedge, since 55 does not divide 360 evenly. In fact, $360 = 6 \cdot 55 + 30$. Thus, the view into a hinged mirror at an angle of 55° should make 6 copies of the wedge with a 30° slice left over. Below is a 55° wedge with a lizard drawn in it. Viewed in the mirror assembly, you should see 6 lizards and the leftover 30° piece. Of course, this means that you have a fraction of a lizard, but it is very important, at least to the lizard, to know which part of the lizard survives. If just the tail is cut off, the lizard will eventually grow a new one, but if it's the head that is cut off, then you've got a dead lizard.

▷ **Exercise 14.** Place your hinged mirror assembly on the picture on the previous page, with the hinge at the dot and the mirrors lined up on the dotted lines. It is possible to see both of the images below. Describe how this can be done.

Thus, a 55° hinged mirror gives a distorted and ambiguous view of the lizard: you do not know whether there are 6 heads and 7 tails or 7 heads and 6 tails. Only certain angles are appropriate for use with hinged mirrors if one wants an undistorted and unambiguous view.

▶ **Exercise 15.** Below is a lizard with lines forming angles of 60°, 72°, 90°, 108°, 120°, and 150°. For each angle, place your hinged mirrors so that the edges lie along the relevant dotted lines and the hinge is at the dot. Describe what you see. Is there any ambiguity when the image is viewed from either side?

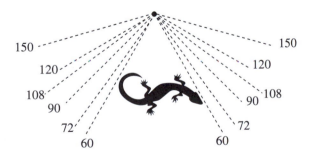

▷ **Exercise 16.** Which angles give you a whole number of (undistorted) lizards? Are there an equal number of right-handed and left-handed lizards? How many lizards are formed? Give general formulae for the angles that work and the relationship between the angle and the number of lizards. You may wish to experiment with your mirrors and a protractor to answer this question.

▶ **Exercise 17.** Below is a symmetrical figure with lines forming angles of 60°, 72°, 90°, 108°, 120°, and 150°. For each angle, place your hinged mirrors so that the edges lie along the relevant dotted lines and the hinge is at the dot. Describe what you see. Is there any ambiguity when the image is viewed from either side?

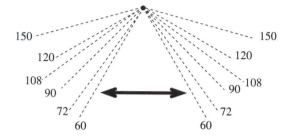

▷ **Exercise 18.** Which angles give you a whole number of arrows? How many arrows are formed? Give general formulae for the angles that work and the relationship between the angle and the number of arrows. Explain any differences between the results of this exercise and those of Exercise 16.

▷ **Exercise 19.** Place your protractor on the picture below, centered at the dot, then put your hinged mirror on top so that you can measure the angles formed by the mirrors. Be sure to keep the angle symmetric. Which angles give you regular polygons? Which regular polygons can you form? Give general formulae for the angles that work and the relationship between the angle and the number of sides in the regular polygon.

•

Using Three and Four Mirrors

One can assemble several mirrors in the shape of a prism. The reflections of the base of the prism, where the mirrors hit the sheet of paper, will form a tiling, or tesselation, of the plane. First, assuming that your mirrors are all the same size, tape three of them together, forming an equilateral triangle with the mirrored surfaces on the inside.

▷ **Exercise 20.** Draw an asymmetric scribble (so that any symmetry will be due to the action of the mirrors) and place your 3-fold mirror over this. Draw what you see.

▷ **Exercise 21.** Assuming that your mirrors are rectangular, tape them together to form an isosceles triangle by standing two on their sides and the third on end. Consider the views looking into one of the short wide mirrors and into the taller narrower mirror. Do you have a tiling? Draw what you see.

The experiments above should have showed you that only some configurations of mirrors give an unambiguous view of the plane. Exercise 16 showed that only angles of the form $\frac{\pi}{k} = \frac{180°}{k}$ with integer $k \geq 2$ do not distort the view of asymmetric objects. Therefore, if one builds a polygon from mirrors that does not distort the plane, all of the angles must be of this form. A polygon, not necessarily regular, such that each angle is of this form is called a *Coxeter polygon*. It is somewhat surprising to find that there are only four Coxeter polygons. For example, consider the 30°-60°-90° triangle. Since $30° = \frac{180°}{6}, 60° = \frac{180°}{3}$, and $90° = \frac{180°}{2}$, this is a Coxeter triangle.

▶ **Exercise 22.** There are only two other Coxeter triangles. Find these, listing their angles and showing that each satisfies the condition above.

▷ **Exercise 23.** There is only one Coxeter quadrilateral (though the lengths and proportions of the sides may vary). Find it.

Kaleidoscopes

The kaleidoscope was invented by Sir David Brewster in 1813 and named by him from the Greek χαλος (beautiful), ειδος (form), and σχοπεω (to view), thus meaning "beautiful form viewer." In general, the basic anatomy of a kaleidoscope consists of four parts: the eyepiece, usually with a clear lens, which serves only to keep extraneous light and dust out; the mirror assembly, which is the heart of the kaleidoscope and which we will discuss in detail later; the tube, which houses the mirror assembly and is often very decorative; and the object chamber, which holds the objects to be viewed, often bits of colored glass. The ideal focal length for the human eye is 6 to 8 inches, so that is usually how long the tube is. The object chamber, also called the vivenda case, is capped by a piece of frosted glass to let in light but to obscure the view of the surroundings. It often rotates, so that the view through the kaleidoscope changes as the objects rattle around. Some kaleidoscopes do not have object chambers but are capped by a clear lens, usually curved, and one views whatever the kaleidoscope is pointed towards. These are called *teleidoscopes*, or "distant form viewers."

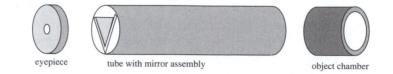

eyepiece | tube with mirror assembly | object chamber

Kaleidoscopes come in two basic versions: those with two mirrors and those with three. The view through a 2-mirror kaleidoscope is like the view in your hinged double mirror above: symmetric reflections surrounding a point. The mirrors are commonly set at an angle of either 60° or 45°, though any angle of the form $\frac{180°}{k}$ where k is an integer could be used. The mirrors in a 3-mirror kaleidoscope, called a polycentral kaleidoscope by Brewster, are most commonly set in an equilateral triangle, but more rarely one finds the other Coxeter triangles, and in all cases the view is of an infinite tiling.

A primitive kaleidoscope can easily be built using a tube such as those for paper towels or wrapping paper. Lenses can be cut from sheets of clear rigid thin plastic (from an artists' supply stores) to fit the ends of the tube, or if you are on a strict budget, you can use plastic film. Cut a piece of paper with a half-inch hole in it to fit the viewing end. Tape the paper and one of the lenses to one end of the tube. Two or three mirrors of glass or rigid plastic should be cut to fit neatly inside the tube at a 60° angle to each other (or use another valid angle). Alternatively, one can glue reflective Mylar (also available at artists' supply stores) to pieces of cardboard or matboard to form the mirrors, though care must be taken to make these as perfectly smooth as possible. Remember that these mirrors are the most important part of the kaleidoscope, and they should fit as precisely as possible. Cut the mirrors a half inch shorter than the tube, and insert them into the tube. Then cut another round piece of clear plastic that just fits inside the tube. Insert this into the tube so that it rests on the ends of the mirrors. If you cut this carefully enough, it will hold in place without any tape or glue. The half-inch space between this and the lens on the far end forms a chamber in which you can place the objects to be viewed. The best objects to use in a kaleidoscope are translucent and colored: glass beads work nicely. Insert any such objects and finish with the end lens. A circular cover of tracing paper over the far lens will give a frosted finish.

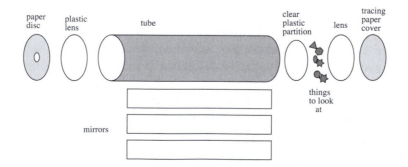

paper disc | plastic lens | tube | clear plastic partition | lens | tracing paper cover

things to look at

mirrors

▷ **Exercise 24.** Build a kaleidoscope.

SUPPLIES

1. Both Thom Boswell's *The Kaleidoscope Book* and Gary Newlin's *Simple Kaleidoscopes* give detailed instructions for making a variety of kaleidoscopes, ranging from very easy to quite professional.
2. Plastic mirrors (for those who cannot be trusted with glass) are available from Dale Seymour Publications and from Creative Publications. Cuttable plastic mirrors are also available from Dale Seymour.
3. Clear rigid plastic and reflective Mylar are available at most artists' or architects' supply stores. One of the biggest mail-order suppliers is Dick Blick (www.dickblick.com).
4. A CD-ROM called *KaleidoMania* by Kevin Lee includes (in the Macintosh version) an interactive activity called 3 Mirror Experiment that allows the user to change the angles of reflection to simulate a three-mirror kaleidoscope. It is published by Key Curriculum Press.

SUGGESTED READINGS

W.W. Rouse Ball and H.S.M. Coxeter, *Mathematical Recreations and Essays*, Dover, New York, 1987.

Thom Boswell, *The Kaleidoscope Book*, Sterling, New York, 1992.

Sir David Brewster, *The Kaleidoscope*, Van Cort Publishers, Holyoke, MA, 1987 (reprint of original text of 1819, Constable & Sons).

H.S.M. Coxeter, *Regular Polytopes*, Dover, New York, 1973.

Joe Kennedy and Diane Thomas, *Kaleidoscope Math*, Creative Publ., Sunnyvale, 1989.

Gary Newlin, *Simple Kaleidoscopes*, Sterling, New York, 1996.

Ian Stewart, *The Magical Maze: Seeing the World Through Mathematical Eyes*, Wiley & Sons, 1997.

E.B. Vinberg, "On Kaleidoscopes," *Quantum* 7, 1997.

5. Two-Dimensional Symmetry

◆ 5.2. ROSETTE GROUPS: POINT SYMMETRY

SUPPLIES
> tracing paper
> colored pencils
> mirror

All regular and some irregular polygons exhibit some symmetry. For example, an equilateral triangle *ABC* as pictured below can be rotated counterclockwise through $\frac{360°}{3} = 120°$ to get an identical triangle with *A* where *B* used to be, *B* where *C* used to be, and *C* where *A* used to be. We say that the equilateral triangle has *rotational symmetry*, with a rotation of 120°, or a *3-fold rotation*, since three turns will return it to its original position.

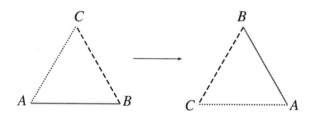

If we label the original configuration of the triangle as 1 and the configuration after a 120° counterclockwise rotation as *R*, then it makes sense to think of a 240° counterclockwise rotation as two successive 120° counterclockwise rotations, or R^2. Then it is clear that $R^3 = 1$, since three counterclockwise 120° rotations is the same as a 360° rotation, leaving the figure the same as it was in the beginning. Mathematicians tend to rotate everything counterclockwise, though everybody else in the world rotates clockwise. You can do whichever you want, as long as you are consistent.

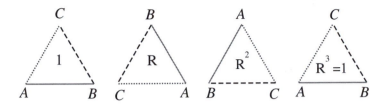

▶ **Exercise 1.** One could also rotate the triangle 120° clockwise, but this is the same as which of the operations already pictured?

This triangle also has three mirror lines, or lines of reflection, as you can see by placing your mirror along the lines below:

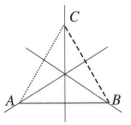

Let us denote the reflection across the vertical line by F (for Flip). Two reflections across the vertical line returns every point to where it came from, so $F^2 = 1$.

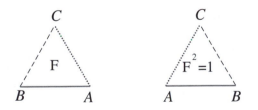

These operations, rotation and reflection, act on the triangle to give an identical figure with the points rearranged. You can keep track of where the points move by labeling the vertices, though I find it easier to color the sides. Another operation is the *identity*, which leaves the triangle untouched: This is the operation we have been denoting by 1.

One can also combine rotations and reflections. For example, RF means first rotate the triangle by 120° counterclockwise, then reflect it across the vertical line:

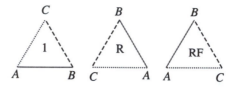

Note that the final configuration is different from those already pictured above. Similarly, we can draw R^2F:

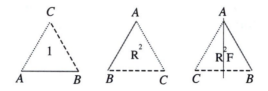

To summarize, we have found the following six configurations, or re-arrangements, of the triangle:

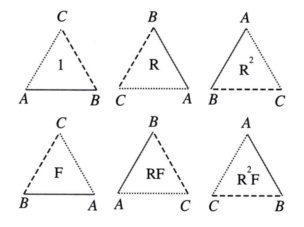

Next, we try to show that these are indeed all of the possible configurations. For example, we have reflected the original triangle across the vertical mirror line, but there are two other diagonal lines of reflection noted above.

▷ **Exercise 2.** Reflecting the original triangle across the diagonal line that goes through the lower left vertex A turns out to be the same as one of the above configurations. Which one? [Note that the easiest way to do this exercise is with tracing paper. What happens if you use a mirror instead?]

▷ **Exercise 3.** Reflecting the triangle across the diagonal line that goes through the lower right vertex B also turns out to be the same as one of the above configurations. Which one?

▶ **Exercise 4.** Recall that *RF* means to rotate the original triangle by 120° counterclockwise and then reflect across the vertical line. Thus, *FR* would mean to reflect the original triangle across the vertical line first and then rotate 120° counterclockwise. Which configuration of those pictured above is the same as *FR*? *FR²*? *RFR*?

I have chosen F to designate flipping the triangle across the vertical line. This decision was arbitrary, since one flip can be rotated to another, but one must be consistent. Thus there are only 6 different configurations of this triangle, 1, R, R^2, F, RF, and R^2F. The choice always to rotate first, then flip, was arbitrary, but again one must be consistent. This list of operations is called the *rosette group* of the triangle, and is denoted by D_3, the *dihedral group with a 3-fold rotation*. Thus $D_3 = \{1, R, R^2, F, RF, R^2F\}$. The term "dihedral" and the letter D are used to indicate that the figure has a reflection line, and the subscript 3 in D_3 indicates that there is a 3-fold rotation. We can build a multiplication table for the group D_3. Be careful, and remember that $RF \neq FR$. Note that each entry in the table is one of our six configurations: 1, R, R^2, F, RF, or R^2F. To use the table, for example to figure out which configuration is given by FRF, we write $FRF = F \cdot RF$ so that it is a product of two of our chosen entries. Then find the row that begins with F and the column headed by RF. The intersection of the F row and the RF column is $FRF = R^2$ from the table. Similarly to find RFR^2 using the table, we write $RFR^2 = RF \cdot R^2$ and look up the RF row and the R^2 column. Finding the intersection of this row and column gives us $RFR^2 = R^2F$.

D_3: Symmetries of an Equilateral Triangle

	1	R	R²	F	RF	R²F
1	1	R	R²	F	RF	R²F
R	R	R²	1	RF	R²F	F
R²	R²	1	R	R²F	F	RF
F	F	R²F	RF	1	R²	R
RF	RF	F	R²F	R	1	R²
R²F	R²F	RF	F	R²	R	1

Notice that each of our 6 configurations occurs exactly once in each row and column. After you are about halfway through filling in the table, you can use what you already know to figure out the rest. For example, once you know that $FR = R^2F$, then when you need to find RFR, you can calculate $RFR = R(FR) = R(R^2F) = R^3F = 1F = F$. I chose R to designate a rotation by 120° counterclockwise. As above, one can also rotate clockwise by 120°, which we can designate by R^{-1} to indicate a backwards rotation, and this is the same as rotating counterclockwise by 240°, so $R^{-1} = R^2$.

▶ **Exercise 5.** For the symmetries of the equilateral triangle, use the table above to simplify each of the following to one of the elements of $D_3 = \{1, R, R^2, F, RF, R^2F\}$.
(a) FR^2FR
(b) $RFRFRFRF$
(c) FR^5FR^{10}
(d) $R^{-2}FR^2$
(e) $R^2FR^{-3}FR^2$

The table for D_3 and the rules we have found are true only for the equilateral triangle and any figures that have precisely the same symmetries as an equilateral triangle: a rotation by 120° counterclockwise that takes the figure back onto itself and a reflection across a mirror line. Actually, if there is a 3-fold rotation and one mirror line, there must be three reflection lines, since the rotations must take a mirror line to another mirror line.

All rosette groups are of two types: cyclic, written C_n, for a figure with an n-fold rotation and no reflectional symmetries, and dihedral, written D_n, for a figure with an n-fold rotation and a reflection across a line. This fact is called *Leonardo's theorem*, after Leonardo da Vinci, who first showed that these are the only possibilities.

My favorite way of keeping track of all of the different configurations is to color the figure, making each leg a different color. I trace the figure, with the color coding, on a small piece of tracing paper. Then I can rotate the piece of tracing paper until it aligns with the original figure. I then know what R looks like, so I make a little sketch beside the original, with the colors shifted as on my tracing, and label it R. Do this for each configuration: all the rotations, flips, and combinations of rotations and flips. It is then easy to see whether $RF = FR$ or $RF = FR^2$, and to fill in the table.

▷ **Exercise 6.** Draw another figure that has the same symmetries, and so the same rosette group D_3 and the same multiplication table as the equilateral triangle.

Of course, the next one you have to do by yourself, but this one is easier than the triangle above. Consider the figure below:

▷ **Exercise 7.** This figure has rotational symmetry. What is the angle of rotation? How many rotations will return the figure to its original position?

The figure above has no reflectional symmetry, so the multiplication table is very easy. For this figure, the rosette group is denoted by C_6, where the C stands for *cyclic* and the 6 for the six rotations of the figure.

▷ **Exercise 8.** List the elements of the rosette group C_6 of the figure above and make the multiplication table for this group.

▶ **Exercise 9.** Draw another figure that has the same symmetries, and so the same rosette group C_6 and the same multiplication table as the one above.

▷ **Exercise 10.** Draw a figure that has 8-fold rotational symmetry but no lines of reflection, and thus rosette group C_8.

▷ **Exercise 11.** Draw a figure that has reflectional symmetry and an 8-fold rotation. This will have rosette group D_8 since there is a reflection. How many lines of reflection does your figure have? Mark them on your drawing. Can you think of another figure with the same rotational symmetry, but with a different number of lines of reflection?

▶ **Exercise 12.** Draw a figure that has a line of reflection, but no rotational symmetry. List the elements of the rosette group D_1 of the figure and make the multiplication table for this group.

▷ **Exercise 13.** Draw a figure that has rosette group C_1.

▷ **Exercise 14.** Draw a figure that has rosette group C_2.

▷ **Exercise 15.** Draw a figure that has rosette group D_2.

▷ **Exercise 16.** Describe in words the symmetries of a square.

▷ **Exercise 17.** List the elements of the rosette group D_4 of the square and make the multiplication table for this group.

▷ **Exercise 18.** Draw another figure that has the same symmetries, and so the same rosette group D_4 and multiplication table, as the square.

▶ **Exercise 19.** Describe the symmetries and give the multiplication table for the rosette groups of each of the following patterns. Identify each rosette group as D_n or C_n.

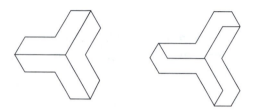

▷ **Exercise 20.** Describe the symmetries and give the multiplication table for the rosette groups of each of the following patterns. Identify each rosette group as D_n or C_n.

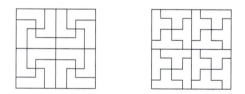

Patterns with rosette symmetry can be made by folding and cutting paper. A figure with reflectional symmetry only, and thus group D_1, is made by taking a piece of paper and folding it in half along the dotted line and then cutting a pattern as shown below. One must cut an asymmetric pattern so that any symmetry is due to the folds and not to the cuts. Since some figures will require a number of folds, it is best to use thin paper, such as tissue paper or tracing paper.

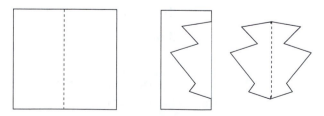

▷ **Exercise 21.** Take a square of tracing paper and fold along the dotted lines as shown below. Cut an asymmetric pattern. Unfold and identify the rosette group of the final figure.

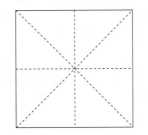

Note that the square paper used in Exercise 21 dictates certain symmetries. For the remaining exercises, use paper discs, which are appropriate for any symmetry. Each time you fold the paper, you are adding a line of reflection.

▷ **Exercise 22.** Figure out how to fold the paper to cut a pattern with rosette group D_3. Fold and cut such a pattern.

To cut a figure with no reflectional symmetry, you must slit the paper to the center and form a cone. Wrap the slit paper into a cone so that the slit edges line up with each other as precisely as possible and the cone is two layers of paper thick to get a figure with rosette group C_2, as pictured below:

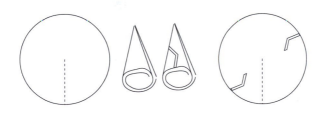

▷ **Exercise 23.** Slit a piece of paper to the center point and wrap it into a cone so that the cut edges line up and the cone is three layers thick. Cut an asymmetric figure. Unwind the paper and identify the rosette group of the result.

▷ **Exercise 24.** Figure out how to wrap the paper to cut a pattern with rosette group C_4. Create such a pattern.

SOFTWARE

1. *Kali*, by Jeffrey Weeks, of The Geometry Center at the University of Minnesota is a freeware program for Macintosh that generates rosette, frieze, and wallpaper patterns, and is well worth investigating, but it uses John Conway's orbifold notation, so you will need to translate (ftp://geom.umn.edu/pub/software/Kali). It is also available as a Java version by Mark Phillips, for any Java-enabled web browser at www.geom.umn.edu.
2. *KaleidoMania*, by Kevin Lee, also generates rosette, frieze, and wallpaper patterns, using either crystallographic or Conway's notation. It includes a very nice tutorial, and is available from Key Curriculum Press.

SUGGESTED READINGS

David W. Farmer, *Groups and Symmetry*, AMS, Providence, 1991.

A.V. Shubnikov and V.A. Koptsik, *Symmetry in Science and Art*, translated by G.D. Archard, Plenum Press, New York, 1974.

Peter S. Stevens, *Handbook of Regular Patterns*, MIT Press, Cambridge, 1980.

5. Two-Dimensional Symmetry

◆ 5.3. FRIEZE PATTERNS: LINE SYMMETRY

SUPPLIES
> tracing paper
> paper cash register tape
> scissors

Frieze, or border, patterns are formed by repetitions of a motif along a line. Think of a pattern made with a rubber stamp. Each repetition of the basic motif must be exactly the same in size and shape as the original figure. The allowed transformations or moves are called *rigid motions* or *isometries*. Isometry means "same measure": these are movements that do not change the motif in length, angle measure, or area. Two of these isometries are rotation and reflection. Another isometry is translation: shifting the motif along a straight line so that the copy is exactly parallel to the original.

Rotation (about a center point) moves the motif through a given angle. If the center of rotation is in the middle of the motif, it will rotate in place. In the picture below, the center of rotation, marked by a small circle, lies halfway between the toes of the two feet. Note that both footprints were made by the left foot.

Reflection across a line gives the mirror image, as if the mirror were placed with its edge on the line. If you are still thinking about rubber stamps, then you have to imagine that your stamp is two-sided. Reflection turns the stamp over. In the picture below, there is one left footprint and one right.

147

Translation along a line moves the motif to a position parallel to the original. Again, both footprints were made by the left foot.

There is one other isometry. **Glide reflection** in a line gives the mirror image, as if the mirror were placed with its edge on the line, and then translates it along the same line. Again, you have to imagine that your rubber stamp has been turned over. In the picture below, there is one left footprint and one right.

These four, rotation, reflection, translation, and glide reflection, are the only four rigid motions on the plane: the only four transformations that preserve size and shape without distortion.

If we want to create a border or strip pattern from these four moves, then we can only use them in ways compatible with the linear character of the strip. Imagine a strip of paper (cash register tape will do), with a dotted line running lengthwise down the middle. Use a motif without any symmetry of its own, so that any symmetry comes from applying one of the four isometries.

If we are going to apply a translation to a motif on the strip of paper, we must translate in the direction of the dotted line. This gives a frieze pattern that we name *hop*:

Reflecting our motif across the center line gives us two feet, a left and a right, and then we can translate this doubled motif along the center line to get another frieze pattern called *jump*:

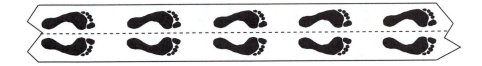

If, instead of reflecting across the center, we reflect in a vertical line perpendicular to the dotted line, and then translate the doubled motif along the strip, we get *sidle* or sidestep:

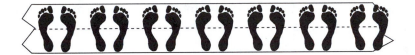

A glide reflection along the center line gives the frieze pattern called *step*:

So far we have used translations, reflections, and glide reflections, but no rotations. To respect the linear symmetry of the strip itself, we only allow rotations by 180°, or 2-fold rotations around points on the dotted center line. A fifth pattern is given by rotating the motif by 180° and then translating to get *spinning hop*. Note that there are only left feet.

Spinning jump is generated by reflecting the original foot print across the center line, then rotating the doubled motif by 180° and translating:

The last frieze pattern is generated by reflecting the original motif across a line perpendicular to the center line, then rotating the doubled motif by 180° and translating to give *spinning sidle*:

All frieze or border patterns can be classified as one of these seven types. They are generated by the four isometries, alone or in combination. A little experimenting will convince you that we have covered all combinations. For Exercises 1 through 6 below, give the relationship between the original motif and the end result of the two given isometries, ignoring the intermediate copy of the motif. Phrase your answers as one of these isometries: translation, rotation by 180°, reflection across the center line or a vertical line, or glide reflection.

▶ **Exercise 1.** What happens to a motif if you perform two successive glide reflections across the center line?

▶ **Exercise 2.** What happens if you reflect the original motif first across the center line, then across a line perpendicular to the center line?

▷ **Exercise 3.** What happens if you reflect the original motif first across the center line, then apply a glide reflection along the center line?

▷ **Exercise 4.** What happens if you reflect the original motif first across a line perpendicular to the center line, then apply a glide reflection along the center line?

▷ **Exercise 5.** What happens if you reflect the original motif first across the center line, then rotate by 180° about a point on the center line?

▶ **Exercise 6.** What other pairs of the allowable motions (rotation by 180°, reflection across the center line, reflection perpendicular to the center line, glide reflection along the strip, and translation along the strip) are there? Can you generate any results that cannot be classified as one of the four isometries?

▷ **Exercise 7.** Everyone in the class form a line and demonstrate the seven patterns by moving down the hall in the prescribed manners. (This is my favorite exercise in the book: it is hard to grade but fun to watch. The only trouble is that after the seventh pattern, the students are all way down at the end of the hall and tend to disappear down the stairs before I can catch them. Note that this exercise is justified by well-documented pedagogical theory, which asserts that doing this will imprint the patterns firmly in your mind.)

The hop-skip-jump terminology above comes from John Conway. Patterns like these have also been studied extensively by crystallographers, chemists who analyze how crystals are shaped, and they have their own more intimidating notation:

hop or $p111$	→　→　→　→　→　→　→
spinning hop or $p112$	←　→　→　→　→　→　→
step or $p1g1$	→　→　→　→　→　→　→
jump or $p1m1$	⇒⇒⇒⇒⇒⇒⇒⇒⇒
spinning jump or $pmm2$	⇒⇐⇒⇐⇒⇐⇒⇐
sidle or $pm11$	⇑ ⇑ ⇑ ⇑ ⇑ ⇑ ⇑ ⇑
spinning sidle or $pmg2$	⇑ ⇓ ⇑ ⇓ ⇑ ⇓ ⇑ ⇓

The crystallographic format is p＿＿＿. The p stands for primitive cell, but just ignore that for now. The first blank is filled in by m if there is a vertical mirror, and 1 otherwise. The second blank is filled in by m if there is a horizontal mirror, g if there is a glide reflection, and 1 otherwise. The last blank is filled by 2 if there is a 180° (or 2-fold) rotation, and 1 otherwise. Since jump has no vertical mirror, a horizontal mirror, and no rotation, its symbol is $p1m1$. Spinning sidle has a vertical mirror, no horizontal mirror (but it does have a glide reflection) and a 2-fold rotation, so its symbol is $pmg2$. In classifying a frieze pattern with this notation, use the following outline:

1. The first character is always **p**.
2. The second character is:
   m if there is a mirror line perpendicular to the center line of the strip.
   1 if there is no mirror line perpendicular to the center line.
3. The third character is:
   m if there is a mirror line along the center line of the strip.
   g if there is not a mirror line but there is a glide reflection along the center line of the strip.
   1 if there is no reflection or glide reflection along the center line.
4. The fourth character is
   2 if there is a 180° rotation.
   1 if there is no 180° rotation.

In all cases, "1," designates no to the question asked for each position. One easy way to determine whether a motif is reflected or rotated is to trace the pattern on a bit of tracing paper, then reflect (turn the paper over, being careful to distinguish between a horizontal or vertical reflection) or rotate and see whether the traced pattern lines up with the original. Alternatively, you can use a mirror to check for reflection lines.

▶ **Exercise 8.** Classify the following patterns:
(a) DDDDDDDDDDDDD
(b) XXXXXXXXXXXXX
(c) ZZZZZZZZZZZZ
(d) YYYYYYYYYYYYY
(e) ∪∩∪∩∪∩∪∩∪∩∪∩
(f) GGGGGGGGGGGGG
(g) bpbpbpbpbpbp

▶ **Exercise 9.** Classify the following patterns:
(a)

(b)

(c)

(d)

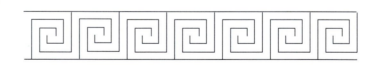

(e)

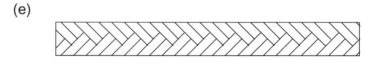

(f)

(g)

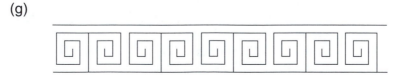

(h)

► **Exercise 10.** Classify the following patterns:

(a)

(b)

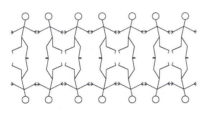

(c)

(d)

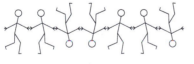

(e)

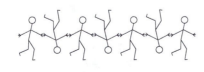

(f)

(g)

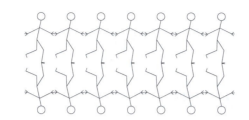

▷ **Exercise 11.** Draw your own examples of each of the seven frieze patterns.

▷ **Exercise 12.** How would you classify the pattern formed by the footprints of someone skipping?

► **Exercise 13.** Fold a long strip of paper in a zigzag and cut out an asymmetric motif. Unfold and classify the pattern.

▷ **Exercise 14.** Roll a long strip of paper around several times into a cylinder. Do not flatten the cylinder, but cut through one side of it, cutting out an asymmetric motif. Unroll and classify the pattern.

▷ **Exercise 15.** Fold a long strip of paper lengthwise and then roll it around several times into a cylinder and cut out an asymmetric motif. Unroll and classify the pattern.

► **Exercise 16.** Roll a long strip of paper around several times into a Möbius band (a cylinder with a half twist) as shown below and cut out an asymmetric motif. Unroll and classify the pattern.

▷ **Exercise 17.** In Exercises 13–16, you should have created four of the seven frieze patterns. Figure out how to fold and/or roll a strip of paper to make the other three. Cut out one of each. The easiest way to figure these out is to lightly sketch a simple pattern of the desired type on the strip of paper (space the motifs well apart), then find a way to fold the paper so that the motifs land on top of each other. Then cut out your pattern.

SOFTWARE

1. *Kali*, by Jeffrey Weeks, of The Geometry Center at the University of Minnesota, is a freeware program for Macintosh that generates rosette, frieze, and wallpaper patterns, and is well worth investigating, but it uses John Conway's orbifold notation, so you will need to translate (ftp://geom.umn.edu/pub/software/Kali). It is also available as a Java version by Mark Phillips, for any Java-enabled web browser at www.geom.umn.edu.

2. *KaleidoMania* by Kevin Lee also generates rosette, frieze, and wallpaper patterns, using either crystallographic or Conway's notation. It includes a very nice tutorial, and is available from Key Curriculum Press.

3. It is also an interesting exercise to create frieze patterns with a drawing program such as *Macromedia FreeHand* or *Adobe Illustrator*. Such programs have reflection, rotation, and alignment transformations, which allow one to manipulate the motifs as needed.

SUGGESTED READINGS

John H. Conway, "The Orbifold Notation for Surface Groups," in *Groups, Combinatorics, and Geometry*, ed. Liebeck and Saxl, Cambridge University Press, New York, 1992.

Donald W. Crowe and Dorothy K. Washburn, *Symmetries of Culture*, University of Washington Press, Seattle, 1988.

David W. Farmer, *Groups and Symmetry*, AMS, Providence, 1991.

Peter S. Stevens, *Handbook of Regular Patterns*, MIT Press, Cambridge, 1980.

5. Two-Dimensional Symmetry

◆ 5.4. WALLPAPER PATTERNS: PLANE SYMMETRY

SUPPLIES
> tracing paper
> colored pencils

Wallpaper patterns are formed by repetitions of a motif in such a way as to cover the plane. As with frieze patterns, they are generated by the four planar isometries: translation, rotation, reflection, and glide reflection. There are exactly 17 different wallpaper patterns. We do not expect you to memorize all of these, but to learn the principles that allow you to design and classify them.

All wallpaper patterns are generated by translation of a basic building block, or *cell*. This cell may or may not have symmetry of its own by rotation, reflection or glide reflection, and may or may not be translated in such a way as to give an overall pattern with symmetry. Think of the cell as a woodblock, which is then inked and copies are stamped out to cover a roll of blank wallpaper. To make sure that no gaps are made, the copies must fit together edge to edge and vertex to vertex, forming a *lattice*, or grid, for the wallpaper pattern, and each region outlined by this lattice is an exact duplicate. The lines laying out the lattice form two parallel families. Translation takes one line, and one cell, to the next. We indicate the translation by arrows or vectors, and a typical cell is shaded. Note that it does not really matter where a translation starts, only its direction and how long it is are important: If one point gets translated 1 inch northwest, then all points get translated 1 inch northwest.

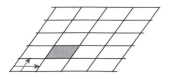

Let us start with three examples of wallpaper patterns, exhibiting various of the isometries in action. The simplest wallpaper pattern is generated by translations alone. Begin with a simple motif that has no symmetry (i.e., no mirror lines or rotations), so that it will be clear which symmetries come from applying an isometry and which came from the original motif. Translating the motif in the direction of a line gives a frieze pattern (the hop pattern). You can then translate this strip in the direction of another line to fill in the whole plane.

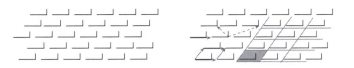

The picture on the right is the same wallpaper pattern but with arrows showing the direction and length of the translations. I have chosen not to make these lines perpendicular, to emphasize that they do not have to be. The dotted lines show other possible choices for the translations, but two directions are enough to fill the plane, so we do not need these. The gray lines in the direction of the translation vectors divide the plane into a lattice of parallelograms. The shaded parallelogram outlined by the lattice of translation lines is called the cell for the pattern, and each parallelogram of the gridded wallpaper is a copy of this cell. This wallpaper pattern is designated **p111**, or **p1** for short, using the standard crystallographic notation, which we will explain in depth later. It must be noted that although this notation looks very much like that used to describe the frieze patterns, there are essential differences.

For our next example we consider **p2mg**, which is a wallpaper pattern with lines of reflection and glide reflection meeting at right angles. We use dashed lines to indicate lines of reflection and dotted lines for lines of glide reflection, but strongly recommend that you use color instead, perhaps red for reflection and blue for glide reflection.

▷ **Exercise 1.** Find the images of the original motif above after a reflection across the dashed line, and after a glide reflection along the dotted line. How are the images of the reflection and the glide reflection related?

Here is a simple **p2mg** pattern. The original motif has no symmetry of its own, but it is reflected and glide reflected to give symmetry to the overall pattern. The basic cell that generates the wallpaper by translation must then contain repetitions of the motif for each of the orientations of the motif that appear in the wallpaper. To preserve the lines of reflection and glide reflection, we are forced to use a rectangular grid of cells.

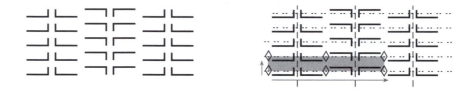

▷ **Exercise 2.** In the **p2mg** pattern pictured above, which frieze pattern is made by one horizontal row?

The third example I have chosen is called **p611** or **p6** and contains only rotational symmetry. The center of each six-part motif is a sixfold center of rotation not only for the motif, but for the overall wallpaper pattern. This pattern is said to have a hexagonal grid.

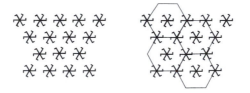

A regular hexagon can be divided into six equilateral triangles, and to make the basic cell as simple as possible, note that the tiling of hexagons can be replaced by a cell consisting of two of these equilateral triangles forming a rhombus.

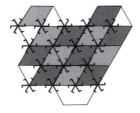

The sixfold motif and the hexagonal grid induce other points of rotation: sixfold centers (marked with a ◯) at the center of the motif, 3-fold centers (△), and 2-fold centers (◇). We mark only those in the basic cell.

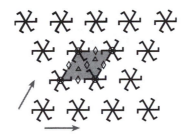

▷ **Exercise 3.** Below is a triangle and its reflection. Draw the mirror line. State a general rule for finding the mirror line of a motif and its reflection.

▶ **Exercise 4.** Below is a triangle and its image after a 180° rotation. Mark the center of rotation. State a general rule for finding the center of rotation for a motif and its image.

▷ **Exercise 5.** Below is a triangle and its images after a 5-fold rotation. Mark the center of rotation. State a general rule for finding the center of rotation for a motif and its images.

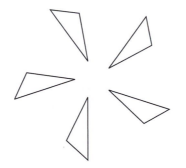

The basic cell and the pattern of reflection lines, glide reflection lines, and rotation centers of various orders completely determine the wallpaper pattern. Once you have the symmetries of the basic cell figured out, the overall pattern is easy to draw using tracing paper to copy translations of the basic cell. Throughout, we use the following standard symbols to indicate elements of symmetry:

direction and length of translation = ⟶

line of reflection = — — — — — — ·

line of glide reflection = -- -- -- -- -- -- ·

center of a 2-fold rotation (180°) = ◆ if it occurs on a line of reflection and ◇ if not

center of a 3-fold rotation (120°) = ▲ if it occurs on a line of reflection and △ if not

center of a 4-fold rotation (90°) = ■ if it occurs on a line of reflection and □ if not

center of a 6-fold rotation (60°) = ⬣ if it occurs on a line of reflection and ⬡ if not

One startling fact about wallpaper patterns is that the only rotations possible are 2-, 3-, 4- and 6-fold. To see that this is true, suppose that we had a 5-fold rotation point at point A. There will, of course, be infinitely many other 5-fold rotation points, either within the same cell or from translating A. Let B be another 5-fold rotation point so that A and B are the closest two points of all the 5-fold rotation centers. Of course, there will be others equally close, the translates of A and B, but there should be no pair that is closer together. If B is rotated by $\frac{360}{5} = 72°$ about A, then we have another 5-fold rotation point D with $AB = AD$, since the wallpaper pattern is symmetric by rotation about A. Similarly, if A is rotated 72° around B, we get another 5-fold rotation point at C with $BC = AB$. But then C and D are closer to each other than A and B were. Something goes wrong whenever we assume that a 5-fold center exist, so there cannot be any such points.

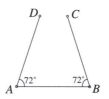

▶ **Exercise 6.** Show that there cannot be a 7-fold rotation point.

Similar arguments eliminate the possibility of rotation points for all orders greater than 6. The argument does not work for rotations of order 2, 3, 4, and 6. For example, if point A, B, and C as above are rotation centers of order 6, then $D = C$, and the points lie equally spaced as pictured below, and so any wallpaper pattern with a rotation center of order 6 must conform to a grid of equilateral triangles (although traditionally they are described as having a hexagonal lattice).

▷ **Exercise 7.** Adapt the explanation in the last paragraph to show that the possibility of rotation points of orders 2, 3, and 4 cannot be eliminated by an argument similar to the one above showing that there cannot be a rotation center of order 5.

In our three examples we used lattices of skew parallelograms for **p111**, rectangles for **p2mg**, and hexagons (or pieces of hexagons) for **p611**. Since we always have translations in two directions, all lattices (even the hexagonal lattice, once we divide it up into equilateral triangles forming rhombi) are parallelograms of some sort. When a reflection is involved, only certain types of parallelograms are possible. For example, consider the motif below, with the indicated translations and reflection line.

Motif *a* is reflected across line 1 to get motif *b*. The doubled motif *ab* is translated in one direction to *cd*, and in the other direction to *ef*, which is then translated to *ij*. But the mirror line 1 is also translated to line 2, and then to line 3, etc. Motif *ab* is reflected across line 2 to *gh*. Thus, there are copies of the motif directly above one another. Whenever a horizontal reflection is present, the lattice must preserve this property that motifs are aligned vertically.

▶ **Exercise 8.** Show that if a wallpaper pattern contains a horizontal glide reflection, then copies of the motif will lie along a vertical line.

Whenever there is a reflection or glide reflection, the lattice must have the property that its corners or vertices must make two sets of parallel lines that are perpendicular. Obvious choices for such lattices are rectangles and squares. The arrangement of two equilateral triangles shown previously for the hexagonal lattice also has this property. The only other choice is a lattice of rhombuses, since it can be shown that the diagonals of a rhombus are perpendicular.

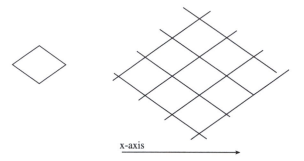

x-axis

We now know that the choices for lattices are parallelograms, rectangles, squares, rhombuses, or hexagons, and the only possible rotation centers are of order 2, 3, 4, or 6. The 17 wallpaper patterns are classified, as were the frieze patterns of the previous section, using standard international crystallographic notation. In analyzing a pattern, one looks for lines of reflection (mark these in red or with dashed lines), lines of glide reflection (blue or dotted lines), and rotation centers, marked appropriately with ◆, ▲, ■ or ⬣ if the rotation center falls on a red (dashed) line, and ◇, △, □ or ◯ if not. Below is a list of all 17 forms. Some books use the short forms given.

The 17 Wallpaper Patterns

Long form	Short form	Lattice
p111	p1	Parallelogram
p211	p2	Parallelogram
p1m1	pm	Rectangle
p1g1	pg	Rectangle
c1m1	cm	Rhombus
p2mm	pmm	Rectangle
p2mg	pmg	Rectangle
p2gg	pgg	Rectangle
c2mm	cmm	Rhombus
p411	p4	Square
p4mm	p4m	Square
p4gm	p4g	Square
p311	p3	Hexagon
p3m1	p3m1	Hexagon
p31m	p31m	Hexagon
p611	p6	Hexagon
p6mm	p6m	Hexagon

The international crystallographic notation is _ _ _ _ .

1. The first blank is filled by **p** or **c**, for the type of cell. Lattices of parallelograms, rectangles, squares, and hexagons have **p** (for primitive), while a rhombic lattice gets **c** (for centered). There are only two wallpaper patterns with type **c** cells, as we shall see later. The cell should have the centers of highest rotation order at its vertices. Primitive (type **p**) cells should have one side along the x-axis and centered cells (type **c**) should have one diagonal of the rhombus cell on the x-axis. Rotate the whole picture if necessary.

2. The second blank is filled in with **n** for the highest order of rotation: this will be 1 (no rotation), 2 when there is a 2-fold rotation, 3 for a 3-fold rotation, 4 for a 4-fold rotation, or 6 for a 6-fold rotation. As we saw in **p611** there are often several centers of rotation with different orders.

3. The third blank is filled by **m** if there is a vertical mirror (perpendicular to the x-axis), **g** if there is no mirror but there is a glide reflection perpendicular to the x-axis, and **1** otherwise.

4. The fourth blank denotes a line of symmetry at angle α to the x-axis where α depends on the order **n** of the second blank. Look at lines at angle $\alpha = 180°$ if **n** = 1 or 2, $\alpha = 45°$ if **n** = 4, $\alpha = 60°$ if **n** = 3 or 6. If this line at angle α is a line of reflection, then there is an **m** in this blank. If it is not a mirror but is a line of glide reflection, then put a **g**. Otherwise, put a **1**.

In all cases, **1** designates no to the question asked for each position. These crystallographic symbols are often abbreviated:

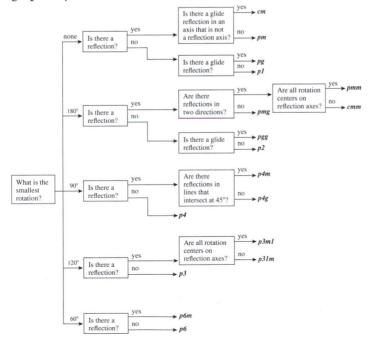

Flow chart for the seventeen one-color, two-dimensional patterns.

We have examined the patterns **p111, p2mg,** and **p611** as introductory examples. We next work through the remaining 14 patterns. Generally, if one can fill in the basic cell, it is easy to generate the wallpaper: Just think of the cell as a rubber stamp, and cover the plane with copies of the cell and its motifs.

▷ **Exercise 9.** Here is a partial basic cell for a **p2** or **p211** wallpaper pattern, with the rotation centers marked. Fill in the rest of the basic cell (you will need one more copy of the motif within the cell) and then draw a section of the wallpaper, four cells wide and three cells tall.

▶ **Exercise 10.** Here is a basic cell for a **p1g1** wallpaper pattern, with the lines of glide reflection marked. Because the motifs do not line up, note that one copy of the motif is split in two, but when you draw the wallpaper, these pieces should fit together. Draw a section of the wallpaper, four cells wide and three cells tall.

▷ **Exercise 11.** Here is a section of **p1m1** wallpaper. Draw two arrows indicating the directions of translation. Draw lines of reflection in red. Outline and shade the basic cell (it will contain two repetitions of the motif, one each of the two orientations).

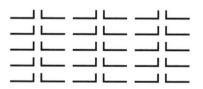

▷ **Exercise 12.** Which frieze pattern is formed by a single horizontal row from the wallpaper **p1m1** pictured above?

▶ **Exercise 13.** Here is a section of **p2gg** wallpaper. Draw two arrows indicating the directions of translation. Draw lines of glide reflection in blue. Put ◇ marks at the 2-fold rotation points. Outline and shade the basic cell, remembering that the corners should be centers of rotation and that the cell should contain copies of the motif in each of its configurations.

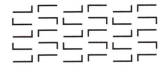

▷ **Exercise 14.** Which frieze pattern is formed by a single horizontal row from the wallpaper **p2gg** pictured above?

▷ **Exercise 15.** Here is the original motif, a shaded partial basic cell, and the lines of reflection and 2-fold rotation centers for the pattern **p2mm**. Note that ◆ is used, since the rotation centers are on the lines of reflection. Fill in the pattern, first in the basic cell, and then in the rest of the grid shown.

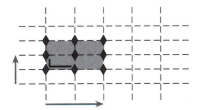

At this point, we have studied the two patterns on parallelogram grids; **p1** and **p2**, and five rectangular patterns; **p2mg**, **p1g1**, **p1m1**, **p2mm**, and **p2gg**. There are two wallpaper patterns, **c1m1** and **c2mm**, that use a centered rhombic lattice. It is necessary to use a rhombus to comply with the mandate that the corners of the cell should be centers of rotation. One of these patterns is **c2mm**, which will have to have rotation centers of order 2, mirror lines perpendicular to the x-axis, and another set of mirror lines parallel to the x-axis.

▷ **Exercise 16.** Below is a piece of a **c2mm** pattern, with the centered cell shaded. Draw reflection lines in red and glide reflection lines in blue. Indicate the 2-fold centers of rotation with a ◆ if on a mirror line, ◇ if not (this pattern has both).

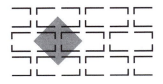

▶ **Exercise 17.** Here are two copies of the motif, a shaded partial basic cell, and the dashed lines of reflection and dotted lines of glide reflection for the pattern **c1m1**. As in the previous patterns

c2mm and **p1g1**, the basic cell contains fragments instead of complete motifs, but these fragments fit together to make the pattern. Fill in the pattern, first in the basic cell, and then to cover the area shown.

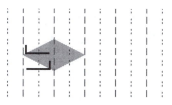

Patterns with 4-fold symmetry must have a lattice with 4-fold symmetry, so the next group of wallpaper patterns use a lattice of squares.

▷ **Exercise 18.** Here is the original motif and a shaded partial basic cell with the central 4-fold center of rotation marked with a □ for the pattern **p4**. Fill in the pattern first in the basic cell, then for a piece of wallpaper four cells wide and three cells tall. There are other rotation centers on the border of the cell. Mark them with the appropriate symbols.

▷ **Exercise 19.** Here is a **p4gm** pattern. Draw the reflection lines in red and the glide reflection lines in blue. You should find families of glide reflection lines in 4 directions. Indicate the centers of rotation appropriately. Shade a square basic cell.

▷ **Exercise 20.** Show what happens when you reflect the motif on the facing page across the dashed lines, which meet at a 45° angle. Be sure to get all the repetitions of the figure.

▶ **Exercise 21.** A **p4mm** will have a square lattice, 4-fold centers of rotation, and mirror lines perpendicular to the *x*-axis and at an angle of 45° to the *x*-axis. It can also be described by reflecting a motif through each of the sides of a right isosceles triangle. Draw such a pattern.

The remaining wallpaper patterns, **p3, p3m1, p31m**, and **p6mm**, use hexagonal lattices, though we use a basic cell consisting of two equilateral triangles forming a rhombus, as explained above.

▷ **Exercise 22.** Below is a motif and a cell for **p3** showing the rotation centers. Fill in the pattern first in the basic cell, then for a piece of wallpaper four cells wide and three cells tall.

▶ **Exercise 23.** The patterns **p3m1** and **p31m** are often confused, even in some textbooks. Below are examples of each. Draw the reflection lines in red and the glide reflection lines in blue. Indicate the centers of rotation appropriately. One key difference between the two is that **p3m1** has all rotation centers on the red lines, while **p31m** has some on the red lines and some not. Figure out which is which and shade the basic cells:

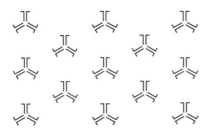

▷ **Exercise 24.** Show what happens when you reflect the motif across the dashed lines, which meet at a 30° angle. Be sure to get all the repetitions of the figure.

▷ **Exercise 25.** A **p6mm** will have a hexagonal lattice, 6-fold centers of rotation, and mirror lines perpendicular to the x-axis and at an angle of 60° to the x-axis. It can also be described by reflecting a motif through each of the sides of a 30°-60°-90° right triangle. Draw such a pattern.

▶ **Exercise 26.** Classify the patterns below:

▷ **Exercise 27.** Draw your own examples of each of the seventeen wallpaper patterns.

SOFTWARE

1. *Kali*, by Jeffrey Weeks, of The Geometry Center at the University of Minnesota, is a freeware program for Macintosh that generates rosette, frieze, and wallpaper patterns, and is well worth investigating, but it uses John Conway's orbifold notation, so you will need to translate (ftp://geom.umn.edu/pub/software/Kali). It is also available as a Java version by Mark Phillips, for any Java-enabled web browser at www.geom.umn.edu.

2. *KaleidoMania*, by Kevin Lee, also generates rosette, frieze, and wallpaper patterns, using either crystallographic or Conway's notation. It includes a very nice tutorial, and is available from Key Curriculum Press.

3. It is also an interesting exercise to create all the wallpaper patterns with a drawing program such as *Macromedia FreeHand* or *Adobe Illustrator*. Such programs have reflection, rotation, and alignment transformations, which allow one to manipulate the motifs as needed.

SUGGESTED READINGS

John H. Conway, "The Orbifold Notation for Surface Groups," in *Groups, Combinatorics, and Geometry*, ed. Liebeck and Saxl, Cambridge University Press, New York, 1992.

H.S.M. Coxeter, *Introduction to Geometry*, Wiley & Sons, New York, 1969.

Donald W. Crowe and Dorothy K. Washburn, *Symmetries of Culture*, University of Washington Press, Seattle, 1988.

David W. Farmer, *Groups and Symmetry*, AMS, Providence, 1991.

Branko Grünbaum and G.C. Shephard, *Tilings and Patterns*, W.H. Freeman, New York, 1987.

Pamela K. McCracken and William S. Huff, "Wallpapers precisely 17: an eye-opening confirmation," *Symmetry: Culture and Science* 3, 1992.

Doris Schattschneider, "The Plane Symmetry Groups: Their Recognition and Notation," *American Mathematical Monthly* 85, 1978.

Peter S. Stevens, *Handbook of Regular Patterns*, MIT Press, Cambridge, 1980.

Hermann Weyl, *Symmetry*, Princeton University Press, Princeton, 1952.

5. Two-Dimensional Symmetry

◆ 5.5. ISLAMIC LATTICE PATTERNS

SUPPLIES
> tracing paper
> non-photo blue pencil

One of the characteristic features of Islamic art is the use of the *girih*, beautifully intricate latticework or interlaced designs based on polygons and star polygons. One theory for the prevalence of such designs is that since the Islamic religion forbids figurative art, craftsmen used such patterns to indicate the perfection and infinite presence of Allah. Another theory is that such patterns were favored because of their adaptability to the most common building materials, brick, plaster, and tile, in a region where wood was scarce.

Such patterns are constructed using well-defined grids and rules, and can be classified as wallpaper patterns. In this section we will attempt to draw some of the simpler ones, following a method explained by A.K. Dewdney in his column in *Scientific American*.

The process of creating a latticework pattern starts with choosing a grid of squares, triangles, or hexagons. For our first example, I have chosen a square grid. At each vertex of the grid, draw a circle. Place equally spaced dots around the circle. To keep the drawing simple, I have not shown the grid, only the circles with eight dots each.

Draw lines bouncing from one circle to another. This can be thought of as playing billiards on a table with circular obstacles, or as pinball without gravity. Remember that the angle of incidence must be the same as the angle of reflection. All reflections off the construction circles should use the same angle, so that a star polygon is formed about the circle. Here I have chosen to use a 45° angle of reflection, creating a 90° angle between the line approaching the circle and the line leaving. We will refer to this drawing as the *design* for the final interlaced figure.

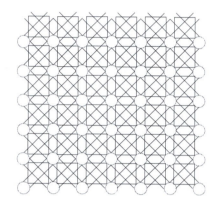

Finally, thicken the lines, and weave them, alternating overpasses and underpasses. Erase the construction circles, or if you draw them lightly with a non-photo blue pencil, they should not show when the finished drawing is photocopied. There are two different sizes of squares appearing in the pattern, each formed by a single strand.

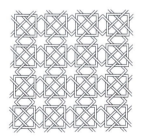

▶ **Exercise 1.** Classify both the design and the interlaced pattern above as wallpaper patterns. Explain what difference the crossings make in the classification.

Here is another example, using a triangular grid of circles with six dots each.

At each circle, the lines approaching and leaving the circle form a 120° angle. We have made one of the strands below heavier to emphasize this point.

The final interlaced lattice pattern exhibits a sixfold symmetry. All of the strands are infinite, unlike the previous example. At first glance, there appear to be three types of strands: sort of slanted steps descending from left to right, their mirror images ascending from left to right, and vertical zigzags. A closer look reveals that all of these are congruent and they are rotations of each other.

▷ **Exercise 2.** Classify both the design and the interlaced pattern above as wallpaper patterns.

In designing an Islamic latticework pattern, one must choose which grid to use, how big a circle to use on the grid, how many dots to place on each circle, the angle that the lines will form at the circle, and the final choice of over- and underpasses. Let us examine each of these choices.

With a triangular grid, the final pattern will have to be one of **p3, p3m1, p31m, p6,** or **p6mm,** with either 3-fold or 6-fold symmetry. A square grid allows the wallpaper patterns **p4, p4mm,** and **p4gm,** all with four-fold symmetry. These are the most common grids for using the method outlined above, though more complicated arrangements of circles use rectangular, rhombic, or parallelogram grids.

Once the grid is chosen and the circles drawn, the number of dots placed on each circle determines the type of star polygon featured in the final design. Since the grid partially determines the wallpaper pattern, the type of star polygon must be compatible with the order of rotation. In general, the more dots, the more satisfyingly complex the final pattern will be.

▷ **Exercise 3.** For a square grid, list several choices that could be made for the number of dots on each circle. Repeat for a triangular grid.

The angle of reflection is determined when you draw the first line connecting a dot on one circle to a dot on another. This choice then dictates all the remaining lines, since the angle of incidence must equal the angle of reflection, and star polygons are formed only when the same angle is applied at each dot.

In the final step of interlacing, once one chooses to make one crossing into an overpass, that decision determines all of the other crossings. Recall that in the section on Celtic knots, we showed that it is always possible to alternate over- and underpasses. Thus, each design for a latticework pattern gives rise to only two possible interlacings.

▷ **Exercise 4.** Draw the interlace pattern started below, in which each strand forms a hexagon:

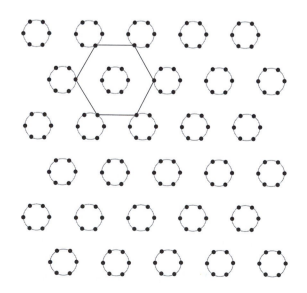

▶ **Exercise 5.** Classify the design and the final interlaced pattern in Exercise 4 as wallpaper patterns.

You may have noticed that I have not yet addressed the question of the size of the circles. This is to some degree an aesthetic decision, but note the following two exercises.

▷ **Exercise 6.** Draw the interlace pattern for the design started below, in which each strand forms a hexagon:

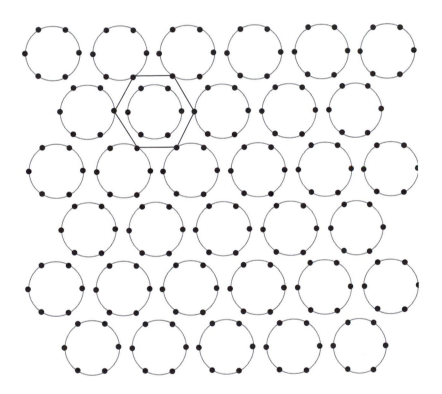

▷ **Exercise 7.** Classify both the design and the final interlaced pattern in Exercise 6 as wallpaper patterns.

The designs of Exercises 4 and 6 use the same grid, the same number of dots per circle, and even the same angles of reflection. Furthermore, each strand forms a hexagon. The only difference is in the size of the circles. Following is yet another design using the same inputs, but varying the size of the circles. In this design three hexagons meet at a point, so they cannot be interlaced by a simple over and under rule.

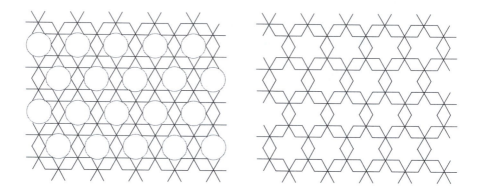

▷ **Exercise 8.** Classify the design above as a wallpaper pattern.

▷ **Exercise 9.** Figure out the relationship between the size of the circles (the radius) and their spacing in the example above. Compare your finding with the same data from Exercises 4 and 6.

▷ **Exercise 10.** Draw the interlace pattern associated with the design started below:

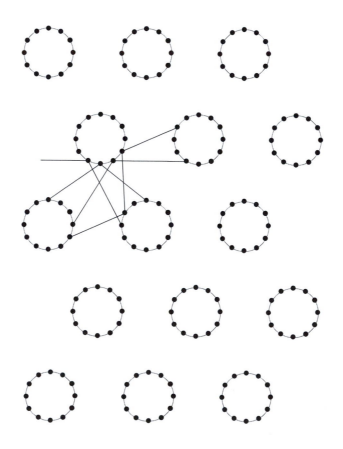

One can also use more complicated grids, such as the one below. This one has circles of two different sizes, laid out in a symmetric pattern.

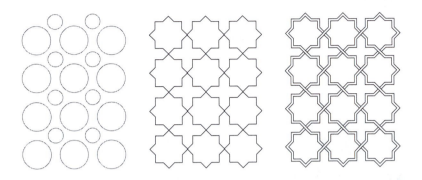

▷ **Exercise 11.** Classify both the design and the interlaced pattern above as wallpaper patterns.

▶ **Exercise 12.** What semiregular tiling from Section 4.1 is this arrangement of circles related to?

▶ **Exercise 13.** Below is a grid of two sizes of circles, with 2, 3, and 12 dots each. I have drawn the two different types of strands. Finish the interlace pattern.

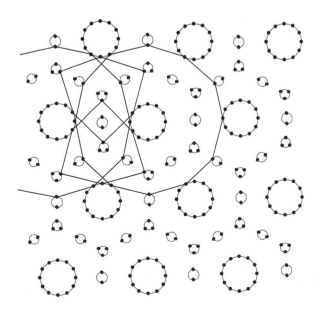

▷ **Exercise 14.** Classify the interlaced pattern in Exercise 13 as a wallpaper pattern.

Our next example has circles and ellipses (ovals) for the bounces:

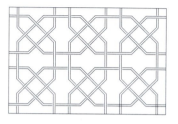

▷ **Exercise 15.** Show that circles alone cannot generate the pattern above, unless one allows them to overlap.

▷ **Exercise 16.** Classify the interlaced pattern above as a wallpaper pattern.

Although no wallpaper can have five- or tenfold symmetry, pentagons and decagons are surprisingly common in interlace patterns. This requires some irregularity in the construction. Below is one such pattern:

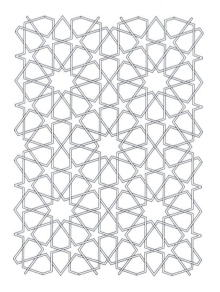

▷ **Exercise 17.** Classify the interlaced pattern above as a wallpaper pattern.

SUGGESTED READINGS

Syed Jan Abas and Amer Shaker Salman, *Symmetries of Islamic Geometric Patterns*, World Scientific Press, Singapore, 1995.

J. Bourgoin, *Arabic Geometrical Pattern and Design*, Dover, New York, 1973.

Titus Burckhardt, *Art of Islam: Language and Meaning*, World of Islam Festival Publishing Co., 1976.

Keith Critchlow, *Islamic Patterns*, Thames and Hudson, New York, 1976.

A.K. Dewdney, "Computer Recreations: Imagination meets geometry in the crystalline realm of latticeworks," *Scientific American* 258, June 1988.

Issam El-Said and Ayşe Parman, *Geometric Concepts in Islamic Art*, Dale Seymour Publ., Palo Alto, 1976.

Branko Grünbaum and G.C. Shephard, "Interlace Patterns in Islamic and Moorish Art," *Leonardo* 25, 1992.

Gülru Necipoğlu, *The Topkapi Scroll—Geometry and Ornament in Islamic Architecture*, The Getty Center for the History of Art and the Humanities, Santa Monica, 1995.

Peter S. Stevens, *Handbook of Regular Patterns*, MIT Press, Cambridge, 1980.

6. Other Dimensions, Other Worlds

◆ 6.1. FLATLANDS

Two-Dimensional Literature

There are several, admittedly minor, literary works that are set in 2-dimensional worlds. The first and definitive such work is Edwin Abbott's *Flatland*, written in 1884. The others, lineal descendants, extend his ideas and aims in various ways. In all of these books, the geometry of the surroundings and its effects on the lives of the inhabitants is a paramount consideration. All of these attempt to make the reader think about dimensionality and its consequences.

Edwin Abbott's *Flatland*

Flatland was written in 1884 by Edwin Abbott, the headmaster of a noted English public school. His intent in writing his book is made clear in the dedication:

> To the Inhabitants of SPACE IN GENERAL and H.C. IN PARTICULAR this Work is Dedicated by a Humble Native of Flatland in the Hope that even as he was Initiated into the Mysteries of THREE DIMENSIONS having been previously conversant with ONLY TWO so the Citizens of that Celestial Region may aspire yet higher and higher to the Secrets of FOUR FIVE OR EVEN SIX Dimensions thereby contributing to the Enlargement of the IMAGINATION and the possible Development of that most rare and excellent Gift of Modesty among the Superior Races of SOLID HUMANITY.

Thus, it appears that his intent was twofold: to get us to imagine higher dimensions by thinking in analogy with his two-dimensional characters as they come to comprehend a third dimension and also to comment on the social conditions of the time.

▷ **Exercise 1.** Read Abbott's *Flatland*. (It is available from Dover Press for only $1.00 and it's quite short.)

First a brief overview of Flatland and its inhabitants: Flatland is pictured as a plane, inhabited by polygons. Women of all classes are straight lines, but for men the number of sides designates their social class. Isosceles triangles are generally of lower class, and the closer to equilateral the more status a triangular man would have. Isosceles triangles with very short bases, and thus very sharp vertices, are generally soldiers, while equilateral triangles are tradesmen. Squares and pentagons form the professional class. The nobility are regular polygons with six or more sides, culminating in the Polygonal class, with a great number of sides. The priestly class, called Circles out of courtesy, consists of polygons with so many very short sides that they are almost circular. In general, offspring of isosceles triangles are closer to equilateral than their fathers, and offspring of regular polygons gain one additional side in each generation. The narrator of the book is A Square, a respectable professional gentleman and amateur mathematician. A Square is generally in complete conformity with the society of which he is a member, and his frequent and unquestioning praise of the wisdom of the laws and conventions of this society make his eventual disgrace, for pursuing completely theoretical but heretical knowledge, more poignant.

In Flatland, women are straight lines and are therefore objects of fear, since they can easily pierce and kill other figures, at considerable hazard to themselves. They are depicted as incapable of thought: fragile, dangerous, ineducable, unpredictable. They are referred to as the Thinner Sex or the Frailer Sex, and there are a number of laws intended to lessen the danger they pose. It must be noted that Abbott was in fact an outspoken advocate for opportunities for education for women, as well as for members of all social classes. At the time *Flatland* was written, women had only recently gained quite limited permission to attend classes at Oxford and Cambridge. His depiction of women, the restrictions placed on them, ostensibly for their protection, and their quite artificial relationship with the males are thus intended as satire.

Another aspect of Flatland also has clear satiric intent: the rigid emphasis on conformity. Abbott had among his students a number from the lower classes and religious minorities who clearly expressed a deep sense of gratitude to him for making their education possible. Nonconformity is persecuted in Flatland, since its existence questions the foundations of the rigid social hierarchy. The inconvenience of not being able to determine the number of sides of an irregular pentagon from a single angle is cited as an insuperable burden to society. In Flatland, the life of an Irregular is described:

The Irregular is from his birth scouted by his parents, derided by his brothers and sisters, neglected by the domestics, scorned and suspected by society, and excluded from all posts of responsibility, trust, and useful activity. His every movement is jealously watched by the police till he comes of age and presents himself for inspection; then he is either destroyed, if he is found to exceed the fixed margin of deviation, or else immured in a Government Office as a clerk of the seventh class; prevented from marriage; forced to drudge at an uninteresting occupation for a miserable stipend; obliged to live and board at the office, and to take even his vacation under close supervision; what wonder that human nature, even in the best and purest, is embittered and perverted by such surroundings.

The upper classes in Flatland appear self-interested and vicious, suppressing all threats to the status quo, which enshrines them by the accident of birth endowing them with a respectable number of sides. They seem obsessed with the threat posed by the lower classes.

▷ **Exercise 2.** Give several examples from *Flatland* of the ways in which the minority nobility and priestly classes limit the threat of revolution from the lower but more numerous isosceles classes.

Abbott gives fairly short shrift to the physics of his two-dimensional world. He introduces a convenient constant Fog, by means of which distance can be deduced by making farther points less distinct than closer ones. Since the Flatland plane seems to lack physical features, it might seem to be difficult to figure out where one is. Abbott explains that there is a constant gravitational attraction to the south, while rain, when it comes, always falls from the north. This provides two ways of orienting oneself in Flatland. However, A Square states,

I must omit many matters of which the explanation would not, I flatter myself, be without interest for my Readers: as for example, our method of propelling and stopping ourselves, although destitute of feet; the means by which we give fixity to structures of wood, stone, or brick, although of course we have no hands, nor can we lay foundations as you can, nor avail ourselves of the lateral pressure of the earth; the manner in which the rain originates in the intervals between our various zones, so that the northern regions do not intercept the moisture from falling on the southern; the nature of our hills and mines, our trees and vegetables, our seasons and harvests; our Alphabet and method of writing, adapted to our linear tablets; these and a hundred other details of our physical existence I must pass over, nor do I mention them now except to indicate to my readers that their omission proceeds not from forgetfulness on the part of the author, but from his regard for the time of the Reader.

▷ **Exercise 3.** Figure out and describe a way for the inhabitants of Flatland to propel and stop themselves.

▶ **Exercise 4.** Figure out and describe a way for the inhabitants of Flatland to fix their houses in place.

▷ **Exercise 5.** Figure out and describe an alphabet and linear system of writing for the inhabitants of Flatland.

Abbott's investigation into dimensionality begins with a description of a dream that A Square had on December 30, 1999, a date that probably seemed sufficiently far in the future when the book was written. In this dream he visits another land, Lineland, which consists of a single line inhabited by lines (men) and points (women). He tries in vain to convince the king of Lineland of the existence of a second dimension.

▷ **Exercise 6.** As part of A Square's argument with the king of Lineland, he passes through the line. What did the king see as he did this?

The next day, December 31, 1999, A Square has a conversation with his grandson, a "most promising young Hexagon of unusual brilliance and perfect angularity: 'I began to show the boy how a Point by moving through a length of three inches makes a Line of three inches, which may be represented by 3; and how a Line of three inches, moving parallel to itself through a length of three inches, makes a Square of three inches every way, which may be represented by 3^2 [square inches].'"

▶ **Exercise 7.** Generalize to give a geometric significance to the quantity 3^3, answering a question posed by the young Hexagon.

That night, the Square is visited by a Sphere from Space, who tries, at first vainly, to convince him of the existence of a third dimension. The Sphere's first argument is that from the third dimension he could see into the house from above, describing all of the inhabitants.

▷ **Exercise 8.** In a paragraph, give an analogous argument about sight from a four-dimensional being to convince a 3-dimensional being of the existence of another dimension. Include details to make your argument as convincing and vivid as possible. Be creative.

The Sphere then passes through the plane, appearing first as a point, then as circles of increasing size, then diminishing to a point and disappearing.

▷ **Exercise 9.** What would a 4-dimensional sphere look like as it passes through 3-dimensional space?

▷ **Exercise 10.** What would a 4-dimensional cube look like as it passes through 3-dimensional space?

The Sphere then tries an argument by analogy: Begin with a single point; then move the point to produce a line segment, which will have 2 vertices (the initial position of the point and the final position of the point) with the line segment connecting the two; next move the line segment parallel to itself to form a square.

▶ **Exercise 11.** Illustrate the process described above, showing a point, the line segment formed by moving the point, and the square formed by moving the line segment. How many vertices, edges, and 2-dimensional faces does the square have and how are they formed?

▷ **Exercise 12.** Now think about moving the square parallel to itself. What figure is formed? How many vertices, edges, 2-dimensional faces, and 3-dimensional solids are there and how are they formed?

▷ **Exercise 13.** Now think about moving the figure of Exercise 12 parallel to itself. The result is called a 4-dimensional *hypercube*. How many vertices, edges, 2-dimensional faces, 3-dimensional solids, and 4-dimensional regions are there and how are they formed?

▷ **Exercise 14.** Now think about moving the figure of Exercise 13 parallel to itself to form a 5-dimensional *hyperhypercube*. How many vertices, edges, 2-dimensional faces, 3-dimensional solids, 4-dimensional regions, and 5-dimensional regions are there and how are they formed?

The Sphere's next argument is to actually descend into a locked cupboard and to remove a tablet without opening the door.

▶ **Exercise 15.** In a paragraph, give an analogous argument about movement from a four-dimensional being to convince a 3-dimensional being of the existence of another dimension. Include details to make your argument as convincing as possible.

It must be particularly noted that the fourth dimension that Abbott is encouraging us to envision is a purely spatial one, just like the three spatial dimensions where we have lived all of our lives. At the time this book was written, the idea of relativity and the concept of time as a fourth dimension lay in the future.

The Sphere's last, and finally convincing, argument is to take A Square with him into Space. Completely converted and enraptured at the possibilities, A Square eventually asks to see the fourth dimension. The Sphere vehemently denies any possibility of such a dimension and sends him back to Flatland. There A Square tries to convince others of the possibility of a third dimension, is arrested, imprisoned for life, and writes his memoir *Flatland*.

Suggested essay questions after reading *Flatland*:

a. Write your own continuation or variation on life in Flatland.
b. Give a description of the ways in which a 2-dimensional universe constrains the environment and its inhabitants.
c. Write a speculation on a fourth spatial dimension.
d. Describe the ways in which Abbott uses irony to relate life in Flatland with specific social problems of the era in which the book was written (1884).

Dionys Burger's *Sphereland*

Sphereland, written in the 1960s by a Dutch mathematician, is a straightforward continuation of *Flatland*. The narrator is A Hexagon, the grandson of A Square. He has a brief appearance in *Flatland* as the boy whose questions set A Square off on his voyage of discovery.

Burger takes it on himself to "correct" some of the faults in *Flatland*, indicating that Flatland has become more progressive since the narrator's grandfather's time. That Abbott's satiric intent is frequently misunderstood by many others besides our own students is clear from the introduction, written by Isaac Asimov: "The Victorian convention of women as a quite inferior form of life was accepted without question. In the twentieth century, this view, both insulting and injurious, to say nothing of being untrue, could not be accepted, and Dionys Burger . . . went to some pains to neutralize it." These pains include statements in *Sphereland* such as "The notion that a female is a stupid creature because of her small brain area has turned out to be false. Some women now even study and get degrees from a university. Man no longer has a monopoly over science, even though nature did create woman first of all for marriage. It is she who by virtue of her great gifts of love and devotion has been destined to raise children and to dedicate herself to homemaking. This takes up such a large part of her mental life that there is usually no room left in her intellect for study of science. The really important inventions and discoveries will undoubtedly continue to be made by men." A doubtful improvement at best. Similarly, the rigid class structure is somewhat relaxed, though still fairly obvious. These "improvements" blunt the impact of the novel, and make it much less interesting and significant than the original.

Our main interest in the book is the extension of the geometric analogy. Flatland is undergoing a period of exploration and scientific discovery. An explorer sets off to travel westward as far as he can, using a gravity meter to ensure the correct direction. After a lengthy journey, he returns to his starting point with the realization that their world is in fact disc-shaped.

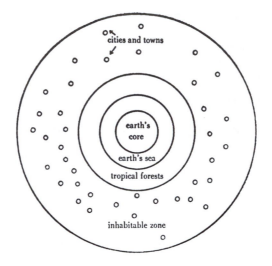

This theory gradually gains acceptance. Further explorations include using a catapult to send a female explorer up into the atmosphere. Once above the mist prevalent throughout Flatland, this explorer saw other world-discs, analogous to the other solar systems in our universe.

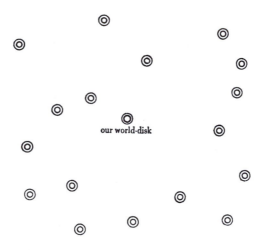

On a New Year's Eve, the Sphere who visited A Square reappears to his grandson and tells of his encounter with a four-dimensional sphere, called an Over-Sphere (perhaps a faulty translation for the mathematical term hyper-sphere), and the refusal of his society to accept such a possibility. This annual visit from the Sphere to A Hexagon and his family becomes a tradition, leading them to greater insights into the geometric nature of their universe.

For example, the inhabitants of Flatland have now domesticated wolves as pets, who occur in two variations: mongrel and pedigreed, according to their orientation.

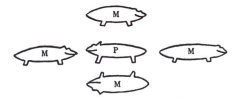

▷ **Exercise 16.** The Sphere kindly changes A Hexagon's daughter's mongrel dog into a pedigreed one. Explain how this was done.

Things get slightly more interesting when the narrator becomes friendly with Mr. Puncto, the recently dismissed director of the Trigonometric Service, an institute established to map Flatland by triangulating their world and accurately measuring the angles and sides of these triangles. He has lost his job when, in the course of his duties, he discovered that the angles in large triangles add up to more than 180°, and that this discrepancy could not be explained by inaccuracies in measurement. For smaller triangles, this discrepancy lay within the margin of error for the angle measurements, but recently the Trigonometric Service had begun to measure larger triangles. The deviation in the sum of the angles could no longer be dismissed, though the discoverer could be. In trying to resolve this problem, A Hexagon's grandson, a promising young octagon, proposes the following triangle:

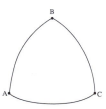

▷ **Exercise 17.** Measure the angles of the triangle above with a protractor and find their sum.

The most obvious objection to this proposal is the fact that the edges of this triangle appear curved. One can get around this objection by noting that a line is defined to be the shortest path between two points; its straightness is an accident due to euclidean geometry where the shortest path between two points is indeed straight. A further objection is shown in the following picture if all triangles have angle sum more than 180°, then consider adjacent triangles $\triangle ABC$ and $\triangle BCD$. In order for angles $\angle ABC$ and $\angle ACB$ to be a bit too fat, the edge BC must curve to the right. But at the same time, in order for $\angle DCB$ and $\angle DBC$ to be a bit too fat, edge BC must curve to the left.

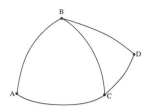

▷ **Exercise 18.** The inhabitants of Flatland found it very hard to imagine how to bend *BC* so that it curves in a direction away from vertices *A* and *D* simultaneously. How can this be done?

▷ **Exercise 19.** Describe a geometric shape where it is natural for triangles to have the properties described above. Explain your answer.

Thus, A Hexagon and Mr. Puncto come to the realization that their world is Sphereland, not Flatland. At this time, space stations have been built in the upper atmosphere, and Mr. Puncto finds employment as director of a telemetry project, intended to map the other visible world-disks. In the process of doing this, he once again finds a discrepancy; in checking the distance to distant objects, he finds that they seem to have moved since an earlier measurement. This allows Burger to introduce the idea of an expanding universe, and another opportunity for martyrdom for the enlightened Sphereland scientists.

C.H. Hinton's *An Episode in Flatland*

A young mathematician, C.H. Hinton, may have influenced Abbott to write *Flatland* by publishing a popular article on the fourth dimension in 1880. It is not known whether Abbott and Hinton ever met, though they had several mutual acquaintances. Indeed, at one time Hinton taught science at the same school where Abbott's close friend Howard Candler (the H.C. of the dedication of *Flatland*) taught mathematics, until Hinton was dismissed and jailed for bigamy, after which he left England. Much of C.H. Hinton's work is on the fourth dimension, and in 1907 he wrote a novel called *An Episode in Flatland*. This work is in some ways more ambitious than Abbott's: He adopts a different and more workable model for a two-dimensional world; he goes into far more detail on the logistics of daily existence for his beings; he has a much more extravagant plot with an idealistic socialist bent. On the other hand, he hasn't Abbott's gift for satire and is often inconsistent in preserving the two-dimensional character of his world. Further, hard as this may be to imagine, his characterizations are even flatter than Abbott's. The book is, perhaps deservedly, out of print.

Hinton's characters live on the rim of a disc-shaped world called Astria. They are basically triangular in shape, but possess legs and arms and a single eye.

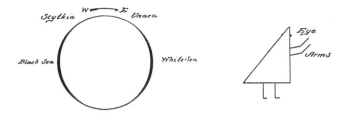

Men are born facing east and women west, and since they are confined to their world these orientations are immutable. While they can bend around to see and reach objects and persons behind them, this is awkward and somewhat difficult, so mirrors are in common use. In traveling, they must climb over the bodies of those they wish to pass. In inhabited lands, all the trees had been felled and the main crop, a springy form of cereal, was planted, but trees became awkward obstacles in primeval territories.

▷ **Exercise 20.** Figure out a procedure for an Astrian to get past a tree, drawing appropriate pictures to illustrate.

Astria is threatened by climatic changes: Summers are getting hotter and winters colder, as its orbit changes into a more eccentric ellipse due to the approach of another planet's gravitational force. This is held secret from the general populace, of course for their own good. A reclusive mathematician, who is the sole believer in a third dimension, holds the key to saving the world. It should be noted that Hinton was a foremost popularizer of the fourth dimension at the time, endowing it with mystical properties. The plot of *An Episode in Flatland* involves fighting bureaucratic forces, uniting true love, and getting all the inhabitants of Astria to unite in willing their world into the third dimension in an act of psychokinesis to escape the gravitational influence of the threatening neighboring planet. We will return to Hinton's beliefs about the fourth dimension in Section 6.2.

A.K. Dewdney's *The Planiverse*

After slogging through Hinton, it is refreshing to turn to Dewdney's *The Planiverse*, published in 1984, the centennial of *Flatland*. Dewdney addresses the big questions that his predecessors left unanswered: What do Flatlanders eat for dinner; how do they reproduce; what do they wear; and do they have pockets?

The framing story in *The Planiverse* is that Dr. Dewdney, a professor of computer science, has the students in his large-scale programming course write a simulation of a two-dimensional universe. Over several semesters the project grows, with different students writing subroutines to incorporate 2-D physics and weather patterns and simple life forms and their predator-prey relationships on a planet the students call Astria, after Hinton's planet. Eventually, creatures called FECs (the initials of three of the students) with rudimentary intelligence are created, whose primary occupation is hunting for smaller creatures called throgs. These FECs can communicate with the programmers through the computer interface in simple dialogue such as (queries from the students are marked >, and replies are marked <):

```
<   FEC HERE—HUNT THROGS.
>   WHY?
<   UNKNOWN: "WHY?"
>   DESCRIBE HUNT.
<   MOVE EAST UNTIL THROG OR SEA.
```

Eventually, a graduate student, Alice Little, who has been involved with the project since its beginnings and who has a slight speech impediment (this seemingly irrelevant bit of information is a hint for the next exercise), calls Dr. Dewdney to tell him that one of the FECs has spoken a word not in the programmed vocabulary; the word is "YNDRD," apparently the name of the FEC. YNDRD, soon renamed Yendred, appears with increasing frequency, and the computer display shows a far more detailed and alien landscape than the one designed and programmed by the students. The form of the FEC also evolves, and it becomes clear that they have established contact with the alien two-dimensional intelligent being Yendred. His world is called Arde, and, like Hinton's Astria, is disc-shaped, with the Ardeans living on the rim. Yendred's first reaction to this contact is, of course, to assume that they are spirits. Over several months, they remain in frequent contact with Yendred, exchanging information about their worlds. Yendred soon sets off on a long journey to the mystic east of his world in search of a deeper spiritual existence, so the humans have the opportunity to see much of Arde. Eventually, Yendred dismisses them:

```
<   WE CANNOT TALK AGAIN. TO TALK AGAIN IS OF NO BENEFIT.
>   BUT WE HAVE SO MUCH MORE TO LEARN FROM YOU.
<   YOU CANNOT LEARN FROM ME. NOR I FROM YOU. YOU DO
    NOT HAVE THE KNOWLEDGE.
>   WHAT KNOWLEDGE?
<   THE KNOWLEDGE BEYOND THOUGHT OF THE REALITY BE-
    YOND REALITY.
>   WOULD IT HELP IF WE LEARNED YOUR PHILOSOPHY AND
    RELIGION?
```

< IT HAS NOT TO DO WITH WHAT YOU CALL PHILOSOPHY OR RELIGION. IF YOU FOLLOW ONLY THOUGHT YOU WILL NEVER DISCOVER THE SURPRISE WHICH LIES BEYOND THOUGHT.

They are never able to raise Yendred again, finding only FECs and throgs when they query the program.

▶ **Exercise 21.** Where does the name Yendred come from?

In the course of talking with and observing Yendred, Dewdney and his students learn much about life on Arde. Dewdney has done an amazing job in presenting a fully realized two-dimensional world, complete with entrancing sidebars and appendices on Ardean physics, meteorology, engineering, and biology. For example, below is the illustration of Yendred's house:

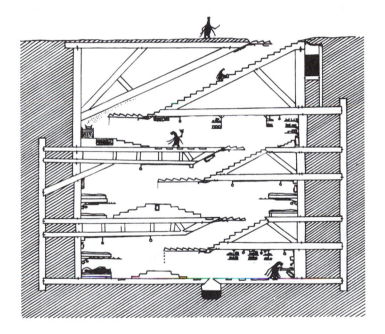

The house is underground, with Yendred standing near the entrance. The entrance is a swinging stair, so that travelers from either direction can step down a few steps then up a few steps to continue their journey. A bar can be inserted to lock the door or block the runoff from rain showers, which otherwise would flood the dwelling. Other swinging stairs connect the levels of the house. On the first level is the cooking and eating area, with recessed stools, a stove, and shelving. The second level contains four sleeping platforms, and the third two sleeping platforms, a desk and book shelves.

In the course of Yendred's journey to the East, between encounters with other Ardeans and adventures, Dewdney describes many simple 2-dimensional contrivances: For example, a piece of string can be used as a bag, a raincoat, or a balloon. More complicated machinery is described and illustrated: printing presses, boats, rockets, steel mills, steam engines, blimps. Two-dimensional games based on volleyball and the oriental game known as Go and musical instruments are described.

▶ **Exercise 22.** How would an Ardean hold onto a string?

▷ **Exercise 23.** Could knots be used in Arde? Explain your answer. What could be used instead of knots?

▷ **Exercise 24.** Could knots be used in a 4-dimensional universe? Explain your answer.

▷ **Exercise 25.** How would you move a heavy object in Arde?

▶ **Exercise 26.** Invent a 2-dimensional musical instrument.

Further Readings

The dimensional analogy and the basic idea of Abbott's *Flatland* is used in several mathematical texts, such as Jeffrey Weeks's *The Shape of Space* and Rudi Rucker's *The Fourth Dimension*. For example, in *The Shape of Space*, Weeks describes a further adventure of A Square: He travels due east, trailing a spool of red string, and comes back to the point where he started. This adventure is similar to the trip in *Sphereland*, part of the evidence that Flatland is really spherical. However, Weeks describes a further journey where A Square travels due north, trailing a spool of blue thread, and again returns to his starting point.

▷ **Exercise 27.** Show that on a sphere, the red and blue paths would intersect somewhere along the path.

▷ **Exercise 28.** In *The Shape of Space*, both the red and blue threads form circular paths but do *not* intersect, except at the starting point. Figure out a possible shape for this 2-dimensional world.

In Rucker's *The Fourth Dimension*, the idea of a sphere visiting Flatland and its higher-dimensional analogy helps the reader imagine the effects of a fourth dimension on 3-space. We will return to this discussion in Section 6.2.

Pointland

The most extreme case of a world in another dimension is to try to imagine living in a single point. In *Flatland*, after visiting Space in the company of the Sphere, the Sphere and A Square visit Pointland in a further vision.

Pointland is inhabited by a single very contented Point, who says, regarding himself: "It fills all Space, and what It fills, It is. What It thinks, that It utters; and what it Utters, that It hears; and It itself is Thinker, Utterer, Hearer, Thought, Word, Audition; it is the One, and yet the All in All. Ah, the happiness, ah, the happiness of Being!" The Sphere, however, denigrates this mindless happiness, stressing the need to strive to overcome such blissful ignorance.

In 1965, the Italian writer Italo Calvino revisited Pointland in a short story that begins with the Big Bang theory: "Through the calculations begun by Edwin P. Hubble on the galaxies' velocity of recession, we can establish there exists a moment when all the universe's matter was concentrated in a single point, before it began to expand in space." Calvino plays with the idea of infinitely many characters, some quite obnoxious and some more kindly, simultaneously inhabiting the same space, unable to detach themselves, until the big bang flings them to the far ends of the universe.

SUGGESTED READINGS

Edwin A. Abbott, *Flatland*, Princeton University Press, Princeton, 1991 (reprint of original of 1884, with an introduction by Thomas Banchoff).

Thomas Banchoff, *From Flatland to Hypergraphics: Interacting with Higher Dimensions*, http://www.geom.umn.edu/~banchoff/ISR/ISR.html

Dionys Burger, *Sphereland*, translated by Cornelie Rheinbolt, HarperCollins, New York, 1983.

Italo Calvino, *All at One Point*, in *Cosmicomics*, translated by William Weaver, Harcourt Brace, New York, 1968.

A.K. Dewdney, *The Planiverse*, Poseidon Press, New York, 1984.

Martin Gardner, "The Wonders of a Planiverse" in *The Last Recreations*, Springer-Verlag, New York, 1997.

Martin Gardner, "Flatlands" in *The Unexpected Hanging and Other Mathematical Diversions*, Simon & Schuster, New York, 1969.

Charles Howard Hinton, *An Episode in Flatland*, Swan Sonnenschein & Co., London, 1907.

Rudy Rucker, *The Fourth Dimension: Toward a Geometry of Higher Reality*, Houghton Mifflin, Boston, 1984.

Jeffrey R. Weeks, *The Shape of Space*, Marcel Dekker, New York, 1985.

6. Other Dimensions, Other Worlds

◆ 6.2. THE FOURTH DIMENSION

The fourth dimension that we wish to discuss in this section is a fourth spatial dimension, not time regarded as a fourth dimension. The interpretation of time as a fourth dimension became common with the widespread dissemination of Einstein's theory of relativity, but that breakthrough was 20 years in the future when Abbott was writing *Flatland*. Long before then, mathematicians and a few philosophers had meditated on a fourth spatial dimension, and the idea became a popular topic of speculation during the period between 1880 and 1910.

In mathematics, the standard view of three-dimensional space is that of a set of points with coordinates. Each point is given coordinates (x, y, z), where x gives the distance from the origin, or center, running forward and backward, y the distance from the origin running left and right, and z the distance up and down. Below is a cube drawn in this coordinate system:

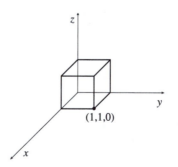

If you are in a rectangular room, stand in one corner so that you have walls behind you and to the right: to find the point $(1, 1, 1)$ step one foot forward, one foot left and place your finger 1 foot from the floor. The point $(1, 1, -1)$ will be one foot forward, one foot left, and one foot below the floor, hovering near the ceiling of the room below you.

▷ **Exercise 1.** Describe the location of the following points relative to your classroom, after you have agreed on where the origin and axes are.
a) (5, 5 ,5)
b) (−5, 5, 5)
c) (5, −5, 5)
d) (5, 5, −5)
e) (−5, −5, 5)

To build up four dimensions, let us consider the progression that leads us to the three we are familiar with: start with a single point (Pointland), which we consider as the origin, or center, of the universe. Then walk back and forth, forming an axis called the x-axis. Turn at right angles and walk in the other direction, forming the y-axis and a plane spanned between the two axes. The direction perpendicular to both the x- and y-axes is the z-axis, forming a three-dimensional space. To find a fourth dimension all we need is to find another direction perpendicular to all of the previous directions. If x denotes back and forth, y left and right, and z up and down, we need words to describe movement in the fourth dimension. C.H. Hinton coined the words *ana* and *kata* for these movements, so for lack of better, we will use these. Four-dimensional space is called *hyperspace*. The 1-dimensional analogue of the unit cube is a line segment 1-inch long lying on the x-axis. Stacking many copies of this line segment on top of each other in the y-direction builds a square. Stacking squares together in the z-direction, like stacking sheets of paper, builds up a cube. In four dimensions, stacking cubes together in the *ana* direction would build a *hypercube*.

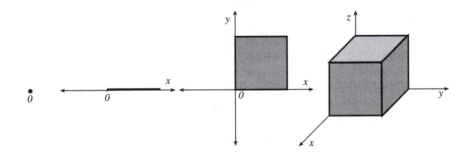

▶ **Exercise 2.** Note that the origin cuts the x-axis in half, and the x-axis cuts the plane in half.
(a) Can you generalize these space-cutting ideas to the third dimension?
(b) Can you generalize these space-cutting ideas to the fourth dimension?

▷ **Exercise 3.** In the plane note that any two nonparallel lines intersect in a point.
   (a) Can you generalize this idea to the intersection of two nonparallel planes in the third dimension?
   (b) Can you generalize this idea to the intersection of two nonparallel 3-dimensional spaces in hyperspace?

▶ **Exercise 4.** In the illustration below, you see that in general, a point and a line on the plane will not intersect at all. The probability that the point will land on the line is, in fact, zero, given that there are so many other points from which to choose. In the second illustration, we see that the most general, or most likely, intersection of a line and a plane in 3-dimensional space is a point. Of course, this does not always occur: the line could be parallel to the plane, or the line could lie on the plane, but these are considered special cases. In each of these cases, jiggling the line a little bit would result in the general case where the line intersects the plane at a single point.

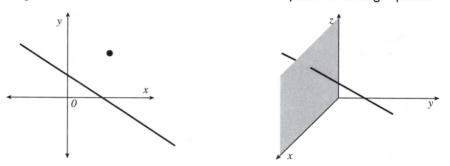

   (a) What is the most general intersection of two lines in the plane?
   (b) What is the most general intersection of two lines in 3-space?
   (c) Find a formula, using the results of this exercise and the previous one, for the dimension of the intersection of a k-dimensional line or plane with an ℓ-dimensional line or plane in n-dimensional space.
   (d) What is the most general intersection of a line and a plane in hyperspace?
   (e) What is the most general intersection of two planes in hyperspace?
   (f) What is the most general intersection of a plane and a 3-space in hyperspace?

To see what is possible in hyperspace, we reason by analogy. First, imagine two points on a line, and let the first point hop up off of the line and back onto the line on the other side of the fixed second point. As a resident of Lineland, the fixed point sees the first point disappear and then reappear on the other side and is completely mystified. To the observer in the plane, it is quite clear what has happened: The first point has taken advantage of the second dimension to perform a maneuver impossible if all action is confined to the line.

Next, imagine a point inside a circle in the Flatland plane. The point can jump up into the third dimension and land back on the plane outside the circle. As a resident of Flatland, the circle sees the point disappear and then reappear outside. To the observer in the third dimension, it is obvious how the point escaped, but the circle is mystified.

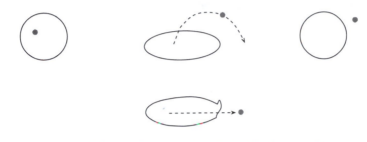

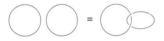

Alternatively, one could imagine that the circle lifted a small section of itself up into the third dimension. To the point in Flatland, this would look as though a gap appeared in the circle, and then the point could make a dash through the gap to escape from the circle.

This sort of maneuver allows us to link two circles: One circle could lift a bit of itself *ana* or *kata* into the fourth dimension, allowing the other to slide through and link. Thus, the following figures are the same in the fourth dimension:

▷ **Exercise 5.** Explain, with a series of pictures, how to untie the knot below. Note exactly where you need to slip into hyperspace.

▷ **Exercise 6.** Explain how an object could be removed from a sealed box in hyperspace.

▷ **Exercise 7.** Explain how a left glove could be changed into a right glove in hyperspace.

From the exercises and examples, we see that if a fourth dimension exists, then beings can walk through walls, disappear and reappear, remove things from sealed containers, untie knots, etc. What sort of beings have these capabilities? Ghosts and spirits!

While mathematicians had been actively studying the fourth dimension and alternative geometries ever since Bernhard Riemann's seminal lecture and essay "On the Hypotheses Which Lie at the Foundation of Geometry" of 1854, the fourth dimension was imprinted on the imagination of the general populace by the London trial for fraud of the American medium Henry Slade in 1877. Mystics and spiritualists were subjects of considerable interest and controversy in Victorian England, but what really captured the attention of the public in this trial was the presence of many eminent physicists and mathematicians (including two who were later to win Nobel prizes) testifying about the possibilities that the existence of a fourth dimension would allow. These prominent scientists were recruited by the German physicist Johann Zöllner, a fervent believer in Slade. Zöllner went on to write a book, *Transcendental Physics*, supporting Slade and the spiritualism now associated with the fourth dimension. In this book, he describes several experiments that he asked Slade to carry out. One was to link two carved wooden rings, another was to reverse the orientation of a spiral snail shell, and a third to tie a knot in a circular loop of gut. While Slade never really performed the stated tasks (instead the rings reappeared around a table leg, the snail shell disappeared and was found under the table, and a hole was burned in the loop), Zöllner managed to maintain his belief in Slade as an agent of the fourth dimension.

▶ **Exercise 8.** How could the rings have ended up on the table leg?

Following the trial of Slade, the fourth dimension as the "realm of spirits" was taken up by many spiritualists, as well as respectable clergymen, such as Abbott, who had a professional interest in the miraculous. C.H. Hinton, the author of *An Episode in Flatland*, also wrote a number of pamphlets and books on the subject. After Hinton's trial and brief imprisonment in England for bigamy, he went on to teach at a school in Japan, and then emigrated to the United States. There he taught at Princeton University (where he seems to have spent most of his time inventing a "baseball gun" — an automatic pitching machine that terrified the batters) and the University of Minnesota. After losing these jobs, he worked at the U.S. Naval Observatory, and finally at the U.S. Patent Office. Coincidentally, during this phase of Hinton's career, Einstein was working at the patent office in Bern, Switzerland, and inventing relativity in his spare time. But Hinton's books were widely read, especially his more philosophical works endowing the fourth dimension with spiritual significance.

Scientific American sponsored a contest in 1909 for essays on the subject of the fourth dimension with a prize of $500, a substantial amount at that time. The best of these essays were collected and edited by Henry Manning, of Brown University, and published as *The Fourth Dimension Simply Explained*. One advocate who helped popularize Hinton's ideas in America was a designer, Claude Bragdon. Typical of the odd mix of mathematics, art, and mysticism that Bragdon adopted is the following illustration:

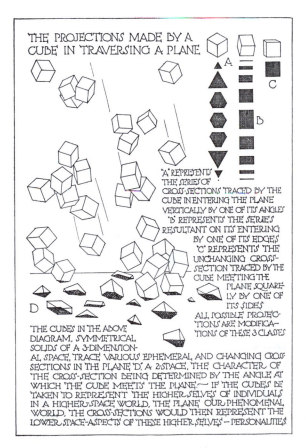

Bragdon went on to urge the adoption of design, which he called "Projective Ornament," based on four-dimensional figures:

> Many sincere workers in the field of art have realized the aesthetic poverty into which the modern world has fallen. Designers are reduced either to dig in the boneyard of dead civilizations, or to develop a purely personal style and method . . . . Is there not some source, some secret spring of fresh beauty undiscovered, to satisfy our thirsty souls?

Bragdon goes on to say:

> Ornament is the outgrowth of no practical necessity, but a striving toward beauty. Our zeal for efficiency has resulted in a corresponding aesthetic infertility. Signs are not lacking that consciousness is now looking in a new direction—away from the contemplation of the facts of materiality towards the mysteries of the supersensuous life. This transfer of attention should give birth to a new aesthetic, expressive of the changing psychological mood. The new direction of consciousness is well suggested in the phrase, *The Fourth Dimension of Space*, and the decorative motifs of the new aesthetic may appropriately be sought in four-dimensional geometry.

Many of Bragdon's designs are based on the *hypercube* or *tesseract*: the four-dimensional analog of the familiar cube. We summarize the results of Exercises 11-14 of Section 6.1, where v denotes the number of vertices or corners, e the number of edges, f the number of faces, s the number of solids, t the number of 4-dimensional regions, and u the number of 5-dimensional regions.

n-dimensional Cubes

Dimension	Figure	v	e	f	s	t	u
1	Line segment	2	1				
2	Square	4	4	1			
3	Cube	8	12	6	1		
4	Hypercube	16	32	24	8	1	
5	Hyperhypercube	32	80	80	40	10	1

▷ **Exercise 9.** Find the Euler characteristic, $\chi = v - e + f - s + t - u$, for each of the figures in the table above.

In the previous section, we discussed how a cube is formed from a square by movement to a position parallel to the original position. A hypercube can be formed by moving a cube *ana* or *kata* through hyperspace to a new position parallel to the original position.

▶ **Exercise 10.** From the table preceding Exercise 9, a hypercube contains eight cubes. Two are illustrated above. Find the other six in the projection below.

Another way of viewing the hypercube is to consider the cube as a wire framework that can be stretched and laid flat as on the left on the next page. One can think of the center square as the bottom face of the cube, and the outer square the upper face. The side faces of the cube are stretched to form the trapezoids surrounding the inner square in this projection. The 4-dimensional analogue of this projection, distorting the hypercube to fit it into 3-dimensional space, is shown on the right.

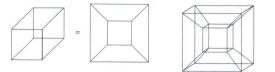

▷ **Exercise 11.** Describe the positions of the eight cubes that make up the hypercube in the projection above.

Yet another way of envisioning the hypercube is by analogy with the formation of a cube from a net or pattern that can be cut up and folded to form a cube. One net for the cube is shown below on the right.

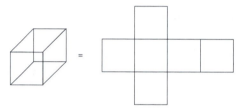

▷ **Exercise 12.** Cut out the net above and tape it together to build a cube.

Similarly, a hypercube could be assembled from eight cubes with their faces or sides glued together in pairs correctly. A net for the hypercube is shown below:

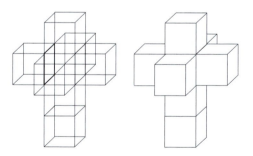

▷ **Exercise 13.** Build the net above (sugar cubes work well). Figure out how the sides of the cubes should be glued together to build the hypercube and color code them.

Salvador Dali used this motif in his 1954 painting *The Crucifixion: Corpus Hypercubicus*:

Another four-dimensional figure can be built up in analogy with the triangle. Start with a single point. In the next generation, add another point above the first and connect the two, getting a line segment. Place the line segment on the floor and add another point above the line segment and connect this point with each of the points on the line segment, obtaining a triangle. Place the triangle flat on the floor and add another point above it; connecting the new point to each of the points in the triangle gives a three-dimensional solid called the tetrahedron or triangular pyramid.

Adding another point in the fourth dimension, *ana* or *kata* the figure, and connecting this point to each point in the tetrahedron gives a figure called the *hypertetrahedron or pentahedroid*.

▷ **Exercise 14.** Complete the table below.

n-dimensional Triangular Figures

Dimension	Figure	v	e	f	s	t
1	Line segment	2	1			
2	Triangle					
3	Tetrahedron					
4	Hypertetrahedron					

The founders of Cubism, Picasso and Braque, were greatly influenced by the idea of the fourth dimension. They had already absorbed the ideas of Cézanne that everything in nature should be interpreted as geometrical solids. A second major influence on the movement was African sculpture, which was also highly geometric in nature. Refusing to be bound by the traditional depiction of three dimensional space in two dimensions by using perspective, the Cubists sought to display multiple aspects of an object at once. A grand example is Pablo Picasso's *Portrait of Daniel-Henry Kahnweiler*, shown below on the left. On the right is a photograph of Kahnweiler taken by Picasso. This painting of Picasso's friend and art dealer was one of his first Cubist portraits.

The views from several angles cross and merge to demonstrate the dynamic nature of reality. This is where the fourth dimension comes into play. The most famous mathematician of the time, Henri Poincaré, wrote several books presenting mathematical ideas to a general audience. In *Science and Hypothesis* of 1902, he writes:

> Well, in the same way that we draw the perspective of a three-dimensional figure in the plane, so we can draw that of a four-dimensional figure on a canvas of three (or two) dimensions. To a geometer this is but child's play. We can even draw several perspectives of the same figure from several different points of view . . . . In this sense we may say that we can represent to ourselves the fourth dimension.

Imagine what would happen if A Square tried to show all sides of a polygon at once and in the same place. The views would overlap, and the object would not be easily discernible. Theoretically, though, more of the object would be portrayed than a traditional one-dimensional Flatland drawing would allow. To see all aspects of a polygon simultaneously with any clarity, we must look down from the third dimension.

While this is a gross simplification, the same concepts apply to Cubist art. To see all aspects of a three-dimensional object clearly and simultaneously requires viewing it from a fourth dimension. The Cubists wanted to show enough of that perspective to portray the inherent truth and reality of the object. The writer and art critic Apollinaire, one of Cubism's most vehement admirers, wrote in *Les Peintres Cubistes* in 1913 (as quoted in Henderson):

> Until now, the three dimensions of Euclid's geometry were sufficient to the restiveness felt by great artists for the infinite.
>
> The new painters do not propose, any more than did their predecessors, to be geometers. But it may be said that geometry is to the plastic arts what grammar is to the art of the writer. Today, scholars no longer limit themselves to the three dimensions of Euclid. The painters have been led quite naturally, one might say by intuition, to preoccupy themselves with the new possibilities of spatial measurement which, in the language of the modern studios, are designated by the term *fourth dimension*.
>
> As it presents itself to the mind, from the plastic point of view, the fourth dimension appears engendered by the three known dimensions: it represents the immensity of space eternalizing itself in all directions at any given moment. It is space itself, the dimension of the infinite; the fourth dimension endows objects with plasticity. It gives objects the proportions which they merit in the work of art, whereas in Greek art, for instance, a somewhat mechanical rhythm constantly destroys the proportions.
>
> Greek art had a purely human conception of beauty. It took man as the measure of perfection. The art of the new painters takes the infinite universe as its ideal, and it is to this ideal that we owe a new norm of the perfect, which

permits the painter to proportion objects in accordance with the degree of plasticity he desires them to have. . . .

Finally, I must point out that the *fourth dimension*—this utopian expression should be analyzed and explained, so that nothing more than historical interest may be attached to it—has come to stand for the aspirations and premonitions of the many young artists who contemplate Egyptian, negro, and oceanic sculptures, meditate on various scientific works, and live in anticipation of a sublime art.

While the portrait shown previously by Picasso shows the multiple viewpoints typical of Cubism, other artists were experimenting with the idea of time as a fourth dimension at about the same time, as witnessed by Marcel Duchamp's *Nude Descending a Staircase, No. 2.*

That artists are still interested in the fourth dimension is shown in the writings and works by Tony Robbin, a contemporary artist greatly inspired by mathematician Tom Banchoff.

SOFTWARE

1. Jeffrey Weeks has written a shareware program, called *Hypercube*, for Macintosh computers that allows one to view different aspects of the hypercube. This is available from the Geometry Center as part of a package called Geometry Games: www.geom.umn.edu.

SUGGESTED READINGS

Thomas Banchoff, *Beyond the Third Dimension*, W.H. Freeman, New York, 1990.

Claude Bragdon, *A Primer of Higher Space*, Manas Press, Rochester, 1913.

Claude Bragdon, *Projective Ornament*, reprint of 1915 edition, Dover, New York, 1992.

Pierre Cabanne, *Duchamp & Co.*, Terrail, Paris, 1997.

John Canady, *Mainstreams of Modern Art*, Holt Rinehart and Winston, New York, 1959.

Ian Chilvers (ed.), *The Concise Oxford Dictionary of Art and Artists*, Oxford University Press, Oxford, 1996.

Edward F. Fry, *Cubism*, McGraw-Hill, New York, 1966.

Christopher Gray, *Cubist Aesthetic Theories*, Johns Hopkins Press, Baltimore, 1953.

Linda Dalrymple Henderson, *The Fourth Dimension and Non-Euclidean Geometry in Modern Art*, Princeton University Press, Princeton, 1983.

Michio Kaku, *Hyperspace*, Oxford University Press, New York, 1994.

Henry P. Manning (ed.), *The Fourth Dimension Simply Explained*, reprint of 1910 edition, Dover, 1960.

Henry P. Manning, *Geometry of Four Dimensions*, reprint of 1914 edition, Dover, 1956.

Henri Poincaré, *Science and Hypothesis*, Dover, New York, 1952.

Tony Robbin, *Fourfield: Computers, Art, and the Fourth Dimension*, Little, Brown and Company, Boston, 1992.

Rudy Rucker, *The Fourth Dimension: Toward a Geometry of Higher Reality*, Houghton Mifflin, Boston, 1984.

Jeffrey R. Weeks, *The Shape of Space*, Marcel Dekker, New York, 1985.

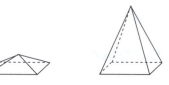

7. Polyhedra

◆ 7.1. PYRAMIDS, PRISMS, AND ANTIPRISMS

SUPPLIES

cardboard polygons (To do all the exercises in this section, you will need 20 equilateral triangles and 3 squares, all having an edge length of 1 inch.)

scissors

tape

A polyhedron (plural: polyhedra) is a three-dimensional figure whose *faces*, or sides, are polygons. A (hollow) cube is an example of a polyhedron. There are many types of polyhedra, and we shall explore some of these in this and the next section.

Pyramids

The Egyptian pyramids were built on square bases, but we can generalize the idea of a pyramid to any base. The classical pyramid is called a *right square pyramid*, which means that the base is a square and the top vertex (or *apex*) is directly above the point in the middle of the square. In the example pictured on the next page, the triangles forming the sides have been chosen to be equilateral. The five faces of this pyramid are thus the square and four equilateral triangles, all regular polygons. The pyramid has five vertices—the apex and the four vertices around the base—and eight edges. On the right is a diagram called a *net* for this pyramid: a two-dimensional drawing of the faces of the pyramid that can be cut out and taped together to build it. The net has the advantage that all the polygonal faces are presented without the distortion inherent to the perspective drawing on the left.

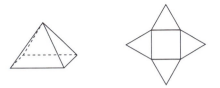

▷ **Exercise 1.** Build a model of a right square pyramid with equilateral sides as pictured above.

▷ **Exercise 2.** If the pyramid pictured above has a square base that is one inch on a side and the sides are equilateral triangles, how tall is it? [Hint: You will need to use the Pythagorean theorem twice.]

▶ **Exercise 3.** Imagine slicing the pyramid above by a horizontal plane. What shape are the cross-sections?

▷ **Exercise 4.** Imagine slicing the pyramid above by a vertical plane parallel to one of the edges of the square. What shape are the cross-sections?

Below are some other pyramids that can also be described as right square pyramids. To distinguish all of these, the one such pyramid with equilateral sides is called the regular right square pyramid. Since in a right pyramid the apex is centered over the square, in each of the pyramids below the side faces must be isosceles triangles with the same base but with different heights for any right pyramid.

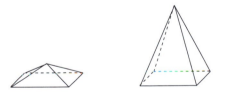

▷ **Exercise 5.** Draw nets for the pyramids above.

▷ **Exercise 6.** While clearly there is no limit for how tall the isosceles triangles forming the side faces of the right square pyramid may be, there is a limit to how short they can be. Assuming that the square base is one inch on the side, find a number that the altitude of the triangular side faces must exceed.

A *skew square pyramid* will have a square for its base, but the apex will not be directly over the center.

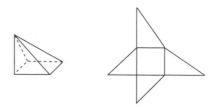

▶ **Exercise 7.** Note that the skew square pyramid pictured above happens to have its apex directly over the front left base vertex. If the square base is 1 inch on each side and the pyramid is 1 inch tall, find the lengths of the other edges of the pyramid.

▷ **Exercise 8.** A *disphenoid* is a skew triangular pyramid all of whose faces are congruent but not equilateral. This can be built by taking any acute-angled triangle. The lines joining the midpoints of the sides form four congruent triangles. Build a model of a disphenoid.

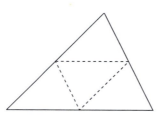

A *right regular-pentagonal pyramid* will have a regular pentagon as the base and a top vertex directly above the center of the pentagon, as pictured below on the left, while a skew irregular-pentagonal pyramid will have an irregular pentagon for the base and the apex off-center, as on the right:

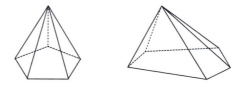

▶ **Exercise 9.** Draw a right regular-hexagonal pyramid.

▷ **Exercise 10.** Draw a skew regular-hexagonal pyramid.

▷ **Exercise 11.** Draw a right irregular-hexagonal pyramid.

▷ **Exercise 12.** Draw a skew irregular-hexagonal pyramid.

▶ **Exercise 13.** If a pyramid has a base with *n* sides, find *f*, the number of faces; *e*, the number of edges; and *v*, the number of vertices.

▷ **Exercise 14.** Find all the pyramids such that all of the faces are regular polygons (including the triangular side faces), and draw a net for each.

Prisms

Another family of polyhedra is the class of *prisms*. A prism has congruent polygonal base and top faces that are parallel, and sides joining corresponding edges of the top and bottom. A *regular-n-sided prism* will have a regular polygonal base and top. A *right prism* will have the side faces perpendicular to the base and the top.

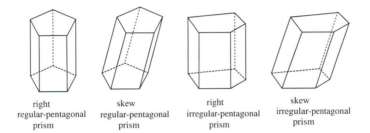

right	skew	right	skew
regular-pentagonal	regular-pentagonal	irregular-pentagonal	irregular-pentagonal
prism	prism	prism	prism

▷ **Exercise 15.** Draw a right regular-hexagonal prism and a net for it.

▶ **Exercise 16.** Draw a skew irregular-hexagonal prism.

▷ **Exercise 17.** If a prism has a base with *n* sides, find *f*, the number of faces; *e*, the number of edges; and *v*, the number of vertices.

▷ **Exercise 18.** How many prisms are there such that all of the faces are regular polygons (including the side faces)?

In general, we are more interested in regular figures than the irregular and skew pyramids and prisms pictured above. The examples studied thus far are also all *convex*. A polyhedron is defined to be convex if for any two points chosen on the surface, the line segment connecting these two points lies inside or on the polyhedron. A nonconvex irregular polyhedron is pictured below with a line segment connecting two points on the prism that lies outside of the solid:

Note that at vertex A, two rectangles and the irregular base of the prism meet. Thus, the polygon angles meeting at A add up to more than 360°. Whenever this occurs, the polyhedron has to bend in such a way as to be nonconvex.

Volume Formulae

In Section 1.1 we gave the formula for the area of a parallelogram, $A = bh$, where b is the length of the base and h the height, explaining that any parallelogram will have area equal to that of the rectangle on the same base with the same height.

You are probably familiar with the formula for the volume of a rectangular box: $V = lwh$, where l is the length of the box, w its width, and h its height. We wish to extend this formula to arbitrary prisms.

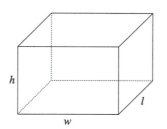

▶ **Exercise 19.** Give a formula for the volume of a right triangular prism with base a right triangle. Explain why your formula works. [Hint: Consider the pictures below.]

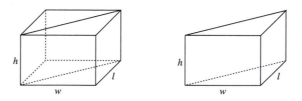

▷ **Exercise 20.** Give a formula for the volume of a skew rectangular prism as pictured on the facing page. Explain why your formula works.

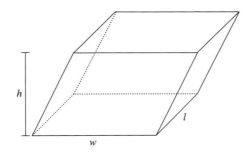

In the formula $V = lwh$ for the box, note that the product $l·w$ gives the area of the rectangular base of the box. Thus this formula could be restated as

$$V(\text{prism}) = hA(\text{base}),$$

and this formula holds true for any prism, skew or right, no matter what the base. To help understand why this is true, one can cut a dozen identical shapes from cardboard and stack them up. If the stack is pushed sideways, like a deck of cards, then the resulting stack is the same height and has the same base. Both stacks clearly have the same volume.

▷ **Exercise 21.** Cut a dozen similar figures of carefully graduated sizes from cardboard and stack them up by size. What can you conclude about the volume of a pyramid?

To find the formula for the volume of a pyramid, let us start with a pyramid with a triangular base. We label this pyramid by its vertices as $A - BCD$, where A is the top vertex and $\triangle BCD$ is the base. Attach at vertex A another triangle $\triangle AEF$ congruent and parallel to $\triangle BCD$, as shown on the next page. A skew prism is formed with $\triangle BCD$ on the base and $\triangle AEF$ on the top. The pyramids $A - BCD$ and $D - AEF$ have congruent bases and equal heights, so they will have equal volumes: $V(A - BCD) = V(D - AEF)$. Pyramid $D - AEF$ is the same as pyramid $A - DEF$, since this is just a change in which vertex is the top and which form the base.

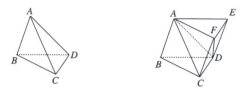

Now consider the parallelogram *CDEF*. The diagonal *DF* cuts this into two congruent triangles, △*CDF* and △*DEF*. The pyramids *A − DEF* and *A − CDF* thus have congruent bases and the same height, and so have equal volumes: $V(A − DEF) = V(A − CDF)$. We have cut the prism up into three pyramids, each of which have the same volume as the original pyramid. Therefore, the pyramid has volume equal to one-third of the volume of the prism with the same base and the same height.

$$V(\text{pyramid}) = \frac{1}{3}hA(\text{base}):$$

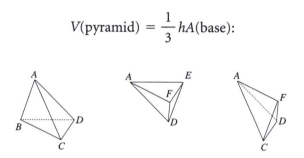

▷ **Exercise 22.** Prove that the volume formula above works for any pyramid by explaining how to cut up a pyramid with an *n*-sided base into triangular pyramids, each with the same height.

Antiprisms

Another class of polyhedra are the antiprisms: polyhedra that have congruent parallel polygons for the base and top polygons with the top polygon rotated so that the vertices lie over the midpoints of the sides of the lower polygon, joined by a band of triangles around the sides.

▷ **Exercise 23.** Build a model of the antiprism with equilateral triangles for base and top faces, and equilateral triangles for all the side faces.

▷ **Exercise 24.** Build a model of the antiprism with squares for base and top faces, and equilateral triangles for all the side faces.

▷ **Exercise 25.** If an antiprism has a base with n sides, find f, the number of faces; e, the number of edges; and v, the number of vertices.

SUGGESTED READINGS

H.S.M. Coxeter, *Regular Polytopes*, Dover, New York, 1973.

H.M. Cundy and A.P. Rollett, *Mathematical Models*, Oxford University Press, New York, 1961.

Alan Holden, *Shapes, Space, and Symmetry*, Dover, New York, 1971.

David Gay, *Geometry by Discovery*, Wiley & Sons, New York, 1998.

Anthony Pugh, *Polyhedra: A Visual Approach*, University of California Press, Berkeley, 1976.

Hermann Weyl, *Symmetry*, Princeton University Press, Princeton, 1952.

7. Polyhedra

◆ 7.2. THE PLATONIC SOLIDS

SUPPLIES
> cardboard polygons (To do all the exercises in this section, you will need 90 equilateral triangles, 6 squares, and 12 pentagons, all having an edge length of 1 inch.)
> scissors
> tape

Platonic Solids

Regular polyhedra are those whose faces are congruent regular polygons, and whose *vertex configurations* (the number and types of polygons meeting at each vertex) are the same for every vertex. Thus, all the faces of a regular polyhedron will be identical and will be assembled in such a way that the vertices are also identical. Polyhedra are named by the number of faces they have, using the Greek prefixes from Section 1.2. Thus, a *tetrahedron* should have four sides. A cube, which has, of course, six sides, is sometimes called a *hexahedron*. Since a cube has three squares meeting at each vertex, we denote it by the Schläfli symbol 4.4.4, analogous to the notation we used to describe tilings. Note that the sum of the angles at each vertex of the cube is $3 \times 90° = 270°$, since each square contributes 90°.

You will need to cut sets of congruent equilateral triangles, squares, regular pentagons, etc., out of cardboard to use in the following exercises. Note that if you tape two identical polygons together, you get a sort of sandwich, which we do not consider truly three-dimensional. Therefore, any polyhedron must have at least three polygons meeting at each vertex.

We begin with equilateral triangles. The comment above indicates that each vertex must have at least three triangles meeting at each vertex, while we also know that six equilateral triangles at a vertex form the tiling 3.3.3.3.3.3. Since the angle sum at each vertex is then 360°, this tiling lies flat instead of building a solid figure. Therefore, any polyhedron formed by equilateral triangles must have at least three, but no more than five, triangles at each vertex. We are also only interested, for the moment, in finite polyhedra: those whose shapes are generally spherical. Such polyhedra will close back on themselves, enclosing a spherical space.

▶ **Exercise 1.** Form polyhedra from equilateral triangles;
  (a) so that three meet at each vertex. How many faces does the solid have? What should we call it?
  (b) so that four meet at each vertex. How many faces does the solid have? What should we call it?
  (c) so that five meet at each vertex. How many faces does the solid have? What should we call it?

▷ **Exercise 2.** What is the greatest number of squares that can be used at each vertex to build a polyhedron? There is only one regular polyhedron formed by squares. Build it and name it.

▷ **Exercise 3.** What is the greatest number of regular pentagons that can be used at each vertex to build a polyhedron? There is only one regular polyhedron formed by regular pentagons. Build it and name it.

▷ **Exercise 4.** What happens if you tape three regular hexagons at each vertex?

At this point, you have found and named the five *Platonic solids*, the only regular polyhedra. Some of these we met in the previous section. For example, the cube is a regular right square prism.

▶ **Exercise 5.** Which of the Platonic solids can be classed as a pyramid?

▷ **Exercise 6.** Which of the Platonic solids can be classed as an antiprism?

▷ **Exercise 7.** Fill in the table below, giving the type of faces (triangles, squares, etc.), the Schläfli symbol (such as 4.4.4 for the cube), and the numbers of faces, edges, and vertices.

Platonic Solids

Polyhedron	Face type	Symbol	f	e	v
Tetrahedron					
Cube	Squares	4.4.4			
Octahedron					
Dodecahedron					
Icosahedron					

You will note some similarities in the numbers in the table of Exercise 7. For example, the numbers of faces, edges, and vertices for the cube are the same as the numbers for the vertices, edges, and faces for the octahedron. Thus, the octahedron has a vertex for every face on the cube, the same number of edges, and a face for every vertex of the cube. The cube and the octahedron are *duals*. If one takes a cube and places a vertex at the center of each face, one gets an octahedron neatly embedded inside the cube. Similarly, if one places a new vertex in the center of each triangular face of an octahedron and connects these new vertices, one gets a cube embedded inside the octahedron.

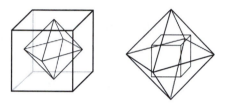

▷ **Exercise 8.** Fill in the table below.

Duals

Polyhedron	Dual
Tetrahedron	
Cube	
Octahedron	
Dodecahedron	
Icosahedron	

▶ **Exercise 9.** Draw the dual of the right square pyramid with equilateral sides.

▷ **Exercise 10.** Draw the dual of the right hexagonal prism with square sides.

▶ **Exercise 11.** What is the dual of the antiprism with equilateral triangles for bottom and top faces, and with equilateral triangles for all of the side faces?

The ideal image of a polyhedron is, of course, a carefully built 3-dimensional model, but they are awkward to carry around with you. Nets have the advantage of representing the polygonal faces without distortion, but need to be assembled before you can see and study the vertex configurations. Below is a net for the cube:

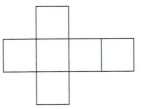

▷ **Exercise 12.** Determine which of the following are nets for the cube:

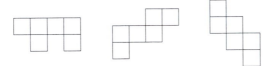

▷ **Exercise 13.** Determine which of the following are nets for the tetrahedron:

▷ **Exercise 14.** Draw nets for the octahedron, icosahedron, and dodecahedron.

Another way of representing a polyhedron is by a *Schlegel diagram:* If one imagines either a wire frame model or a model with clear faces of the polyhedron, imagine a light suspended directly over the center of one of the faces of a wire frame or transparent-sided model, and consider the shadow cast by the edges.

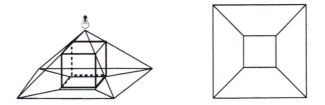

This representation has the advantage that one can see the vertex configurations on the diagram, though the shape of the faces is distorted. Below are the Schlegel diagrams of the icosahedron on the left and the dodecahedron on the right:

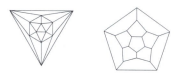

▶ **Exercise 15.** Draw the Schlegel diagram of the tetrahedron.

▷ **Exercise 16.** Draw the Schlegel diagram of the octahedron.

The next group of exercises ask you to visualize cross-sections of the various polyhedra. You can imagine slicing each polyhedron by planes as specified, or dipping the polyhedron into a glass of water and then describing the shape of the water line.

▶ **Exercise 17.** Stand a cube on one of its faces. Imagine slicing it by a horizontal plane. What shape are the cross-sections?

▷ **Exercise 18.** Stand a cube on one of its edges. Imagine slicing it by a horizontal plane. What shape are the cross-sections?

▷ **Exercise 19.** Stand a cube on one of its vertices. Imagine slicing it by a horizontal plane. What shape are the cross-sections?

▶ **Exercise 20.** Stand a tetrahedron on one of its faces. Imagine slicing it by a horizontal plane. What shape are the cross-sections?

▷ **Exercise 21.** Stand a tetrahedron on one of its edges. Imagine slicing it by a horizontal plane. What shape are the cross-sections?

▷ **Exercise 22.** Stand a tetrahedron on one of its vertices. Imagine slicing it by a horizontal plane. What shape are the cross-sections?

All of the Platonic solids, as well as the pyramids, prisms, and antiprisms, satisfy *Euler's formula*:

$$\chi = v - e + f = 2$$

▶ **Exercise 23.** Verify Euler's formula for each of the Platonic solids.

To see why Euler's formula is true for any of the convex solids, consider an arbitrary spherical polyhedron P (one that would have the shape of a sphere if we pretend that it is a balloon and blow it up). The general idea is this: First remove one of the faces, leaving the edges and vertices behind to get a new figure P_1.

Since P_1 has one face fewer than P did, we have $\chi(P_1) = \chi(P) - 1$. Next remove the faces one by one. There are three cases to consider.

In the first case, suppose we wish to remove a face that is connected to the rest of the figure along two edges. Leave those edges and their endpoint vertices behind. Thus, after throwing away this triangle, we have left the same number of vertices, one less face, and one fewer edges as in the original. The new Euler characteristic is thus $\chi = v - (e - 1) + (f - 1) = v - e + f$. Therefore, there is no net change in the Euler characteristic.

▷ **Exercise 24.** Suppose we wish to remove a face that is connected to the rest of the figure along one edge. Leave that edge and its endpoint vertices behind. What is the net change in the Euler characteristic χ?

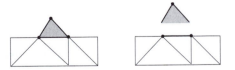

▷ **Exercise 25.** Suppose we wish to remove a face that is connected to the rest of the figure only at a vertex. Leave that vertex behind. What is the net change in the Euler characteristic χ?

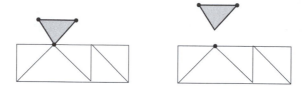

Finally after a finite number of steps, we are left with a single face P_k, with all of its edges and vertices intact. By Exercises 24–26, we know that $\chi(P_k) = \chi(P) - 1$. This face will have $F = 1$, $E = n$, and $V = n$, where n is the number of edges (and vertices) in this polygon. Thus, $\chi(P_k) = n - n + 1 = 1$. Therefore, $\chi(P) = \chi(P_k) + 1 = 2$, so for any polyhedron with spherical shape, Euler's formula is true.

Deltahedra

Irregular polyhedra can have either irregular faces or irregular vertex configurations. One class of irregular polyhedra are the *convex deltahedra* which have only equilateral triangles as faces. The three regular convex deltahedra were found in Exercise 1. For any polyhedron with triangular faces, note that one starts with a pile of triangles. Each of these triangles has three edges. If we let t be the number of triangles, we have $3t$ edges before we start building the model. When we tape the triangles together, we match up the edges in pairs, so that the completed polyhedron must have $\frac{3t}{2}$ edges. Since one cannot have fractional edges, this means that t must be divisible by 2, i.e., t must be even. Thus, all deltahedra must have an even number of triangles. We must have at least three triangles meeting at each vertex to avoid the sandwich effect, but we cannot have more than five and retain convexity. Therefore, the tetrahedron is the smallest (in the sense of having the fewest faces) deltahedron with only four triangles, and the icosahedron the largest with 20 triangles meeting five at each vertex. The octahedron is another regular deltahedron using eight triangles. Since the others must be irregular, they will have, for example, three triangles meeting at one vertex and four at another, and will also require an even number of triangles somewhere between the number needed for the tetrahedron and the icosahedron.

▷ **Exercise 26.** There are five convex irregular deltahedra. Build models of these.
  (a) Build a convex deltahedron that consists of six equilateral triangles. This polyhedron is called the triangular dipyramid.
  (b) Build a convex deltahedron (the pentagonal dipyramid) that consists of ten equilateral triangles.

(c) Build a convex deltahedron (the Siamese dodecahedron or snub disphenoid) that consists of twelve equilateral triangles.

(d) Build a convex deltahedron (the tri-augmented triangular prism) that consists of fourteen equilateral triangles.

(e) Build a convex deltahedron (the gyro-elongated square dipyramid) that consists of sixteen equilateral triangles.

Although the argument above would lead one to think that there would be an irregular convex deltahedron with 18 triangles, it is impossible to assemble such a polyhedron. Try it!

Regularity

The definition of regularity for polyhedra requires that the faces be regular polygons, that these faces are all congruent, and that the vertex configurations be the same for every vertex. All three conditions are necessary: The disphenoid of Section 7.1 has congruent nonregular faces and identical vertex configurations, the deltahedra have congruent regular faces but varying vertex configurations, while some of the prisms and antiprisms of Section 7.1 have only regular polygons as faces and identical vertex configurations. These three conditions guarantee that the polyhedron has rotational symmetry about each vertex and about the center of each face.

SOFTWARE

1. *KaleidoTile*, by Jeffrey Weeks, of the Geometry Center, University of Minnesota, is freeware, for Macintosh only. It can be found at ftp://geom.umn.edu/pub/software/KaleidoTile.

SUGGESTED READINGS

Anatole Beck, Michael N. Bleicher, and Donald W. Crowe, *Excursions into Mathematics*, Worth Publishers, New York, 1969.

H.S.M. Coxeter, *Regular Polytopes*, Dover, New York, 1973.

H.M. Cundy and A.P. Rollett, *Mathematical Models*, Oxford University Press, New York, 1961.

Alan Holden, *Shapes, Space, and Symmetry*, Dover, New York, 1971.

Anthony Pugh, *Polyhedra: A Visual Approach*, University of California Press, Berkeley, 1976.

Hermann Weyl, *Symmetry*, Princeton University Press, Princeton, 1952.

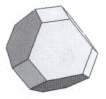

7. Polyhedra

◆ 7.3. ARCHIMEDEAN SOLIDS

SUPPLIES
cardboard polygons
scissors
tape

In Section 4.1, we investigated the three regular tilings, by squares, equilateral triangles, and regular hexagons, and then went on to find the eight semiregular, or Archimedean, tilings. For polyhedra, the five Platonic solids, the tetrahedron, cube, octahedron, icosahedron, and dodecahedron, are the analogues of the regular tilings. We now wish to find the three-dimensional analogues of the semiregular tesselations. Such semiregular, or Archimedean, polyhedra must have the properties that all faces are regular polygons, though more than one type of polygon will be used, and that each vertex must have the same polygons occurring in the same order. While Archimedes apparently knew of all thirteen of the solids that bear his name, his book on the subject was lost many centuries ago. Several of the semiregular polyhedra appear in texts and pictures of the Renaissance, but the complete list was rediscovered by Kepler, who also defined the classes of prisms and antiprisms. We proceed as he did in finding them, as described in Cromwell's *Polyhedra*.

Since we are interested only in convex polyhedra, at each vertex the sum of the angles of the polygons meeting there must be strictly less than 360°. We repeat from Section 4.1 the table of angles for the regular polyhedra that we will be using.

Vertex Angles

Polygon	Sides	Angle
Triangle	3	60°
Square	4	90°

Polygon	Sides	Angle
Pentagon	5	108°
Hexagon	6	120°
Heptagon	7	128.57°
Octagon	8	135°
Nonagon	9	140°
Decagon	10	144°
Dodecagon	12	150°

We start with some basic guidelines for our search.

Rule 1: Every semiregular polyhedron must have the angles of the polygons meeting at each vertex sum to less than 360°.

Rule 2: Every semiregular polyhedron must have at least three but no more than five polygons meeting at each vertex.

▶ **Exercise 1.** Explain why Rule 2 is true.

Rule 3: No semiregular polyhedron can have four different types of polygons meeting at each vertex.

▷ **Exercise 2.** Explain why Rule 3 is true.

A little additional thought gives us further limits on types of vertex configurations that can occur. Note that if we try to arrange one equilateral triangle and two other different polygons at a vertex, to make pattern $3.n.m$, they would be arranged as below:

At vertex A, we have the vertex configuration $3.n.m$. A triangle and an m-sided polygon meet at B, so the face marked with "?" ought to have n sides. But at vertex C there are already a triangle and an n-sided polygon, so the face marked "?" should have m sides. Therefore, we cannot have any vertex configuration of the form $3.n.m$ if $n \neq m$.

▷ **Exercise 3.** Show that one cannot have a vertex configuration of the form 5.*n.m* where *n* ≠ *m*. Generalize the result to any vertex configuration of the form *k.n.m* where *k* is odd and *n* ≠ *m*.

This gives us the following rule:

Rule 4: No semiregular polyhedron can have vertex configuration *k.n.m* where *k* is odd and *n* ≠ *m*.

Another configuration that cannot occur is 3.*k.n.m* unless *k* = *m*, as shown in the illustration below. At vertices *A* and *B* we have the configuration 3.*k.n.m*. This forces the configuration 3.*k.n.k* at vertex *C*. If this is to be a semiregular polyhedron, we must have *k* = *m*.

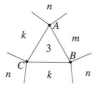

Rule 5: No semiregular polyhedron can have vertex configuration 3.*k.n.m* unless *k* = *m*.

We will investigate the possible semiregular polyhedra systematically by types of vertex configurations, beginning with combinations consisting of triangles and one other type of polygon. From Rule 2, we know that every vertex must have at least 3 polygons, but no more than 5. Five triangles gives us 3.3.3.3.3, the icosahedron. Four triangles and a square gives us 3.3.3.3.4, with an angle sum of 60° + 60° + 60° + 60° + 90° = 330°.

▷ **Exercise 4.** Build a snub cube, 3.3.3.3.4.

Four triangles and a pentagon gives us 3.3.3.3.5, with an angle sum of 60° + 60° + 60° + 60° + 108° = 348°.

▷ **Exercise 5.** Build a snub dodecahedron, 3.3.3.3.5.

Four triangles and a hexagon give us 3.3.3.3.6, with an angle sum of 60° + 60° + 60° + 60° + 120° = 360°, so this would lie flat and make a tiling instead of a convex polyhedron. Next we consider combinations of three triangles and two other polygons. The combination 3.3.3.4.4 gives an angle sum of 360°, and we investigated this tiling pattern in Section 4.1. Any combination of the form $3.3.3.n.m$ with n, $m \geq 4$ will have angle sum more than 360° and so would not form a finite polyhedron. The next exercise addresses other vertex configurations with three triangles.

▶ **Exercise 6.** Explain why the polyhedron given by the vertex configuration $3.3.3.n$ is the n-gon antiprism.

Now we consider patterns with two triangles. Rule 4 forbids forms like $3.3.n$, unless $n = 3$, in which case the figure is the tetrahedron. Rule 5 forbids forms like $3.3.n.n$, but we could rearrange the order of the polygons and get some viable candidates.

▷ **Exercise 7.** Build a cuboctahedron, $3.4.3.4$.

▶ **Exercise 8.** Build an icosidodecahedron, $3.5.3.5$.

▷ **Exercise 9.** What is formed by the pattern $3.6.3.6$?

▷ **Exercise 10.** Show that vertex configurations $3.4.3.5$, $3.4.3.6$, and $3.5.3.6$ cannot be used to build an Archimedean solid.

Next we try vertex configurations with only one triangle. Rule 4 implies that the only possible configurations using three polygons at each vertex must have form $3.n.n$. The configuration $3.4.4$ gives the triangular prism.

▶ **Exercise 11.** By rewriting the configuration $3.5.5$ as $5.3.5$ and applying one of the rules above, show that it cannot give a semiregular polyhedron. Generalize to $3.n.n$ for n odd.

▷ **Exercise 12.** Build a truncated tetrahedron, $3.6.6$.

▶ **Exercise 13.** Build a truncated cube, $3.8.8$.

▷ **Exercise 14.** Build a truncated dodecahedron, $3.10.10$.

▷ **Exercise 15.** Build a rhombicuboctahedron, $3.4.4.4$.

▷ **Exercise 16.** Show that patterns such as $3.4.4.5$ and $3.4.4.6$ do not give semiregular polyhedra.

▷ **Exercise 17.** Build a rhombicosidodecahedron, $3.4.5.4$.

▶ **Exercise 18.** Show that the patterns $3.n.4.n$ and $3.4.n.4$ for $n > 5$ do not give semiregular polyhedra.

We have now exhausted all possibilities that contain triangles, and move on to vertex configurations with no triangles but containing squares. The cube has three squares at each vertex, for symbol $4.4.4$, and the vertex configuration $4.4.n$ gives an n-gon prism with square sides.

▷ **Exercise 19.** Show that the patterns 4.4.*n.n* and 4.4.*n.m* do not give semiregular polyhedra, where *n, m* > 4.

▷ **Exercise 20.** Show that the pattern 4.*n.n* does not give semiregular polyhedra for *n* > 4 and odd.

▷ **Exercise 21.** Build a truncated octahedron, 4.6.6.

▷ **Exercise 22.** Show that 4.8.8 does not give a semiregular polyhedron. Generalize to 4.*n.n* for *n* > 8 and *n* even.

▷ **Exercise 23.** Show that 4.*n.m* does not give a semiregular polyhedron for *n* ≠ *m* and either *n* or *m* odd.

After the above series of exercises, we only need to investigate combinations of the form 4.*n.m* where *n* ≠ *m* and *n* and *m* are both even.

▶ **Exercise 24.** Build a great rhombicuboctahedron, 4.6.8.

▷ **Exercise 25.** Build a great rhombicosidodecahedron, 4.6.10.

▷ **Exercise 26.** Show that 4.6.12 and 4.8.10 do not give semiregular polyhedra.

That disposes of polyhedra containing squares, so we move on to polyhedra with pentagons as their smallest polygon. The dodecahedron is 5.5.5. By Rule 4, semiregular polyhedra containing pentagons as their smallest polygon must have form 5.*n.n*.

▷ **Exercise 27.** Build a truncated icosahedron, 5.6.6.

▷ **Exercise 28.** Show that 5.*n.n* does not give a semiregular polyhedron for *n* > 6.

You should now have models of all of the thirteen Archimedean solids. This will be studied in more detail in the following sections of this chapter and the next so don't squish them.

SOFTWARE AND SUPPLIES

1. *KaleidoTile*, by Jeffrey Weeks, of the Geometry Center, University of Minnesota, is freeware, for Macintosh only. It not only draws the regular and semiregular polyhedra, but speaks their names and allows vertex and edge truncations. It can be found at ftp://geom.umn.edu/pub/software/KaleidoTile.

2. A particularly nice and well-designed set of cardboard polyhedron models are the *Geodazzlers*, made by Design Science Toys, 1362 Rte. 9, Tivoli, NY 12583.

SUGGESTED READINGS

Peter R. Cromwell, *Polyhedra*, Cambridge University Press, New York, 1997.

H.M. Cundy and A.P. Rollett, *Mathematical Models*, Oxford University Press, New York, 1961.

Alan Holden, *Shapes, Space, and Symmetry*, Dover, New York, 1971.

Anthony Pugh, *Polyhedra: A Visual Approach*, University of California Press, Berkeley, 1976.

Duncan Stuart, *Polyhedral and Mosaic Transformations*, North Carolina State University School of Design Publ., Raleigh, 1963.

7. Polyhedra

◆ **7.4. POLYHEDRAL TRANSFORMATIONS**

SUPPLIES
    Models of platonic and archimedean solids
    clay
    knife

In this section, we will investigate other ways of deriving the Archimedean solids and also explain their nomenclature.

One way of getting new polyhedra from known ones is *truncation*. For example, take a cube and slice off each of the vertices. Cutting off the vertices introduces equilateral triangular faces where each vertex was, and cuts the corners off the squares to turn them into octagons.

Gradually enlarge the triangles until the octagons are regular (all the edges are equal).

230

At this point, one has a semiregular polyhedron: a polyhedron whose faces are all regular polygons of more than one type, and such that each vertex has the same configuration. The polyhedron below is called a *truncated cube* and has two regular octagons and an equilateral triangle meeting at each vertex, and so is denoted by 3.8.8:

The cube had 6 faces, 12 edges, and 8 vertices. In truncation, the 8 vertices are replaced by 8 triangles, so the truncated cube has $f = 14$, with 6 octagonal faces and 8 triangular faces. The truncated cube has the 12 edges from the original cube, plus three new edges around each of the triangles, for a total of $e = 12 + 3 \cdot 8 = 36$. None of the vertices of the original cube survive, but there are three new vertices at each of the triangles, so $v = 24$. Note that $\chi = 14 - 36 + 24 = 2$.

If we continue to slice away at the truncated cube until the slices meet at the midpoint of the edge of what was once the cube, we get another semiregular polyhedron, the cuboctahedron, with pattern 3.4.3.4:

The cuboctahedron has 14 faces, six squares from the original cube and eight triangles. None of the eight original edges survive, so $e = 24$, three edges each for the triangles. The truncated cube had 24 vertices, but these have merged in pairs in forming the cuboctahedron, so $v = 12$.

▷ **Exercise 1.** Make a cube out of clay, and perform the series of truncations shown above.

▷ **Exercise 2.** Imagine truncating the vertices of an octahedron. Which polygon will replace the vertices? What polygon replaces the faces?

▶ **Exercise 3.** Truncate an octahedron as demonstrated above for the cube. At a point halfway through the truncation process and again where the slices meet, semiregular polyhedra are formed. Which of the Archimedean solids are these?

▷ **Exercise 4.** Give the number of faces, edges, and vertices formed by the Archimedean solids of Exercise 3.

▷ **Exercise 5.** Truncate a tetrahedron as above for the cube. At a point halfway through the truncation process and again where the slices meet, semiregular polyhedra are formed. Which of the Archimedean solids are these?

▶ **Exercise 6.** Give the number of faces, edges, and vertices formed by the Archimedean solids of Exercise 5.

▷ **Exercise 7.** Truncate a dodecahedron as above for the cube. At a point halfway through the truncation process and again where the slices meet, semiregular polyhedra are formed. Which of the Archimedean solids are these?

▷ **Exercise 8.** Give the number of faces, edges, and vertices formed by the Archimedean solids of Exercise 7.

▷ **Exercise 9.** Truncate an icosahedron as above for the cube. At a point half way through the truncation process and again at the time when the slices meet, semiregular polyhedra are formed. Which of the Archimedean solids are these?

▷ **Exercise 10.** Give the number of faces, edges, and vertices formed by the Archimedean solids of Exercise 9.

This process explains how the five Archimedean solids whose names begin with "truncated" got their names.

▶ **Exercise 11.** Explain how the cuboctahedron and the icosidodecahedron got their names.

To return to the transformation of the cube, if we continued to slice away after the point where the cuboctahedron was formed, the triangles would run into each other and change form into irregular hexagons, and the squares would no longer touch each other but would be separated by the hexagons. As you keep slicing, the hexagons enlarge until they are regular, and we have the semiregular polyhedron 4.6.6, the truncated octahedron.

We can continue to enlarge the hexagonal faces until the squares get completely eaten up, ending with an octahedron.

▷ **Exercise 12.** Describe the similar set of operations on a dodecahedron, by first truncating the vertices (which introduces triangular faces which gradually grow, forming two different Archimedean solids in Exercise 7). Then allow these triangular faces to run into each other to form hexagons and form another of the Archimedean solids. Which Archimedean solid is formed? Continue until the original pentagonal faces of the dodecahedron disappear. Which solid is formed?

To get some others of the Archimedean solids, we need to back up a bit. After truncating the cube to get the cuboctahedron, we could have decided to truncate the vertices of the cuboctahedron rather than continuing to enlarge the triangles. If we had done this, we would have gotten a different figure, the *great rhombicuboctahedron* shown below, with vertex configuration 4.6.8.

▶ **Exercise 13.** Find the numbers of faces, edges, and vertices for the great rhombicuboctahedron.

▷ **Exercise 14.** Continue the truncation series above until the squares of the great rhombicuboctahedron meet. Which of the Archimedean solids is formed?

▷ **Exercise 15.** Give the number of faces, edges, and vertices formed by the Archimedean solid of Exercise 14.

▷ **Exercise 16.** Describe the similar set of operations on an icosidodecahedron, by truncating its vertices. Which two Archimedean solids are formed?

▷ **Exercise 17.** Give the number of faces, edges, and vertices formed by the Archimedean solids of Exercise 16.

There are only two of the Archimedean solids left: the snub cube and the snub dodecahedron. To form the snub cube, in the center of each face of the cube place a smaller square, slightly rotated. Surround this smaller square by a belt of equilateral triangles. The original cube had six square faces, so the snub cube will have six square faces. Each square will share an edge with 4 triangles, as shown below. Six of these will fit together to form the snub cube, with the addition of one triangle where each of the original vertices of the cube was:

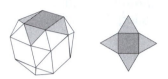

Thus, the snub cube will have $F = 6 + 6 \cdot 4 + 8 = 38$. The vertices are easier: Each of the central square faces has four vertices, and all the triangles share these vertices, so $V = 6 \cdot 4 = 24$. We can use the Euler characteristic to find the number of edges:

$$\chi = 2 = V - E + F,$$
$$2 = 24 - E + 38,$$
$$E = 62 - 2 = 60.$$

The snub dodecahedron is formed similarly.

▷ **Exercise 18.** Find the number of faces, edges, and vertices in a snub dodecahedron.

▷ **Exercise 19.** What happens when you snub a tetrahedron?

SOFTWARE

1. *KaleidoTile*, by Jeffrey Weeks, of the Geometry Center, University of Minnesota, is freeware, for Macintosh only. It not only draws the regular and semiregular polyhedra, but speaks their names and allows vertex and edge truncations. It can be found at ftp://geom.umn.edu/pub/software/KaleidoTile.

SUGGESTED READINGS

Peter R. Cromwell, *Polyhedra*, Cambridge University Press, New York, 1997.

H.M. Cundy and A.P. Rollett, *Mathematical Models*, Oxford University Press, New York, 1961.

Alan Holden, *Shapes, Space, and Symmetry*, Dover, New York, 1971.

Anthony Pugh, *Polyhedra: A Visual Approach*, University of California Press, Berkeley, 1976.

Duncan Stuart, *Polyhedral and Mosaic Transformations*, North Carolina State University School of Design Publ., Raleigh, 1963.

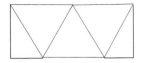

7. Polyhedra

◆ 7.5. MODELS OF POLYHEDRA

This section is meant to provide practical advice for building models of polyhedra with a variety of methods, including:

- cardboard models
- wire-frame models
- origami
- plaited polyhedra
- tetrahedral kites

Each subsection can be read separately. There are no exercises, but you should try to build at least a few examples for each type of model.

Cardboard Models

SUPPLIES
cardboard
scissors
Exacto knife
metal ruler
tape or glue

The Cardboard

The best cardboard models are made from stiff, but not too thick, cardboard. Bristol board, available at artists' supply stores is good, though poster board will do fine. Construction paper tends to be a bit too thin, fades with exposure to light, and goes limp with much handling. In practice, I use the best cardstock that the copier in the printshop will handle. Beautiful models can also be built from metal soldered together or Plexiglas joined with glue.

Choosing a Plan

One way to assemble polyhedra is to use one of the nets from Section 7.2. Cut the net out carefully and score along the fold lines, by putting a metal straightedge along the line and running a knife or the point of a pen over the line. You want to make a groove in the cardboard. Be careful not to cut all the way through. Carefully crease along the scored lines and assemble with tape or glue as described below. In practice, the nets for the larger polyhedra are so complex to cut out and so unwieldy to assemble that it is usually easier to build a polyhedron from individual polygons.

If you choose to use glue to hold your model together, than you may want to leave tabs on each free edge of the net, or each edge if you are using individual polygons.

Cutting Out the Pieces

In Sections 7.2 and 7.3, there is information on how many polygons and of which type are needed for each of the regular and semiregular polyhedra. It is essential that your polygons be completely regular and that all polygons used in a polyhedron have the same edge length. Two inches is a reasonable length to work with, except for the largest figures such as the great rhombi-cosidodecahedron.

One way to make multiple copies of a polygon is to make a template: carefully construct the desired polygon with the desired edge length and then trace as many copies as needed. The disadvantages to this is that each copy will be a tiny bit bigger than the original (but hopefully uniformly so), that it is quite tedious, and that the template, if made of cardboard, will tend to fray, especially at the vertices, so that later copies are less accurate.

Another way to make the polygons is to lay out sheets of each type and then cut off as many as needed. The disadvantage to this method is that pentagons and decagons are awkward, both to lay out and to cut. An advantage to this method is that one can use a copier to make as many duplicates of the sheets as needed. Use the heaviest cardstock that the copier will handle.

Equilateral triangles are easily cut out, by first carefully laying out in pencil a tiling of equilateral triangles (an architect's 30–60–90 triangle is useful for this). Then cut the tiling into strips, preferably using a metal ruler and Exacto knife, since a plastic ruler will soon develop nicks, and scissors do not allow as precise a cut as a ruler and razor blade. However, it should be noted that every architect I know is missing the very tips of their fingers from repeatedly slicing them off. Cut individual triangles off the strip. A tiling of equilateral triangles can also be cut up into hexagons, but in practice, it is easier to use the tiling 3.6.3.6, also given by the grid of equilateral triangles, since one then has longer straight lines to cut.

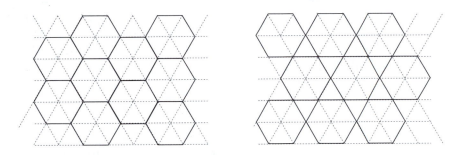

Squares are obviously also easy: just lay out a checkerboard and then cut off strips. Octagons are not too bad: lay out the tiling 4.8.8. It is easier to slice through the squares than to try to save them:

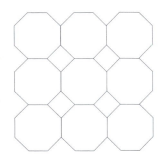

Pentagons are quite tedious, since there is no tiling of the plane with regular pentagons. Perhaps the best one can do is to lay out the drawing below. Decagons are even more troublesome.

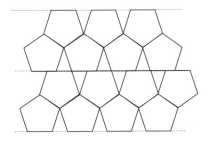

While all of these polygons can be laid out in the traditional manner with ruler and protractor, it is far more efficient to do this on a computer with a drawing program such as Macromedia *Freehand*, Adobe *Illustrator*, or Deneba *Canvas*. Any of these will generate regular polygons, which can then be duplicated and arranged in the manner described above. Take your time with the first row, making sure that the edge length is uniform and of the desired length and checking the alignment of the polygons. Try to arrange them so that they will be easy to cut out later, using tilings whenever possible. Then duplicate one row to fill up a page and copy as needed.

Joining the Pieces with Tape or Glue

Join the polygons in the prescribed manner to make the polyhedron you want using either tape or glue. If you use tape, it looks far better to keep it on the inside of the polyhedron. The tape should run the whole length of each edge. Carefully align neighboring polygons so that the edges and vertices meet cleanly. The last polygon is the most difficult: Try to make sure it is one with as few edges as possible. This last polygon is easily joined to the others along one edge. Attach bits of tape to the other edges and using the taped edge as a hinge, poke it into the polyhedron. Use something long and thin, like a knitting needle or nail file, to lever it into position and smooth the last bits of tape down.

If you prefer to use glue, you can either apply it to the edges or make tabs on each polygon or each free edge of the net. Household glue or contact cement works fine. Superglue is not advised. If you prefer not to use tabs, pour out a puddle of glue and use a toothpick to spread a small amount of glue on a pair of edges. Hold the edges together until the glue begins to set. With tabs, coat each tab with glue, making sure that the glue comes up close to the edge but does not go over it and hold the edges together until set. Clothespins can be used as clamps. The tabs have the additional advantage of strengthening the structure, but they make putting in the last polygon difficult.

Wire-Frame Models

SUPPLIES

> wire, solder, and soldering iron
> drinking straws, string, and 6″ upholstery needle
> toothpicks and marshmallows

The best wire-frame models are, of course, made of wire, soldered together at the vertices. Since it is difficult to solder the wire together at the precise angle needed, it is best to make a cardboard model first, and then build the wire-frame model around that. The wire can be attached to the cardboard at the edges by a drop of glue and then soldered at the vertices. Be sure to use glue that is not waterproof. After the solder has set, soak the polyhedron until the cardboard softens and can be removed.

Rather nice and quite easy models can be made with drinking straws and string: Cut the straws into uniform lengths and thread the string through to pull together the desired edges. Any store selling upholstery supplies will have long straight needles with large eyes that can be used to pull the thread through. If there is an Euler circuit (see Section 12.1), you will only need to pass through each straw once, but in practice there is no harm and added strength by going through each straw more than once. These models are excellent for demonstrating the lack of rigidity of some of the polyhedra, such as the cube.

Low-tech wire-frame models of polyhedra can be made from pipe-cleaners. Really, really low-tech models can be made with toothpick edges stuck into marshmallow vertices. These are surprisingly durable once the marshmallows have hardened, barring dogs, mice, and bugs.

Origami Models

SUPPLIES
   origami paper

We start with examples of origami polyhedra. Be patient and precise. We have chosen examples to show the wide range of origami styles. In order to make the illustrated directions clear and so that you can follow directions in other origami books, we introduce the standard notations used by origamists. Shaded areas in the illustrations indicate the colored side of the paper, white the reverse, usually white, side. Folds occur in two types: *valley folds*, denoted by ----------------, where the paper is folded so that a V is formed pointing away from you, and *mountain folds*, denoted by —··—··—··—··—, where the paper is folded so that a V is formed pointing toward you.

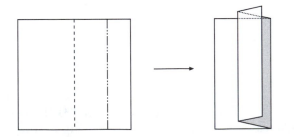

Further symbols used are:

Solid lines show creases made in previous steps. Dotted lines indicate features only visible on the back.

A hooked arrow means fold a flap under and tuck it in as indicated.

An arrow means move the indicated portion in that direction.

An open arrow means you should unfold that portion.

An doubled arrow means that you should form a crease by folding and unfolding.

A looped arrow means you should turn the paper over.

Tetrahedron

This origami model was invented by Kazuo Haga (of Haga's theorem from Section 3.3) and is based on the idea of folding a net for a tetrahedron and then tucking all the extra bits in to hold the model together:

(1) Take a large sheet of origami paper and fold in half along the length and unfold.
(2) Fold again to form a crease at one-fourth of the width.
(3) Fold so that the bottom left corner lies on the crease dividing the paper in fourths and that the dots meet, while folding through the midpoint of the bottom edge.

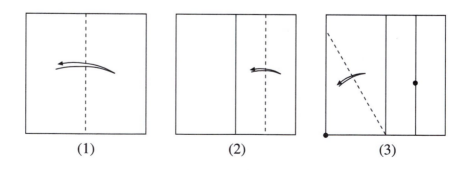

(1)   (2)   (3)

(4) Fold the lower right corner to overlap the triangle made in (3).
(5) Fold the top down from the points formed in (3) and (4).
(6) Unfold all creases, and fold the bottom edge up to meet the fold of (5).

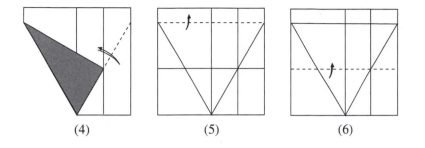

(4)　　　　　　　　(5)　　　　　　　　(6)

(7) Fold so that a crease is formed joining the bottom left corner and the point formed by the intersection of the creases shown by the dot. Fold another crease through the bottom right corner and the same dot point. Unfold.

(8) Fold so the marked points intersect and unfold. Repeat on the right side.

(9) Fold in top corners.

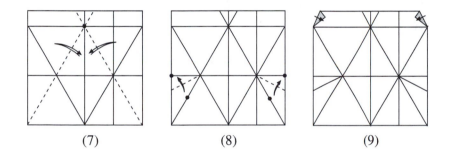

(7)　　　　　　　　(8)　　　　　　　　(9)

(10) You should now have a sheet divided into equilateral triangles. The shaded triangles will form a net for the completed tetrahedron, while the others will get tucked inside.

(11) Remember that we are making a three-dimensional object. Fold as shown, and then holding the point marked × down on the table, fold • up to the point ∘, so that triangles B and C are no longer in the original plane.

(12) Fold triangle A up the meet triangle B, and tuck the long top strip in to hold in place.

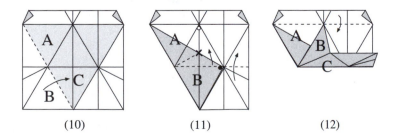

(10)　　　　　　　　(11)　　　　　　　　(12)

(13) Bring point • to point ∘, folding point × under the flap formed by the long top strip.

(14) Tuck point • under the long flap at point ∘.

(15) Bring point • up to point ∘, tucking the long flap in to complete the figure.

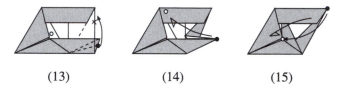

(13)   (14)   (15)

Traditional Origami Inflatable Cube

(1) Take a piece of square origami paper and crease along both diagonals so that the colored side is outside, and unfold.

(2) Crease along both center lines, so that the reverse side is outside, and unfold.

(3) Fold along the creases formed to make a triangle with pleated sides.

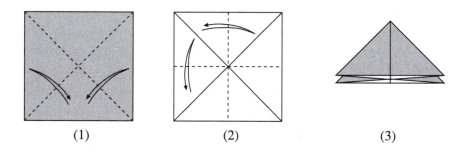

(1)   (2)   (3)

(4) Fold each of the base vertices of the triangle up to the top vertex. Do this on both sides to make a square.

(5) Fold the right and left corners of the square to the center point. Do this on both sides.

(6) Fold the tabs and slide into the pockets formed in (5). Do this on both sides.

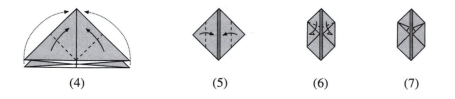

(4)   (5)   (6)   (7)

(7) One end, the lower vertex in the last figure above, has a small hole. Gently inflate.

(8) If you crease between each set of vertices, it becomes a cube.

Octahedron

This model is the invention of Tomoko Fusé and is an example of a belt modular approach. It will take 5 sheets of origami paper. It also builds on the initial net formed for the tetrahedron above.

(1) Take a sheet of origami paper and fold into equilateral triangles as for the tetrahedron up to Step (7).

(2) Fold down the long strip at the top.

(3) Fold through the midpoint on the left side so that the upper left corner meets the line as shown. Repeat for the right side and the lower right corner.

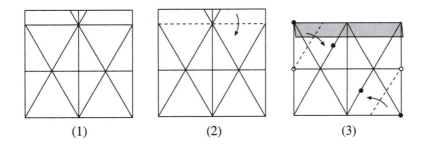

(1)   (2)   (3)

(4) Fold the sides in along the crease lines as shown.

(5) It should now look like this.

(6) Turn over, and crease as shown.

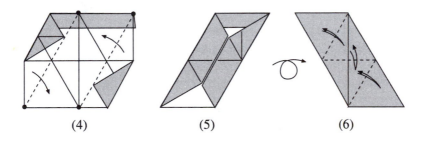

(4)   (5)   (6)

(7) Make three units as above, and tuck each into the next as shown.

(8) Tuck the last unit into the first to make a belt, or cylinder, with the slit sides outward.

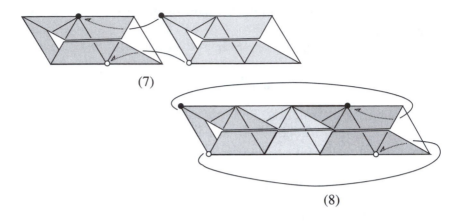

(7)

(8)

(9) Fold two more pieces of origami paper as in Step (1) and then fold on the dotted lines (note the mountain folds).

(10) These two last pieces, which should look like tetrahedra, will form the top and bottom of the octahedron, while the cylindrical shape of (8) forms the belt of triangles around the middle. Tuck the upper and lower triangles into the slits in the belt as shown.

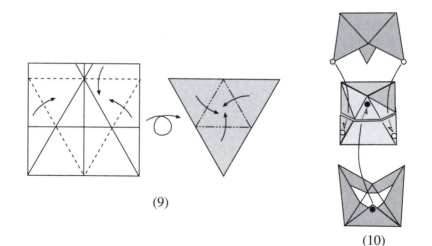

(9)

(10)

Icosahedron

This model is an example of the modular approach: Each face of the polyhedron is a separate module, and small paper tabs join the faces along the edges. Thus, for the icosahedron, you will need 20 faces and 30 joining tabs. Luckily, each component is quite easy to fold.

(1) Take a sheet of origami paper and fold in half lengthwise, then fold through the lower left corner so that the lower right corner meets the center line.
(2) Fold along the edge of the triangle formed in (1).
(3) Unfold the triangle formed in (1), and fold an equilateral triangle by folding a crease between the marked points.

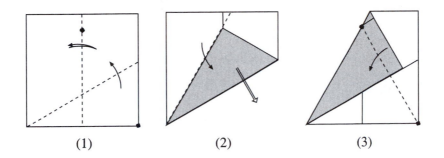

(1)   (2)   (3)

(4) Fold the triangle in half as shown, with the top part folded behind the lower part.
(5) Rotate the paper 180° and fold the leftmost triangle down over the center triangle. Also fold in the small flap at the right.
(6) Tuck the rightmost triangle and the rear flap into the center triangle.
(7) Your completed module should be an equilateral triangle with a pocket in each of the three sides. Make 20 modules.

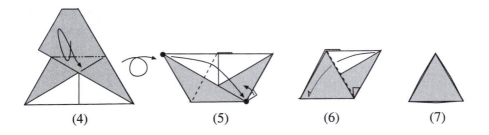

(4)   (5)   (6)   (7)

(8) Each of the joining tabs takes one quarter sheet of paper of the same size as used for the modules above. Cut a sheet of origami paper into four smaller squares. Fold in half.
(9) Fold the four corners in to meet at the center point.
(10) Here is the completed joining tab. Make 30 of these.

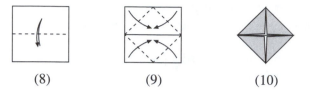

(8)   (9)   (10)

(11) Join triangle modules with the joining tabs so that 5 triangles meet at each vertex. You may want to put a spot of glue on each tab to make a more stable model.

Plaited Polyhedra

SUPPLIES
stiff paper tape

The idea of braiding polyhedra seems to go back to a long out-of-print book, *Plaited Crystal Models*, written by a chemist, John Gorham, in 1888. The idea was picked up by A.R. Pargeter in the article listed in the references and the illustrations below are redrawn from those in that article.

We will begin with a cube. Trace the pattern below onto graph paper, making each square about an inch on a side. Cut along the heavy lines. Fold a valley crease along each of the lines. First, fold the pattern up so that the two squares marked "O" (over) and "U" (under) are aligned, with the "O" face over the "U" face. Then braid the model, bearing in mind that you are trying to end up with a cube, that in a cube three squares must meet at each vertex, and that braiding the strands requires that strands alternately pass over and under each other. Note that one of the squares is slightly tapered: this should be the last to fall into place and is tapered to make tucking into the obvious flap easier.

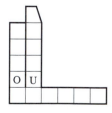

The same principles apply to the next model, which results in an octahedron, with four triangles meeting at each vertex.

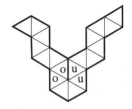

The pattern below results in an icosahedron, with five triangles meeting at each vertex. At the end there will be two loose ends to be tucked in: for both use two triangles for this tab instead of one. This makes a more secure fastening.

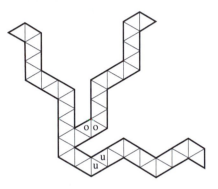

A variation on the cube is to braid it on the diagonal. The pattern below gives such a cube. Fold along the black lines. The paler grey lines are to help you draw the plan on graph paper and will also help you align the pieces. Each of the triangles forms one half of a face of the cube, but they are braided so that each face looks like the picture below on the right:

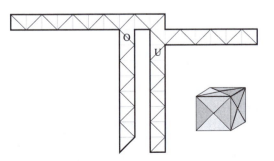

A similar construction can be used to build a tetrahedron, with each triangular face divided into three sectors.

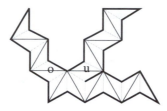

Peter Hilton and Jean Pedersen have extended the idea of braiding polyhedra to make polyhedra out of several straight strips of paper tape, appropriately folded. We will discuss only the cube. First mark a strip of paper tape into five squares. Make three such strips and fold along the lines. Wrap one of the strips to form the square walls of a box without roof or floor, letting the first and last squares overlap. Secure with a paper clip or spot of glue. Wrap the next strip of paper around the cube formed, making sure to overlap the two ends of the previous strip and that the overlapping ends of this second strip cover one of the holes (the roof or the floor). Now weave the last strip over the second strip and under the first, tucking the ends to form a cube.

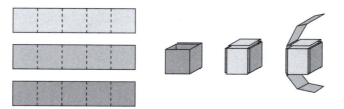

Alexander Graham Bell's Tetrahedral Kite

SUPPLIES

    drinking straws
    6-inch upholstery needle
    Mylar wrapping paper or plastic grocery bags
    kite string

Between 1880 and 1910, Alexander Graham Bell engaged in a series of experiments on kites, his solution to the inquiry on the feasibility of manned flight. These experiments culminated with the flight, on December 6, 1907, of the *Cygnus I*, a compound kite made of 3,393 tetrahedral modules. This giant kite was towed aloft by a steamer and flew for 7 minutes carrying a man, Thomas Selfridge. Bell's research group, the Aerial Experiment Association, went on to develop a glider and an airplane, which was the first plane in America to fly farther than a kilometer.

At the same time, of course, the Wright brothers were experimenting with the manned flight problem. They also used kites to test possible wing configurations. The famous first flight of the Wright brothers' plane took place on December 17, 1903 at Kill Devil Hills in Kitty Hawk, North Carolina. Earlier the same year, on April 23, Bell presented the results of his kite studies to the National Geographic Society:

> Since then I have been continuously at work with experiments relating to kites. Why, I do not know, excepting perhaps because of the intimate connection of the subject with the flying-machine problem.
>
> We are all of us interested in aerial locomotion; and I am sure that no one who has observed with attention the flight of birds can doubt for one moment the possibility of aerial flight by bodies specifically heavier than the air. In the words of an old writer, 'We cannot consider as impossible that which has already been accomplished.'
>
> I have had the feeling that a properly constructed flying-machine should be capable of being flown as a kite; and conversely, a kite should be capable of use as a flying-machine when driven by its own propellers. I am not so sure, however, of the truth of the former proposition as I am of the latter.

Bell takes as his starting point the recent innovation of the box kite, but points out the inherent instability of a rectangular structure without cross-braces.

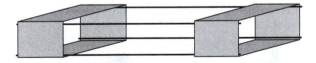

He thus moves on to a triangular box kite. These have the advantage not only of stability, but also of ease in joining to make compound kite structures. However, this stability is only in one direction: lengthwise, rectangles, with all their tendency to shear, are still being formed.

Bell came to realize that the simplest strong structure was the tetrahedron, and all of his later kites are based on the tetrahedral cell: Each face is an equilateral triangle, and two faces are covered with material while the other two faces are open:

These units could be assembled to make larger modular structures, in a variety of configurations. Below is pictured a four-celled tetrahedral kite, and four four-celled kites can be joined to form a 16-celled kite:

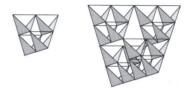

These bigger modules could also be assembled to make kites. Here is a picture from Bell's article showing two of his assistants holding a 64-celled kite:

A simple tetrahedral kite can be made easily from drinking straws and mylar gift-wrapping paper, or plastic grocery bags. This procedure is adapted from Anthony Thyssen's kite web page. First, using nylon kite string (because it's springy and knots well) and an upholstery needle, thread drinking straws together to form several tetrahedral cells. Thread the cells together by running the string through the straws again. While a 4-cell kite flies, it is not too stable in gusty conditions, and the 16-cell kite illustrated above is recommended. The drinking straws are not really strong enough to support anything bigger than 16 cells, but dowels could be used instead for a larger and more permanent kite. Separate 4-cell units can be joined by tying them together at the vertices. After building the framework, cut out a piece of mylar or plastic for each cell. This must cover two adjacent tetrahedral faces, so use a pattern made of five straws assembled as shown on the left and cut the material to overhang the pattern by at least one inch in each direction, and leave cutouts at the vertices.

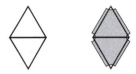

Wrap the extra material around the edges and glue it to itself, once for each cell. Attach the kite string at the center of the lower edge of the first tetrahedral cell, and fly!

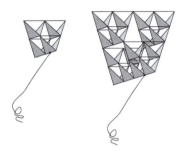

Below is Alexander Graham Bell flying one of his creations:

KITS AND SUPPLIES

1. Zometool makes the best construction kit for building models of polyhedra.
2. A simpler construction kit, using magnetic rods and steel balls for the connectors, is called Roger's Connection.
3. Plans for tetrahedral kites can be purchased from TetraLite, at www.aol.com/TetraLite
4. Tetrahedral kites in a variety of configurations can be purchased from Tetrahedral Pterygoids, P.O. Box 993442, Redding, CA 96099, or www.c-zone.net/tetra.

SUGGESTED READINGS

Alexander G. Bell, "The Tetrahedral Principle in Kite Structure," *National Geographic Magazine* XIV(6), June 1903.

H.M. Cundy and A.P. Rollett, *Mathematical Models*, Oxford University Press, New York, 1961.

Tomoko Fusé, *Unit Origami*, Japan Publ., Tokyo, 1990.

Rona Gurkewitz and Bennett Arnstein, *3-D Geometric Origami*, Dover, New York, 1995.

Peter Hilton and Jean Pedersen, *Build Your Own Polyhedra*, Addison Wesley, Menlo Park, 1994.

Alan Holden, *Shapes, Space, and Symmetry*, Dover, New York, 1971.

Scott Johnson and Hans Walser, "Pop-up Polyhedra," *Mathematics Magazine* 81, 1997.

Kunihiko Kasahara, *Origami Omnibus*, Japan Publ., Tokyo, 1988.

Kunihiko Kasahara and Toshie Takahama, *Origami for the Connoisseur*, Japan Publ., Tokyo, 1987.

David Mitchell, *Mathematical Origami*, Tarquin Publ., Stradbroke, England, 1997.

A.R. Pargeter, "Plaited Polyhedra," *Mathematical Gazette* 43, 1959.

Jean Pedersen, "Some Isonemal Fabrics on Polyhedral Surfaces," in *The Geometric Vein: The Coxeter Festschrift*, ed. C. Davis, B. Grünbaum, and F. Scherk, Springer-Verlag, New York, 1981.

Anthony Pugh, *Polyhedra: A Visual Approach*, University of California Press, Berkeley, 1976.

Anthony Thyssen, *Tetrahedral Kite Using Straws*, http://www.sct.gu.edu.au./~anthony/kites/straw_plan

Magnus Wenninger, *Polyhedron Models*, Cambridge University Press, New York, 1971.

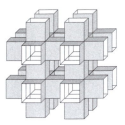

7. Polyhedra

SUPPLIES
    cardboard polygons
    models of regular and semiregular polyhedra, all with the same edge
      length
    tape

In Section 4.1, we discussed the checkerboard tiling, 4.4.4.4 or 4^4, made by putting squares together edge to edge so that four squares meet at each vertex. In Section 7.2, we constructed the cube, 4.4.4 = 4^3, by putting squares together edge to edge so that three squares meet at each vertex. We noted that all tilings must have the sum of the angles at each vertex summing to 360°, while every convex polyhedron must have the sum of the angles meeting at a vertex strictly less than 360°. In this section, we will investigate some structures that occur when the angle sum at a vertex is greater than or equal to 360°.

▶   **Exercise 1.** Build a structure from squares put together edge to edge so that only two edges meet at a time and 5 squares meet at each vertex. There are several ways to do this, but try to build as symmetrical a structure as possible.

The structure you built in Exercise 1 is an example of an infinite polyhedron: You can continue the structure forever. The convex Platonic and Archimedean polyhedra close back on themselves to contain a hollow cavity, but the infinite polyhedra do not. There are, naturally enough, infinitely many infinite polyhedra, so we will not try to build all of them. There are, however, only three completely regular infinite polyhedra, first discovered by Petrie and Coxeter:

One day in 1926, J.F. Petrie told me with much excitement that he had discovered two new regular polyhedra; infinite, but without false vertices. When my incredulity had begun to subside he described them to me: one consisting of squares, six at each vertex and one consisting of hexagons, four at each vertex. It was useless to protest that there is no room for more than four squares round a vertex. The trick is to let the faces go up and down in a kind of zig-zag formation so that the faces that adjoin a given 'horizontal' face lie alternately 'above' and 'below' it. When I understood this, I pointed out a third possibility: hexagons, six at each vertex.

Below is illustrated a section of the infinite regular polyhedron 4^6 that Coxeter describes. It is understood that the structure should go on forever. The polyhedron divides space up into two pieces: the space inside the tunnels of the polyhedron, and the space outside of the polyhedron. Both of these spaces are exactly the same shape! Imagine walking around, first inside the tunnels and then outside, to convince yourself of this.

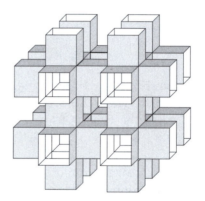

▶ **Exercise 2.** The second regular infinite polyhedron described by Coxeter can be built by making a number of truncated octahedra, but leaving all of the squares off to make holes. Stack the truncated octahedra together so that the square holes match up. Build a section of this infinite polyhedron, using eight truncated octahedra stacked two polyhedra wide, two deep, and two tall. How would you denote this polyhedron? How do the spaces enclosed by the polyhedron and outside the polyhedron compare?

▷ **Exercise 3.** The third regular infinite polyhedron described by Coxeter can be built by making a number of truncated tetrahedra, but leaving all of the triangles off to make holes. Stack the truncated tetrahedra together so that the triangular holes match up. Build a section of this infinite polyhedron, using twelve truncated tetrahedra stacked three polyhedra wide, two deep, and two tall. How would you denote this polyhedron? How do the spaces enclosed by the polyhedron and outside the polyhedron compare?

The first infinite regular polyhedron illustrated above is built from cubes, missing a pair of opposite faces, assembled in a symmetric manner. But another even simpler infinite structure could be built from similar cubes: imagine taking a pile of cubes, missing their tops and bottoms, and stacking them up in an infinite square cylinder.

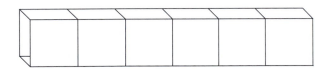

This square cylinder is an infinite polyhedron, but is not considered completely regular, although each vertex has the same configuration, since the fourfold symmetry of each square face is not carried over to the infinite polyhedron.

▷ **Exercise 4.** How would you denote this polyhedron? How do the spaces enclosed by the polyhedron and outside the polyhedron compare?

▷ **Exercise 5.** Draw a sketch of a similar cylindrical polyhedron made by stacking triangular prisms, missing their top and bottom faces. How would you denote this polyhedron?

▷ **Exercise 6.** Draw a sketch of a similar cylindrical polyhedron made by stacking pentagonal prisms, missing their top and bottom faces. How would you denote this polyhedron?

The previous exercises should have shown you two things: First, that there are infinitely many cylindrical prismatic polyhedra, and second, that the notation we have been using for tilings and polyhedra is inadequate to distinguish among the infinite polyhedra. Similar cylindrical structures can be built by stacking antiprisms, without their tops and bottoms:

▷ **Exercise 7.** Build a section of a cylindrical antiprismatic polyhedron by stacking up four square antiprisms missing their top and bottom faces. How would you denote this polyhedron?

▷ **Exercise 8.** Build a section of another cylindrical polyhedron by stacking up alternately square prisms and square antiprisms missing their top and bottom faces. How would you denote this polyhedron?

The simplest way to build the cylindrical square prismatic polyhedron pictured above (or any of the other cylindrical prismatic polyhedra) is to take a sheet of squares and cut out the following net, then crease and tape the sides together.

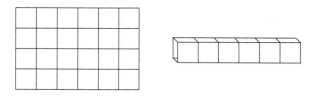

▷ **Exercise 9.** Draw a net for the cylindrical square antiprismatic polyhedron.

Just as strips of squares can be rolled up, so can strips of equilateral triangles, to make a square cylinder.

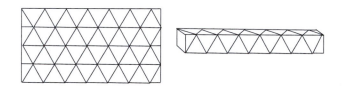

▷ **Exercise 10.** Fold up strips of triangles to make a pentagonal cylinder. How would you denote this polyhedron?

▷ **Exercise 11.** Fold up strips alternately of squares and triangles as shown on the next page to form yet another type of cylindrical polyhedron. How would you denote this polyhedron?

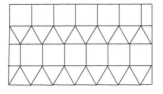

So far we have five types of cylindrical polyhedra: prismatic (Exercises 4 through 6), antiprismatic (Exercises 7 and 9), alternate prism and antiprism (Exercise 8), formed by strips of triangles (Exercise 10), and formed by alternate strips of triangles and squares (Exercise 11). There is one other class of cylindrical polyhedra, and in some ways it is the most interesting.

▷ **Exercise 12.** Cut out the pattern of triangles below and tape the sides together. The seam will spiral around the cylinder that is formed. If you crease the edges of the triangles alternating mountain and valley folds, you will get a polyhedron with flat faces. How would you denote this polyhedron?

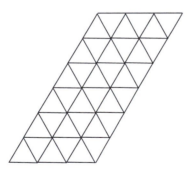

Another group of infinite polyhedra, the double-sided infinite polyhedra, is formed from the tilings we found in Section 4.1. In that section we found the three regular tilings, $4.4.4.4 = 4^4$, $3.3.3.3.3.3 = 3^6$, and $6.6.6 = 6^3$, and eight semiregular tilings, $4.8.8$, $3.12.12$, $4.6.12$, $3.4.6.4$, $3.6.3.6$, $3.3.4.3.4$, $3.3.3.4.4$, and $3.3.3.3.6$. To build a double-sided infinite polyhedron, one takes two copies of one of these tilings and places them parallel to one another. Then punch out copies of one of the tiles, and connect these holes with polygonal tunnels. For example, if we take the tiling $4.8.8$ and replace all of the squares with tunnels made of squares, we get the following infinite polyhedron, which can be denoted (but not uniquely) by $4.4.8.8$ since at every vertex two squares and two octagons meet.

Similarly, we could remove octagons and build octagonal tunnels connecting the two sheets of tilings. We cannot take all the octagons out, or the tiling would fall apart, so we take out every other one to get an infinite polyhedron denoted by 4.4.4.8.

In building infinite polyhedra, we must make sure that the polygons meet edge to edge, and that the edges meet in pairs only. The following sort of intersection, where more than two edges meet, is not allowed:

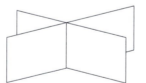

▷ **Exercise 13.** Describe the three different double-layered infinite polyhedra that can be formed from the tiling 4.6.12. How would you denote each of these polyhedra?

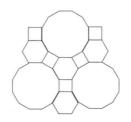

For the tiling 3.4.6.4, two double-sided infinite polyhedra can be formed, one with the triangles removed and replaced by tunnels, and one with the hexagons removed and replaced by tunnels. If you try to take out all of the squares, the tiling is connected only at the vertices, and one is forced to have edges meeting as in the forbidden "x" pattern above. If you try to remove only some of the squares and still make a semiregular polyhedron, you have trouble making every vertex have the same configuration: If you take out every other square around one of the hexagons, then you can't take out every other square around the neighboring hexagons. In the picture below on the right we have tried to remove every other square, but there comes a point where we either have two shaded deleted squares touching, or two unshaded squares to be kept. Thus, we either have bad edge intersections or have nonuniform vertex configurations.

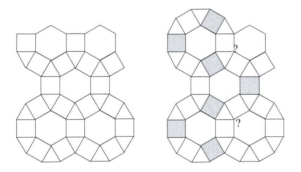

There are nineteen double-layered infinite polyhedra, all derived from the regular and semiregular tilings.

▶ **Exercise 14.** Mark which squares you could replace by tunnels in the tiling 3.3.4.3.4 below to build a double-layered infinite polyhedron, being sure that the result will have only two polygons meeting at each edge, that you remove as many squares as possible, and that each vertex will have the same configuration.

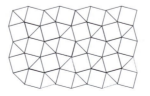

▷ **Exercise 15.** Note that in the following tiling 3.3.3.3.6 the hexagons could be replaced by tunnels to form an infinite polyhedron

3.3.3.3.4.4. Another infinite polyhedron can be formed by replacing some, but not all of the triangles. Mark which triangles you could replace by tunnels in the tiling to be sure that the result will have only one type of vertex, have exactly two polygons meeting at each edge, and is as symmetrical as possible.

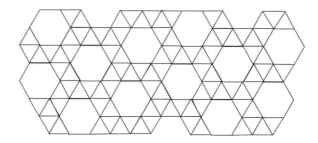

There are also infinite polyhedra with multiple layers, sometimes called *sponges*. Among these are the three regular infinite polyhedra discovered by Coxeter and Petrie discussed at the beginning of this section. One typical example is formed from infinitely many parallel copies of the tiling 4.8.8 with all of the squares removed and replaced by cubical tunnels. These tunnels lead alternately up to the layer above and down to the layer below:

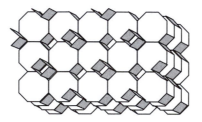

▶ **Exercise 16.** Build a multiple-layered infinite polyhedron from the tiling 4.8.8 by deleting half of the octagons and building tunnels alternately to the layer above and the layer below.

Another way of generating infinite polyhedra is to connect modules, each of which is one of the regular or semiregular polyhedra. For example, an infinite polyhedron consisting of rhombicuboctahedra joined together is shown on the next page:

▷ **Exercise 17.** Connect four rhombicuboctahedra with cubical tunnels, instead of connecting them directly as in the picture above. This is the start of another infinite polyhedron.

Many of this type of infinite polyhedra come from close packings of regular and semiregular polyhedra. A *close packing*, or *space filling*, of polyhedra is the three-dimensional analogue of a tiling on the plane: a way of fitting together polyhedra face to face so that they fill up all of 3-space. There is exactly one close packing using only one type of regular polyhedron; the familiar stacking of cubes.

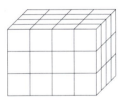

▷ **Exercise 18.** Only one of the semiregular polyhedra close-packs by itself. Which is it?

There are close packings that use more than one type of semiregular polyhedra. For example, the infinite polyhedron formed by the rhombicuboctahedra above leaves spaces that can be filled by cubes and cuboctahedra.

▶ **Exercise 19.** Using the models built in Section 7.3, figure out how to close-pack tetrahedra and octahedra.

▷ **Exercise 20.** Figure out how to close-pack octahedra and truncated cubes.

▷ **Exercise 21.** Figure out how to close-pack rhombicuboctahedra, tetrahedra, and cubes.

▷ **Exercise 22.** Figure out how to close-pack cuboctahedra, truncated octahedra, and truncated tetrahedra.

SUGGESTED READINGS

H.S.M. Coxeter, "Regular skew polyhedra in three and four dimensions, and their topological analogues," *Proceedings of the London Mathematical Society* 43, 1937.

Peter R. Cromwell, *Polyhedra*, Cambridge University Press, New York, 1997.

Alan Holden, *Shapes, Space, and Symmetry*, Dover, New York, 1971.

Anthony Pugh, *Polyhedra: A Visual Approach*, University of California Press, Berkeley, 1976.

A. Wachman, M. Burt, and M. Kleinmann, *Infinite Polyhedra*, Technion, Haifa, Israel, 1974.

Robert Williams, *The Geometric Foundation of Natural Structure*, Dover, New York, 1979.

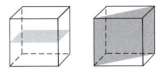

8. Three-Dimensional Symmetry

◆ 8.1. SYMMETRIES OF POLYHEDRA

SUPPLIES
models of Platonic and Archimedean polyhedra

In this section we will investigate the symmetries of 3-dimensional objects, just as in Chapter 5 we investigated symmetries in two dimensions. This is far easier to do with models in hand, so we will assume that you have such models, either constructed in Chapter 7 or provided some other way.

First we'll consider the simplest solid, the tetrahedron, or triangular pyramid. If you set the tetrahedron on its base, and consider the line passing from the top vertex through the center of the base, you have an *axis of rotation*: The tetrahedron can be rotated about this axis by 120° to land back in the same position, though all points except the top vertex and the center point of the base have moved. Below is an illustration of the tetrahedron with this axis of rotation and below that, bird's-eye views of the side faces, shaded to show the effect of three 120° rotations. If you have trouble seeing these rotations, hold a model of the polyhedron with one finger at the apex and your thumb at the center of the base. Use your other hand to rotate the polyhedron about these two points.

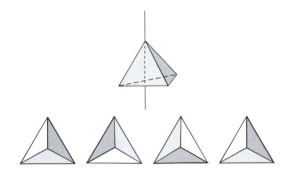

We say that the tetrahedron has a 3-fold axis of rotation, for a rotation angle of $120° = \frac{360°}{3}$.

▶ **Exercise 1.** How many other 3-fold axes of rotation does the tetrahedron have?

The tetrahedron has another type of rotational symmetry. Below is a tetrahedron with an axis passing through the midpoint of one edge and coming out at the midpoint of the opposite edge. This is a 2-fold axis of rotation. Sighting down the axis, you will see only two of the triangular faces, and a 180° rotation will interchange these.

▷ **Exercise 2.** How many other 2-fold axes of rotation does the tetrahedron have?

Next, consider the cube. Below is a cube with a 4-fold axis of rotation passing through the center points of two opposite faces. The Schlegel diagrams below the cube show the effect of 90° rotations around this axis. After 4 rotations through 90°, the cube returns to its original configuration.

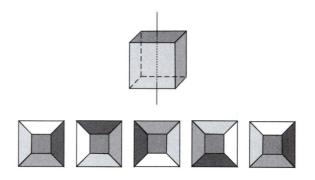

▷ **Exercise 3.** How many other 4-fold axes of rotation does the cube have?

▶ **Exercise 4.** An axis of rotation for the cube is shown on the next page, passing from front left top vertex to the rear right bottom vertex. What is the order of the axis (i.e., n-fold)? What is the angle of rotation? How many other axes of rotation of the same type are there?

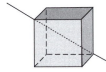

▷ **Exercise 5.** An axis of rotation for the cube is shown below, passing from the midpoint of one edge to the midpoint of the edge diametrically opposite. What is the order of the axis? What is the angle of rotation? How many other axes of rotation of the same type are there?

▷ **Exercise 6.** Fill in the following table with the number of distinct axes of rotation for each regular solid.

Rotations of Platonic Solids

Polyhedron	2-Fold Axes	3-Fold Axes	4-Fold Axes	5-Fold Axes
Tetrahedron		4	0	0
Cube				
Octahedron				
Dodecahedron				
Icosahedron				

In addition to rotational symmetry, 3-dimensional figures can also have reflectional (mirror) symmetry. For example, the tetrahedron is cut into two identical pieces by a plane through one edge and passing through the middles of the two faces opposing that edge.

▷ **Exercise 7.** How many other mirror planes does the tetrahedron have?

The cube can be cut in two ways, by a plane parallel to the faces, or diagonally.

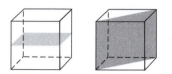

▶ **Exercise 8.** How many mirror planes does a cube have?

▷ **Exercise 9.** Fill in the table below.

Reflections of Platonic Solids

Polyhedron	Mirror planes
Tetrahedron Cube Octahedron Dodecahedron Icosahedron	

▷ **Exercise 10.** You will notice certain coincidences in the information you have collected in Exercises 6 and 9. Explain why some of the figures have the same numbers of lines of rotation and planes of reflection.

▶ **Exercise 11.** What are the symmetries of the truncated forms of the platonic solids?

▷ **Exercise 12.** What are the symmetries of the cuboctahedron and the icosidodecahedron?

▶ **Exercise 13.** What are the symmetries of the snub cube and the snub dodecahedron?

We have discussed two types of symmetry for 3-dimensional objects so far: rotational and reflectional symmetry. These are two of the seven symmetry operations, or *isometries*, in three dimensions. An isometry is a geometric operation which does not change length or angle measure. Thus it does not change the size or shape of an object. One of the other isometries is very easy: the **identity** is the isometry that leaves everything alone. If an isometry fixes four or more noncoplanar points (points that do not happen to lie on a single plane), then it must be the identity. If an isometry fixes three noncollinear points (points that do not lie on a single line), then it must fix the triangle formed by these three points, and so it must fix the plane that this triangle lies on. The only isometry that fixes a plane but moves everything not on that plane is **reflection** through that plane. If an isometry fixes two points, then it must fix all the points on the line connecting these two points. The only isometry that fixes a line and moves everything else is the **rotation** about that axis.

If an isometry fixes exactly one point it is a **rotary reflection** (or rotary inversion in some books), and the fixed point is called the center of symmetry. Illustrated below is a typical rotary reflection, in which point 1 moves to point 2, point 2 to 3, 3 to 4, 4 to 5, 5 to 6, and 6 back to point 1.

A rotary reflection can be thought of as a rotation (in the example above by 60°) followed by a reflection in a plane perpendicular to the axis of rotation (or vice versa). One solid you have seen that has rotary reflectional symmetry is the antiprism.

▷ **Exercise 14.** For the pentagonal antiprism above, draw the axis of rotation and the plane of reflection. What is the angle of rotation?

By a theorem too advanced for this book, any isometry of a finite polyhedron such as we studied in Section 7.1 must have a fixed point. Thus the remaining three isometries of three dimensions that do not fix any points apply only to infinite polyhedra. Two of these will be familiar from our study of frieze and wallpaper patterns: **translation** and **glide reflection**. For example, the infinite polyhedron shown below, which we met in Section 7.6, has translations in three directions.

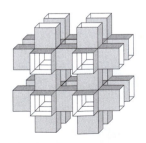

▷ **Exercise 15.** Give an example or build a section of an infinite polyhedron that has translational symmetry in only two directions.

▷ **Exercise 16.** Give an example or build a section of an infinite polyhedron that has glide reflectional symmetry in one direction.

The last of the seven isometries is called a **screw rotation** or screw displacement. This can be thought of as a rotation followed by a translation (or vice versa). A screw has, of course, screw rotational symmetry. One figure that has screw rotational symmetry is the *tetrahelix*, made by gluing tetrahedra together face to face.

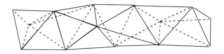

▷ **Exercise 17.** Here is another way to build a section of the tetrahelix. Cut a strip of equilateral triangles and crease along the lines as shown below. Tape the edges together according to the numbering, so that the two edges marked by a "1" get taped together, the edges marked "2" get taped together, etc.

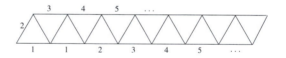

▷ **Exercise 18.** Here is yet another way to build a tetrahelix, due to Buckminster Fuller, as shown to Anthony Pugh. Make a copy of the picture below. Crease with a mountain fold along the lines marked ————————, and with a valley fold along the lines marked ················. Roll up into a cylinder, so that the shaded cells lie under the top row of triangles and then slide and twist so that the triangle marked *A* lies underneath triangle *B*.

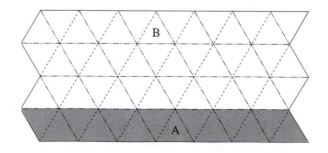

Reflection can be thought of as the most important of the seven isometries, since the other operations can be built up out of reflections. For example, a translation is the result of two reflections in parallel planes. In the illustration below, this principle is shown in the plane by reflecting a motif through two lines. The dashed motif is the first reflection, and the two solid motifs are the original and its translation.

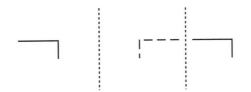

Rotation is reflection in two planes that intersect in a line, which will form the axis of rotation. Again, this can be illustrated in the plane: Reflecting the original motif in the first line gives the dotted motif. A further reflection in the other line gives the final result, which is a rotation of the original.

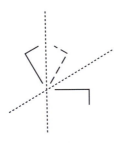

▷ **Exercise 19.** Show that every isometry is the composition of at most 4 reflections.

▶ **Exercise 20.** Show that every isometry with at least one fixed point is the composition of at most three reflections.

SUGGESTED READINGS

H.S.M. Coxeter, *Introduction to Geometry*, Wiley & Sons, New York, 1969.
Alan Holden, *Shapes, Space, and Symmetry*, Dover, New York, 1971.
Anthony Pugh, *Polyhedra: A Visual Approach*, University of California Press, Berkeley, 1976.
Marjorie Senechal, *Crystalline Symmetries*, Adam Hilger, Bristol, 1990.

8. Three-Dimensional Symmetry

◆ **8.2. THREE-DIMENSIONAL KALEIDOSCOPES**

SUPPLIES
> heavy cardboard or foamcore
> reflective Mylar
> model knife
> tape
> glue

In this section, we will experiment with reflections of three-dimensional objects. To start, take three mirrors and place one flat on the table and the other two perpendicular to the table and meeting at an angle of 90°. Thus, a right angle is formed between each pair of mirrors:

Place a small cube (about an inch on each side) in the 3-mirror assembly where the three mirrors meet. You should see a bigger cube in the reflection, with each side twice the length of the original cube.

▷ **Exercise 1.** Cut an equilateral triangle (about an inch on each side) out of cardboard and place it symmetrically in your 3-mirror assembly where the three mirrors meet. What geometric figure do you see?

▷ **Exercise 2.** Copy the figure below and bend along the dotted lines. Place it symmetrically in your 3-mirror assembly where the three mirrors meet, and adjust the three 45°-45°-90° isosceles triangles so that each meets the join of two mirrors at a right angle. What geometric figure do you see?

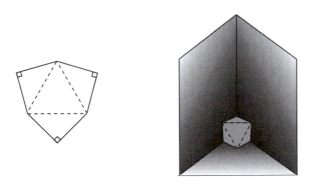

Recall from Section 8.1 that the cube has nine planes of reflection. Illustrated below in the first picture are the three mirror planes that form a 90°-90°-90° assembly. The other 6 mirror planes are illustrated in pairs:

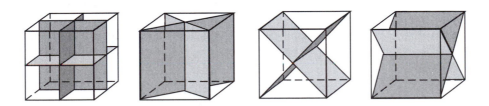

The cube, sliced by all 9 mirror planes, is illustrated below:

Note that the cube is sliced by these mirror planes into 48 congruent wedge-shaped pieces (formally, irregular tetrahedra). We would like to build a mirror assembly in the form of one of these wedges, called an *orthoscheme*.

The first three planes cut the big cube into 8 smaller cubes. If the original cube were 2 inches on a side, then the little cube would be 1 inch on each side. Throw everything out except the upper left front little cube:

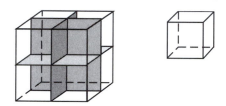

Only one of the next two planes cuts through the upper left front cube, and it slices it diagonally in half:

▷ **Exercise 3.** Find the length of each edge of the wedge-shaped region pictured above, if the small cube is 1 inch on each side.

Only one of the next two planes cuts through the wedge-shaped region, and it slices it diagonally into two pieces:

▶ **Exercise 4.** Find the length of each edge of the triangular wedge pictured above, if the small cube is 1 inch on each side.

Neither of the last two planes cut through the triangular wedge, so this wedge is the final orthoscheme:

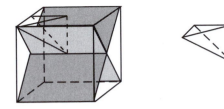

▷ **Exercise 5.** Below is a net for the wedge, or orthoscheme. Using your results from Exercise 4, label the lengths of each unmarked edge:

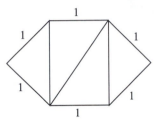

▷ **Exercise 6.** Build the orthoscheme out of cardboard, using 1 inch as your basic unit of measure and the net from Exercise 5.

▷ **Exercise 7.** Build the orthoscheme out of mirrors (with the mirrored surfaces facing in), leaving off the face that formed part of the original cube and using 6 inches or 1 foot for the basic measure instead of 1 inch.

▶ **Exercise 8.** Place the small cardboard orthoscheme so that it nests inside the mirrored orthoscheme. What do you see?

Next we move on to the octahedron. As found in Section 8.1, the octahedron also has nine mirror planes, illustrated below, which divide it up into triangular wedges:

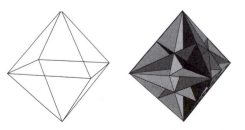

▷ **Exercise 9.** How many triangular wedges are formed by the mirror planes of the octahedron?

▷ **Exercise 10.** Explain why the cube and the octahedron have not only the same number of planes of reflection, but the very same mirror planes.

The first three mirror planes divide the octahedron into eight sectors. Let us concentrate on the front left upper sector:

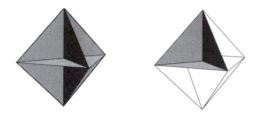

▷ **Exercise 11.** Assuming that each edge of the octahedron is 1 unit long, mark the lengths of each edge of the front left upper sector.

The next pair of mirror planes cut through the top and bottom vertices and slice one pair of opposite sides around the equator in half. This cuts the sector formed by the first three mirror planes in half:

▶ **Exercise 12.** Assuming that each edge of the octahedron is 1 unit long, mark the lengths of each edge of the wedge formed by the first 5 mirror planes.

The next pair of mirror planes cut through the front right and back left vertices and slice one pair of opposite sides in half. This cuts the sector formed by the first five mirror planes in half:

▷ **Exercise 13.** Assuming that each edge of the octahedron is 1 unit long, mark the lengths of each edge of one of the wedges formed by the first seven mirror planes.

The last pair of mirror planes cut through the front left and back right vertices and slice one pair of opposite sides in half. These planes do not affect our chosen wedge:

▷ **Exercise 14.** Draw a net for the orthoscheme of the octahedron.

▶ **Exercise 15.** Cut your net of Exercise 14 out of cardboard and assemble. Place into your mirrored orthoscheme for the cube. What do you see?

The tetrahedron has six planes of symmetry. Each passes through two of the vertices and cuts the edge connecting the other two vertices in half:

The six mirror planes cut the tetrahedron into wedges:

▷ **Exercise 16.** If you take the three mirror planes that cut through the top vertex, the tetrahedron is divided into how many pieces?

A bird's-eye view from the top vertex looks like the illustration below:

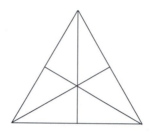

Assuming that each edge of the tetrahedron is 1 unit long, we have marked the lengths of each edge of one of the wedges formed by the first three mirror planes:

The next mirror plane slices through this wedge. Concentrate on the front left bottom wedge:

▶ **Exercise 17.** Assuming that each edge of the tetrahedron is 1 unit long, mark the lengths of each edge of one of the wedges formed by the first four mirror planes.

The next mirror plane does not affect the front left bottom wedge:

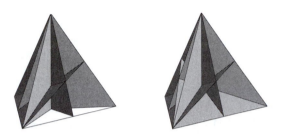

The last mirror plane does not affect the front left bottom wedge:

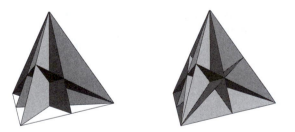

▷ **Exercise 18.** Draw a net for the orthoscheme for the tetrahedron.

▷ **Exercise 19.** Build the orthoscheme out of cardboard, using 1 inch as your basic unit of measure and the net from Exercise 18.

▷ **Exercise 20.** Build the orthoscheme out of mirrors (with the mirrored surfaces facing in), leaving off the face that formed part of the original tetrahedron and using 6 inches or 1 foot for the basic measure instead of 1 inch.

▷ **Exercise 21.** Place the small cardboard orthoscheme inside the larger mirrored one. What do you see?

SUPPLIES

Though you can purchase cuttable mirrors from Dale Seymour, it is far more economical to make (rather imperfect) ones. Get foamcore or stiff matboard and some reflective Mylar (preferably about 0.5 mm thick) from an artists' or architects' supply store. Sometimes you can find the Mylar with a peel-off adhesive backing, but if not, you can mount the Mylar on the foamcore or matboard with spray-on photomount glue. You can thus make large cuttable mirrors.

SUGGESTED READINGS

W.W. Rouse Ball and H.S.M. Coxeter, *Mathematical Recreations and Essays*, Dover, New York, 1987.

H.S.M. Coxeter, *Introduction to Geometry*, Wiley & Sons, New York, 1969.

H.S.M. Coxeter, *Regular Complex Polytopes*, Cambridge University Press, New York, 1991.

H.S.M. Coxeter, *Regular Polytopes*, Dover, New York, 1973.

H.M. Cundy and A.P. Rollett, *Mathematical Models*, Oxford University Press, New York, 1961.

Alan Holden, *Shapes, Space, and Symmetry*, Dover, New York, 1971.

Jay Kappraff, *Connections: The Geometrical Bridge Between Art and Science*, McGraw-Hill, New York, 1991.

Anthony Pugh, *Polyhedra: A Visual Approach*, University of California Press, Berkeley, 1976.

9. Spiral Growth

◆ 9.1. SPIRALS AND HELICES

SUPPLIES
   calculator
   graph paper
   polar graph paper
   string

Generally, a spiral lies in the plane and winds around the origin. The simplest spiral is called the *Archimedean spiral* and is the shape formed by coiling a rope. Since the rope is of constant thickness, the spacing of the coils is regular:

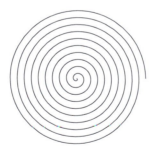

▷ **Exercise 1.** One way to draw an approximation of such a spiral is to take two pencils and a bit of string. Tie the ends of the string to the pencils loosely enough to slip as you rotate the pencil, and wind the string around one of the pencils. Holding the pencil with the string wound on it fixed, let the other pencil trace out the curve given by circling the fixed pencil while letting the string unwind. Draw an Archimedean spiral using this method.

▷ **Exercise 2.** What is the distance between adjacent coils in your spiral constructed in Exercise 1? [Hint: This depends on the size of the fixed pencil.]

One can also easily draw a rectangular Archimedean spiral: again, the coils are equally spaced.

▷ **Exercise 3.** Draw a triangular Archimedean spiral.

Gnomons

A *gnomon* is the term used to describe the region added to a geometrical figure to make a similar larger figure. For example, the white rectangle on the left is twice as long as it is wide. Four different gnomons are shown in the top row of figures. Each, when combined with the original rectangle, makes a larger rectangle of similar proportions, shown in the row below:

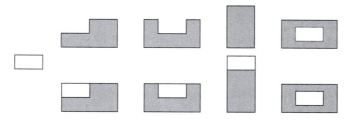

Here are some gnomons for a right triangle. Again, only the shaded figures are gnomons. The gnomon added onto the white triangle makes a new triangle which is similar to, but larger than, the original white triangle:

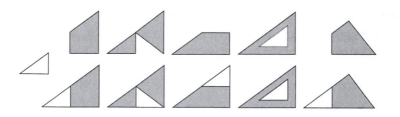

▶ **Exercise 4.** Draw a gnomon for a circle.

▷ **Exercise 5.** Draw a gnomon for the figure below:

▷ **Exercise 6.** Find *x*, where *x* is the length of one side of a rectangle whose other side has length 1, and such that the rectangle is its own gnomon. Thus, two copies of the rectangle make a larger rectangle with similar proportions:

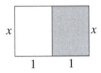

▷ **Exercise 7.** Find *x*, where *x* is the length of one side of a rectangle whose other side has length 1, and such that the rectangle has a square gnomon. Thus, the rectangle plus a square with side 1 make a larger rectangle with similar proportions:

Of great interest to the Greeks was the *golden triangle,* such as $\triangle DCF$ below, and its gnomon $\triangle ADF$, which we met in Section 3.2. There we showed that if $CD = 1$, then $AC = AD = \phi = \frac{1 + \sqrt{5}}{2} = 1.61803398875\ldots$, and that $CF = \frac{1}{\phi} = \phi - 1 = 0.61803398875\ldots$ Since $\triangle CDF$ and $\triangle ADF$ are isosceles, $AF = DF = DC = 1$, and $\alpha = \beta = 36°$:

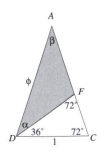

▶ **Exercise 8.** Find *x* and *y* for the gnomon of the 3-4-5 right triangle below:

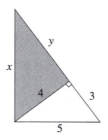

▷ **Exercise 9.** Consider the 30°-60°-90° triangle and its gnomon pictured below. Find the angles marked *α* and *β* and the lengths of the sides marked *x, y,* and *z*:

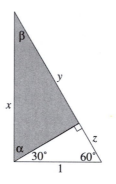

▷ **Exercise 10.** Find a triangle that is its own gnomon. Thus, two copies of this triangle make a larger similar triangle.

Gnomons are of great interest since many living organisms exhibit gnomonic growth. For example, the growth rings of a tree are possible gnomons for Exercise 4. Gnomons can also be used to generate spirals, and these spirals again can be used to describe certain growth patterns. For example, if we join two copies of the self-gnomon rectangle of Exercise 6, we can then attach the gnomon of the resulting rectangle, then attach the gnomon of that composite rectangle, and so on. Connecting corresponding points of these rectangles, we get a polygonal spiral, and rounding off the corners, a spiral curve:

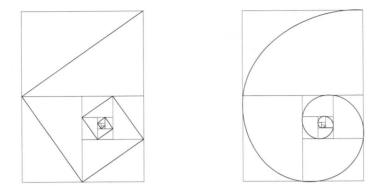

Spirals formed in this way are called *logarithmic* or *equiangular spirals*, since the angles formed by rays from the center point are always the same. This was not true for the Archimedean spiral with which we began this chapter. The coils of a logarithmic spiral are farther and farther apart as one moves away from the center point. Each section of a logarithmic spiral is a gnomon of a previous section. Snail shells are the classic example showing this type of spiral growth, as well as the horns of a ram, an elephant's tusk, or a cat's claw, though of course, these last two examples rarely complete a circuit.

The classic work on mathematical biology is *On Growth and Form* by D'Arcy Thompson, in which he says:

In the great majority of cases, when we consider an organism in part or whole, when we look (for instance) at our own hand or foot, or contemplate an insect or worm, we have no reason (or very little) to consider one part of the existing structure as *older* than another; through and through, the newer particles have been merged and commingled among the old; the outline, such as it is, is due to forces which for the most part are still at work to shape it, and which in shaping it have shaped it as a whole. But the horn, or the snail-shell, is curiously differ-ent; for in these the presently existing structure is, so to speak, partly old and partly new. It has been conformed by successive and continuous increments; and each successive stage of growth, starting from the origin, remains as an in-tegral and unchanging portion of the growing structure.

He further remarks:

In the growth of a shell, we can conceive of no simpler law than this, namely, that it shall widen and lengthen in the same unvarying proportions: and this simplest of laws is that which Nature tends to follow. The shell, like the creature within it, grows in size *but does not change its shape*; and the existence of this constant relativity of growth, or constant similarity of form, is of the essence, and may be made as the basis of a definition, of the equiangular spiral.

A similar process can be carried out for triangular gnomons. Below is pictured the *golden spiral*, based on gnomons for a golden triangle:

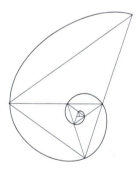

▷ **Exercise 11.** Draw a logarithmic spiral as follows: At the center of a sheet of graph paper (4 or 5 squares per inch) draw a line segment 1 square long, then turn 90° clockwise and draw a seg-ment 2 squares long, then turn 90° clockwise and draw a segment 4 squares long, then turn 90° clockwise and draw a segment 8 squares long, then turn 90° clockwise and draw a segment 16 squares long, to the limits of your sheet of graph paper. Go back to the beginning of your spiral, and turn 90° counterclockwise and draw a segment $\frac{1}{2}$ squares long, then turn 90° counterclockwise and draw a segment $\frac{1}{4}$ squares long, and repeat ad nauseum. Connect the corner points of the polygonal spiral you have thus drawn with as smooth and graceful a curve as possible.

▶ **Exercise 12.** An interesting spiral can be drawn as follows: Near the center of a sheet of graph paper draw 2 small adjacent squares 1 unit on a side, forming a 2 × 1 rectangle. Working your way around the figure, attach a square 2 units on a side to the rectangle, forming a 3 × 2 rectangle. Attach a square 3 units on a side to the rectangle, forming a 5 × 3 rectangle. Attach a square 5 units on a side to the rectangle, forming an 8 × 5 rectangle. Attach a square 8 units on a side to the rectangle, forming a 13 × 8 rectangle. Repeat to the limits of your graph paper. Connect the corner points of the polygonal spiral you have thus drawn with as smooth and graceful a curve as possible.

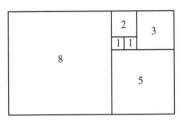

▷ **Exercise 13.** Consider the sequence of squares from Exercise 12: The sides measure 1,1,2,3,5,8, ... This sequence of numbers is called the *Fibonacci sequence*. Find the next 10 terms of the sequence, and give a rule describing how to find the next term of the sequence.

▷ **Exercise 14.** Below is a table giving the ratios of the Fibonacci numbers forming the sides of the rectangles formed in Exercise 12. Fill in the rest of the table and guess the (somewhat familiar) number that these ratios are approaching.

Ratios

$\frac{2}{1}$ = 2.000000
$\frac{3}{2}$ = 1.500000
$\frac{5}{3}$ = 1.666667
$\frac{8}{5}$ = 1.600000

Equations for Spirals

The equations for all spirals are fairly nasty in the more familiar Cartesian coordinates, but relatively simple in *polar coordinates*. Instead of a grid of squares, polar graph paper is marked out by rays from the center, or origin, and concentric circles. Any point on the graph paper is then described by two numbers (r, θ) where r is the radius, giving the distance from the center to the chosen point, and θ is an angle, measured from the horizontal line through the origin. Traditionally, mathematicians measure θ in radians, where the circumference of a circle with radius 1 is 2π radians. If you are more comfortable with angle measurement in degrees, it is easy to convert: Let θ denote the measure of any particular angle in radians and let Θ denote the same angle measured in degrees. Since $2\pi = 360°$, one degree is equal to $\frac{\pi}{180}$ radians, and one radian is equal to $\frac{180°}{\pi}$, so that $\theta = \frac{\pi}{180}\Theta$. Generally, we will give all formulae in both radians (using θ) and degrees (using Θ).

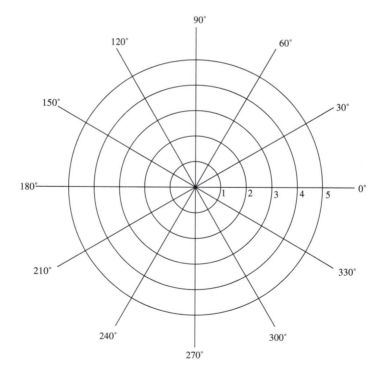

▶ **Exercise 15.** Using polar graph paper, plot the following points:
a. (1, 45°)
b. (2, 90°)

c. (3, 60°)
d. (1, 180°)
e. (1, −90°)
f. (2, 540°)

The polar equation for the Archimedean spiral is $r = k\theta = \frac{k\pi}{180}\Theta$ for some constant k. To draw an Archimedean spiral on polar graph paper, first make up a table of values, then plot the points and connect them using a smooth curve. Here we plot the Archimedean spiral $r = \theta = \frac{\pi}{180}\Theta$, where $k = 1$, using a calculator to compute the values of r for corresponding values of Θ.

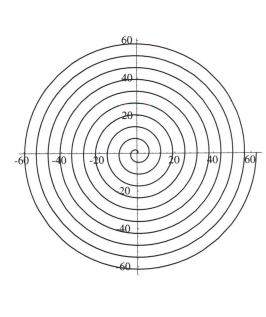

$$r = \theta = \frac{\pi}{180}\Theta$$

Point	Θ	r
1	0°	0
2	30°	0.52
3	60°	1.05
4	90°	1.57
5	120°	2.09
6	150°	2.62
7	180°	3.14
8	225°	3.93
9	270°	4.71
10	315°	5.50
11	360°	6.28
12	450°	7.85
13	540°	9.42
14	630°	11.00
15	720°	12.57

▷ **Exercise 16.** Plot points as above to draw the spiral
$$r = \frac{1}{2}\theta = \frac{\pi}{360}\Theta.$$

As we have seen, the Archimedean spiral begins at the origin and winds around and outward, with successive coils equally spaced. Other spirals may wind inwards, as does the *hyperbolic spiral* on the left at the top of the next page, which has equation $r = \frac{k}{\theta}$. On the right is a *parabolic, or Fermat, spiral*, with equation $r^2 = k\theta$:

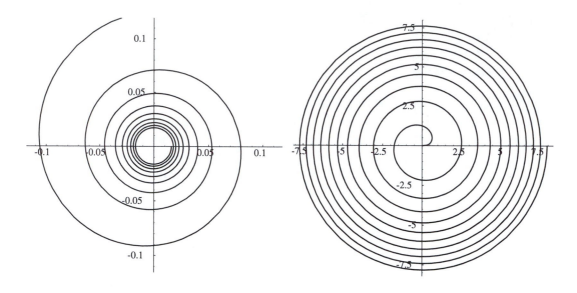

The *logarithmic, or equiangular, spiral* introduced previously has equation $r = k^\theta$, for some number k. Let us plot this spiral for $k = 2$. The magnitude of the numbers we are getting requires us to use a different scale (or a much larger sheet of polar graph paper). Connecting the points in order, we get the curve:

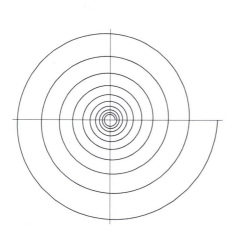

$$r = 2^\theta = 2^{\frac{\pi}{180}\Theta}$$

Point	Θ	r
1	0°	0
2	30°	1.43
3	60°	2.07
4	90°	2.97
5	120°	4.26
6	150°	6.15
7	180°	8.81
8	225°	15.24
9	270°	26.17
10	315°	45.25
11	360°	77.71
12	450°	230.72
13	540°	685.02
14	630°	2048.00
15	720°	6080.61

Helices

All of the spirals we have discussed have been flat, lying in the plane. What is commonly called a spiral is actually a *helix*: the most familiar embodiment being the marvelous Slinky. If you look at a coiled Slinky from directly above, it looks like a circle. From the side, it forms a sine curve.

A spiral helix would look like one of the spirals we have discussed when viewed from above, but would also grow vertically: Below is an Archimedean spiral helix on the left, and a logarithmic spiral helix on the right.

SUGGESTED READINGS

Sir Thomas Cook, *Curves of Life*, reprint of 1914 edition, Dover, New York, 1979.

Philip J. Davis, *Spirals: From Theodorus to Chaos*, A.K. Peters, Wellesley, 1993.

Martin Gardner, *The Unexpected Hanging and Other Mathematical Diversions*, Simon & Schuster, New York, 1969.

Peter S. Stevens, *Patterns in Nature*, Little, Brown and Co., Boston, 1974.

Peter Tannenbaum and Robert Arnold, *Excursions in Modern Mathematics*, Prentice Hall, Upper Saddle River, 1998.

D'Arcy Thompson, *On Growth and Form*, abridged edition, Cambridge University Press, Cambridge, 1961.

9. Spiral Growth

◆ 9.2. FIBONACCI NUMBERS AND PHYLLOTAXIS

SUPPLIES
> an assortment of flowers
> pineapple
> pine cones
> sunflower head
> tree branches
> colored thread
> glue
> polar graph paper

The *Fibonacci numbers*, met briefly in Section 9.1, owe their name to one of the most influential mathematicians of the early Renaissance, Leonardo of Pisa, popularly called Fibonacci. He introduced the use of Arabic numerals to Europe in a book called *Liber Abaci*, an unmitigated blessing, as anyone who has tried to multiply Roman numerals will realize. In that book, there was also a problem that introduced the sequence bearing his name (as quoted in Jacobs):

> How many pairs of rabbits will be generated by one pair in one year? Someone confined a pair of rabbits to an area enclosed by walls from all sides. He wanted to find out how many pairs would be generated, starting with this pair, in one year. It is the nature of rabbits to give birth to one new pair every month; and they start this at the age of two months. The said pair starts proliferating right away. Thus after one month we have two pairs. Of these, the original pair gives birth to another pair next time. This gives a total of three pairs after 2 months. Of these three, two give birth in the next period; five pairs after 3 months . . . and in this fashion we could go on step by step to an arbitrary number of months.

This is the first instance of a mathematical model for population growth. It is, of course, completely unrealistic: Rabbits never die, are infinitely fertile, always give birth to a pair (one male and one female), there must be an unlimited source of food, there are no predators, doesn't anybody worry about all of this inbreeding?

▷ **Exercise 1.** How many rabbits are there after one year?

In Exercise 13 of Section 9.1, you should have found the following *recursive formula*. Let f_n denote the nth Fibonacci number and let $f_1 = 1$ and $f_2 = 1$. Then we can find the nth Fibonacci number by

$$f_n = f_{n-1} + f_{n-2},$$

that is, the next Fibonacci number is the sum of the previous two.

▶ **Exercise 2.** List the first 13 Fibonacci numbers.

▷ **Exercise 3.** Fill in the table below. Find a formula for the sum of the first n Fibonacci numbers.

Sums of Fibonacci Numbers

f_1	1	1
$f_1 + f_2$	1 + 1	2
$f_1 + f_2 + f_3$	1 + 1 + 2	4
$f_1 + f_2 + f_3 + f_4$		
$f_1 + f_2 + f_3 + f_4 + f_5$		
$f_1 + f_2 + f_3 + f_4 + f_5 + f_6$		
$f_1 + f_2 + f_3 + f_4 + f_5 + f_6 + f_7$		
$f_1 + f_2 + f_3 + f_4 + f_5 + f_6 + f_7 + f_8$		
$f_1 + f_2 + f_3 + f_4 + f_5 + f_6 + f_7 + f_8 + f_9$		
$f_1 + f_2 + f_3 + f_4 + f_5 + f_6 + f_7 + f_8 + f_9 + f_{10}$		

We also investigated the ratios of successive Fibonacci numbers in Exercise 14 of Section 9.1, finding that they converge to the golden ratio: In our current notation,

$$\frac{f_n}{f_{n-1}} \to \phi = 1.61803\ldots$$

Here is yet another way of considering the Fibonacci sequence:

▶ **Exercise 4.** A *continued fraction* is a number that can be expressed as a fraction with an infinite number of levels. The simplest such continued fraction is:

$$1 + \cfrac{1}{1 + \cfrac{1}{1 + \cfrac{1}{1 + \cfrac{1}{1 + \cdots}}}}$$

Compute the following:

a) $1 + \frac{1}{1}$

b) $1 + \cfrac{1}{1 + \frac{1}{1}}$

c) $1 + \cfrac{1}{1 + \cfrac{1}{1 + \frac{1}{1}}}$

d) $1 + \cfrac{1}{1 + \cfrac{1}{1 + \cfrac{1}{1 + \frac{1}{1}}}}$

e) What numbers do you get? What number is represented by the infinite continued fraction?

▷ **Exercise 5.** Recall that in Section 3.2, the golden ratio ϕ was defined as the root of the equation $x^2 = x + 1$ or $x = 1 + \frac{1}{x}$. Show that the continued fraction of Exercise 4 also satisfies $x = 1 + \frac{1}{x}$.

Given that the Fibonacci sequence first appears as a completely artificial arithmetic exercise and can be further explored in a purely mathematical setting such as the continued fraction of Exercise 4, it is perfectly astounding to find that this sequence occurs more frequently than any other in nature. Consider the table below of common flowers.

Number of Petals

3	Iris
5	Buttercup, wild rose, geranium, cyclamen
8	Delphinium, cosmos
13	Cineraria, corn marigold
21	Aster, black-eyed susan
34	Pyrethrum daisy
55 or 89	Michaelmas daisy

Of course, there are exceptions, but many of these exceptions fit other Fibonacci-like sequences. Further, many flowers have been intentionally altered by horticulturists from their primitive conformation, most commonly by developing a double form of the original flower with twice as many petals or even more. The one you are most likely to be familiar with is the rose: the common wild rose, considered a rather nasty weed shrub by farmers and those with allergies, has been crossbred for centuries to develop the many-petalled forms common now, though still in disfavor with those with allergies. For the following exercise it is perhaps best to use weed flowers if the season permits.

▷ **Exercise 6.** Count the petals of at least five different types of flowers. Are your numbers part of the Fibonacci sequence?

▷ **Exercise 7.** Another Fibonacci-like sequence is the *Lucas sequence* g_n, whose first few terms are 2, 1, 3, 4, 7, ...
a) Give the next ten terms of the Lucas sequence.
b) Find the limit of the ratio of successive members of the Lucas sequence $\frac{g}{g_{n-1}}$.

Many of the plant patterns which do not fit the Fibonacci sequence will fit the Lucas sequence or other more anomalous sequences which are defined similarly. Further occurrences of the Fibonacci sequence show up in plants such as pineapples, artichokes, pine cones, some cacti, and seed heads such as sunflowers. The study of such plant structures is called *phyllotaxis* (literally, leaf arrangement), though the term is used not only for the study of leaf arrangements but also for the arrangement of petals and seeds and other botanical structures. The point of this section is not that all plant structures involve the Fibonacci sequence, but that a truly surprising number do. The famous geometer Coxeter says:

> Thus we must face the fact that phyllotaxis is really not a universal *law* but only fascinatingly prevalent *tendency*.

For example, the scales on a pineapple are usually roughly hexagonal in form. A closer study of these scale patterns shows clearly that by connecting each scale to one of its neighbors one can see that they are arranged in helices, called *parastichies* by botanists, around the pineapple. There will usually be 3 obvious families of these helices: One family of parallel curves winds around the pineapple in a right-handed manner, while a second family of parallel curves winds around left-handedly, and the third (and usually least obvious) may be either right- or left-handed but more vertical. If one imagines the surface of the pineapple as roughly a cylinder and then cuts the cylinder open and lays it out flat, it would look like the picture below, with the three families of parastichies marked with solid, dashed, and dotted lines:

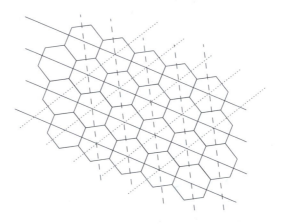

For a small pineapple, there will often be 5 parallel parastichies winding one way, 8 the other way, and 13 the last. For a large pineapple, you may get 8, 13, and 21. Hybrid or genetically altered pineapples may not match the sequence:

▷ **Exercise 8.** Obtain a pineapple and mark these families of parastichies using (nontoxic) glue and colored threads, using one color for the right-hand helices, another for the left-hand helices, and a third for the more vertical ones. How many right-hand parastichies are there? How many left-hand ones? How many vertical ones?

If you are fortunate enough to live in a region with palm trees, look at the trunks: Some varieties show clear leaf scars arranged in helical patterns around the trunk, again with the numbers of helices almost always Fibonacci numbers. Pine cones exhibit similar helical patterns. Since the scales of a pine cone are usually quadrilateral in form, only two families of parastichies are clearly marked. Again, simplifying the pine cone to a cylinder and cutting it open, these look somewhat like the picture below:

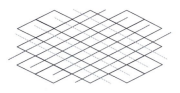

▷ **Exercise 9.** Take pine cones from at least three varieties of conifers and observe the parastichies. Mark these with glue and colored thread. How many right-hand parastichies are there? How many left-hand ones?

▷ **Exercise 10.** What angle is formed between the parastichies on a pine cone?

The most dramatic illustration of this natural growth pattern among commonly available plants is in the seed heads of sunflowers. Sunflower seeds are roughly quadrilateral in form and are arranged on the seed head in beautifully clear parastichies, usually with 34 parallel spirals winding in one direction (clockwise or counterclockwise) and 55 in the other, or for a very large seed head 55 and 89. For a small seed head, one may find only 21 and 34.

▷ **Exercise 11.** Mark one spiral in each direction on a sunflower seed head with glue and colored thread. How many right-hand parastichies are there? How many left-hand ones?

▷ **Exercise 12.** What angle is formed between the parastichies?

Since phyllotaxis means leaf arrangement, no discussion would be complete without some mention of this. Many trees and shrubs also grow according to Fibonacci numbers. Take a twig from a tree and study the arrangement of the leaves along it: Starting from the bottom leaf, consider the clockwise turn to the next leaf along the twig. For an elm tree, these leaves will be placed approximately on opposite sides of the twig: thus, there is a 180° or $\frac{1}{2}$ of a whole turn or clockwise revolution between successive leaves. For a beech tree there is a $\frac{1}{3}$ turn between leaves, for an oak a $\frac{2}{5}$ turn, for a poplar a $\frac{3}{8}$, and for a willow a $\frac{5}{13}$ turn. Not only does each of these involve Fibonacci numbers, each can be rewritten as a ratio of successive terms in the Fibonacci sequence; for example, the $\frac{3}{8}$ clockwise turn between successive leaves of a poplar could also be interpreted as a $\frac{5}{8}$ counterclockwise turn.

▷ **Exercise 13.** Using thread and glue, connect successive leaves on a twig from a tree to form a helix. How much of a complete turn is made between successive leaves?

While these spiral and helical patterns have long been noted, it is only fairly recently that a convincing explanation has been made for the phenomenon. First a few remarks on growth processes and terminology must be made. At the growth tip of a plant is a region of undifferentiated tissue. From this emerges small protrusions, called *primordia*, which will evolve into various features of the plant such as petals, stamens, or leaves. These are gradually pushed farther from the growth center by the emergence of new primordia. Below is a scanning electron micrograph image of a very young sunflower flower head (measuring 2.5 mm.), showing the region of undifferentiated cells and the emerging primordia as they evolve into florets (which will later become the sunflower seeds), already organized into spirals.

The primordia should ideally be well separated from each other around the rim of the growth tip, so that each has as much space to grow as possible. As new ones are generated, old ones move away from the growth region. The following exercise greatly simplifies the situation by considering the growth process as taking place in a plane, rather than in 3 dimensions.

▷ **Exercise 14.** Below are 5 pictures of a growth tip, showing the first primordia. In the first picture primordium #1 has just emerged.

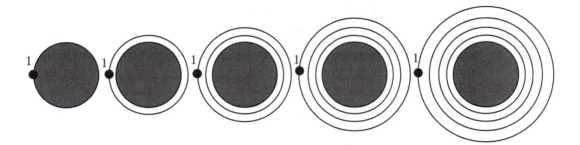

a. In the second picture, primordium #1 has moved away from the rim. Add primordium #2, placing it on the rim of the shaded disc as far as possible from #1.
b. In the third picture, draw primordium #2 on the middle ring as positioned in (a). Add primordium #3, placing it on the rim of the shaded disk as far as possible from #1 and #2.
c. In the fourth picture, draw primordia #2 and #3 on the second and third rings as positioned in (b). Add primordium #4, placing it on the rim of the shaded disc as far as possible from the others.
d. In the last picture, draw primordia #2, #3, and #4 on successive rings as positioned in (c). Add primordium #5, placing it on the rim of the shaded disc as far as possible from the others.
e. Will primordium #5 ever recover from its poor start in life?

Instead of placing each primordium individually in response to the already existing primordia, it makes more sense to have a general rule for where the next one will be inserted. For example, suppose we decide to insert each new primordium at an angle of 60° (measured from the center) from the previous one. Then the florets (or leaves or whatever we are generating) would be positioned as on the next page. Each successive primordium is positioned at an angle of 60° from its predecessor and is a little farther from the center.

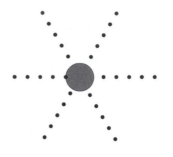

▷ **Exercise 15.** Using polar graph paper, plot 30 primordia emerging at an angle of 0.5(360°) = 180°.

▶ **Exercise 16.** Using polar graph paper, plot 30 primordia emerging at an angle of 0.6(360°) = 216° = −144°.

▷ **Exercise 17.** Using polar graph paper, plot 30 primordia emerging at an angle of 0.7(360°) = 252° = −108°.

These and similar experiments will soon convince you that angles that are rational numbers make the primordia clump up, wasting space, and thus not giving each leaf an optimal share of sunlight and rain, not giving each flower an optimal chance to grow and be fertilized. We need an angle that is irrational, one that does not repeat, so that each primordium has as much space as possible. In some sense, the golden number ϕ is the most irrational of all the irrational numbers, the one that is farthest from any rational approximation. Thus, the ideal angle at which to space the newly emerging primordia is the *golden angle* $\phi \cdot 360° \approx 1.61803(360°) = 582.4908°$ which is equivalent to $222.4908°$ which is equivalent to $−137.5092°$. Below is an illustration (drawn after Prusinkiewicz and Lindenmayer, using *Mathematica*) of the generating patterns for (a) 137.3°, (b) the golden angle, and (c) 137.6°.

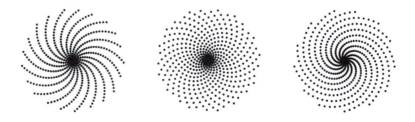

Thus, the appearance of the golden ratio, and its rational approximation as the ratio of successive Fibonacci numbers, is an evolutionary phenomenon for optimal growth.

SUGGESTED READINGS

W.W. Rouse Ball and H.S.M. Coxeter, *Mathematical Recreations and Essays*, Dover, New York, 1987.

H.S.M. Coxeter, *Introduction to Geometry*, Wiley & Sons, New York, 1969.

Stéphane Douady and Yves Couder, "Phyllotaxis as a Self-organized Growth Process," in *Growth Patterns in Physical Sciences and Biology*, ed. J. M. Garcia-Ruiz et al., Plenum Press, New York, 1993.

Konrad Jacobs, *Invitation to Mathematics*, Princeton University Press, Princeton, 1992.

Roger V. Jean, *Phyllotaxis*, Cambridge University Press, Cambridge, England, 1994.

Przemyslaw Prusinkiewicz and Aristid Lindenmayer, *The Algorithmic Beauty of Plants*, Springer-Verlag, New York, 1990.

Peter S. Stevens, *Patterns in Nature*, Little, Brown and Co., Boston, 1974.

Ian Stewart, "Daisy, Daisy, Give Me Your Answer, Do," *Scientific American* 272, January 1995.

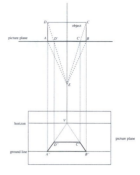

10. Drawing Three Dimensions in Two

♦ **10.1. PERSPECTIVE**

SUPPLIES
ruler

Aside from the far better known *Elements of Geometry*, Euclid also wrote a text called *Optics*. This is based on then current theories of vision and light, including the hypothesis that the eye is filled with luminous rays, and the sight of an object is caused by the reflection of these rays back to the eye. These rays would thus form a cone or pyramid, with the eye at the vertex and the object viewed at the base. Objects that are not within the visual cone of rays are not seen. While physiologically incorrect, these postulates, and the propositions he proves from them, are mostly correct in practice: We now know that the eye is not emitting rays but rather is collecting light rays reflected from the object onto the retina, but the visual cone is a correct image of the field of view when one looks through one eye. Postulate 4 of the *Optics* states that the apparent size of an object depends on the angle it subtends, as shown below.

From this Euclid proves that of two objects of equal size, the one that is farther away looks smaller.

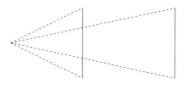

Postulate 7 states that objects viewed by more numerous rays are more distinct, which seems quite mistaken, until one realizes that from this one can derive the physically correct fact that nearer objects will be more distinct, and that if an object is sufficiently far away, it cannot be seen at all.

Proposition 19 uses the fact, which we used in Section 2.1 for billiards, that when light is reflected off a mirror, the angle of incidence is equal to the angle of reflection.

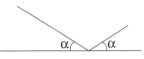

▷ **Exercise 1.** [This exercise is from Pedoe's *Geometry and the Visual Arts*.] Facing a slightly steamy mirror, close one eye and trace the image of your face. Measure the image and verify that it is half the size of your face. Give an explanation, with an appropriate diagram, of why this is true.

▷ **Exercise 2.** How big must a full-length mirror be to reflect the complete image of a person 6 feet tall? Make a drawing showing how the mirror must be hung.

Proposition 6 of the *Optics* gives a proof of the apparent convergence of parallel lines at infinity, like the familiar image of railroad tracks, which we take as a point of departure for our study of perspective. Perspective drawing was invented during the Renaissance, primarily by Brunelleschi, Alberti, Piero della Francesca, Dürer, Leonardo da Vinci, and many other practicing artists, many of whom had a great interest in mathematics. The basic concept of perspective relies on the visual cone or pyramid of Euclid and the picture plane, the plane on which the painting is done. The rays emitted by the eyes, or in modern terms the lines drawn from the eye to the object, intersect the picture plane to give an accurate rendition of what the eye sees. This is illustrated in an engraving from Brook Taylor's *New Principles of Linear Perspective* of 1719.

Many artists gave practical advice and invented various devices to help the artist achieve the effect of perspective. For example, Leonardo da Vinci advised the artist in "How to Portray a Place Accurately" (as quoted in Kemp):

> Obtain a piece of glass as large as a half sheet of royal folio paper and fasten this securely in front of your eyes, that is between your eye and the thing you want to portray. Next, position yourself with your eye at a distance of two-thirds of a *braccio* [a unit of measure, approximately one arm's length] from the glass and fix your head with a device so that you cannot move it at all. Then close or cover one eye, and with the brush or a piece of finely ground red chalk mark on the glass what you see beyond it.

An extension of this idea recommends a screen with fine threads to lay out a grid before the artist, who can then transfer what the eye sees to a gridded sheet of paper, as illustrated in Albrecht Dürer's, *Underweysung der Messung mit dem Zirkel und Richtscheyt* (Nuremberg, 1525), Book 3, Figure 67.

Let us designate by E the point where the eye of the artist is, the *viewing point*. On the next page, the point E and the *object plane* that contains the thing that one wishes to render are illustrated. Perpendicular to the object plane is the *picture plane* on which the image is to be drawn. The *ground line* is the line where the object and picture planes intersect. A line is drawn from E parallel to the object plane and perpendicular to the picture plane, which intersects the picture plane at point V, the *principal vanishing point*. The *vanishing line*, or *horizon*, is on the picture plane parallel to the ground line and through the vanishing point. We will return to the importance of these in a moment. To find the image of a point P on the object plane, imagine a line from E to P. The intersection of this line with the picture plane is P′, the perspective image of P.

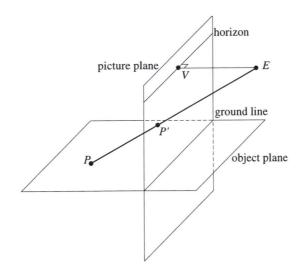

To understand the vanishing point V, consider parallel lines on the object plane, in this case running perpendicular to the ground line. All of the perspective images of all of the points on the line L_1 will lie on the line L_1', where the plane given by L_1 and point E intersects the picture plane. Since L_1 and the line EV are parallel, the line EV will lie on this plane also. Thus the point V must lie on line L_1'. Similarly, one can show that V also lies on the line L_2'. Lines L_1 and L_2 are parallel and never intersect on the object plane, but their perspective images L_1' and L_2' must intersect at the vanishing point V. Thus all lines on the object plane that are perpendicular to the ground line at the intersection of the object and picture planes will have images that pass through the principal vanishing point.

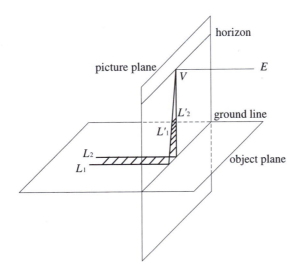

If we consider railroad tracks on the object plane that do not happen to be perpendicular to the ground line, a similar argument shows that the images on the picture plane will intersect at the point V', where EV' is parallel to the lines L_1 and L_2. The point V' must thus lie on the horizon.

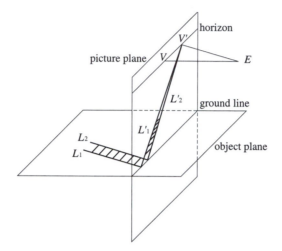

We now have one of the basic rules for perspective drawing: *In general, the images of a set of parallel lines in the object plane will intersect at a unique point on the horizon.*

▷ **Exercise 3.** Are there any exceptions to the general rule stated above?

Note also in both perspective drawings of the railroad tracks that the crossties also seem to get closer and closer together. The rule above can be used to draw the perspective image of a checkerboard or tile floor, and also to find the correct spacing of the horizontal tiles.

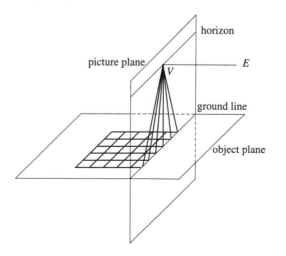

The previous drawing is, of course, an attempt at drawing a perspective view of a perspective view (on the picture plane), and as such is perhaps not as clear as one might like. To see at once both the tile floor on the object plane and the drawing made on the picture plane, we use a technique called *rabattement*, from the French *rabattre* meaning to lower or bring down. The picture plane is rotated and placed in the same plane as the object plane. Below you can see how this is done. The dotted lines connecting the tile floor in the object plane with its image in the picture plane help to keep their respective dimensions the same. The tiling contains four families of parallel lines: vertical lines, horizontal ones, and diagonal lines slanting up to the right and slanting up to the left:

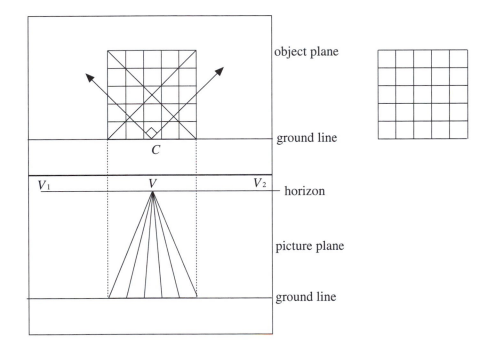

The family of lines corresponding to the vertical lines of the tiling, since they run at right angles to the ground line, will meet at V, the principal vanishing point. The lines corresponding to the diagonals that slant up to the left will meet at another point V_1 on the horizon to the left of V, while the lines corresponding to the diagonals that slant up to the right will meet at a third vanishing point V_2 an equal distance on the other side of V from V_1. To find the precise placement of V_1 and V_2, consider the central point C on the ground line where the object and picture planes intersect. I have drawn arrows from C parallel to the diagonals of the square tiles. These form a right angle at C. Translate the arrows from point C to point E keeping them parallel to the original arrows in the object plane. The points where the translated arrows intersect the picture plane are the points V_1 and V_2.

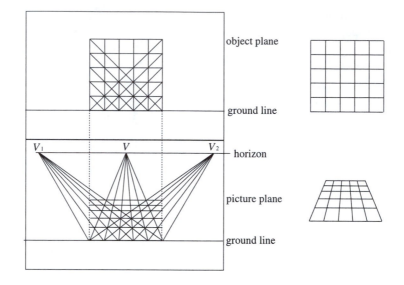

It is clear in the object view of the tiles that the horizontal lines cross where the vertical lines intersect the diagonal lines. Thus, in the perspective view, we place the horizontal cross lines where the corresponding lines intersect. The same techniques, of using a family of parallel diagonal lines intersecting the railroad tracks, is used to find the correct spacing of the crossties in a perspective view of railroad tracks.

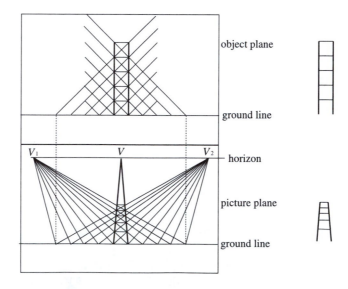

▶ **Exercise 4.** Draw a perspective view of the tiling shown below:

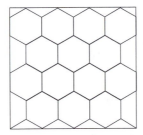

▷ **Exercise 5.** Use the grid below and its intersections with the circle to draw a perspective view of the circle.

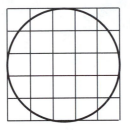

Renaissance artists were so pleased with the technique outlined above that tile floors are quite common in paintings of the era. Piero della Francesca was a prominent artist who wrote a treatise on perspective and made use of it in his *Flagellation of Christ*, ca. 1480. He also used perspective as an artistic metaphor, distancing the casual conversation in the foreground from the rending scene in the back. To the right is a bird's-eye view of the floor, including the footprints of the persons depicted.

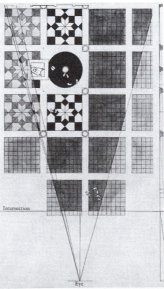

▷ **Exercise 6.** Use tracing paper to find the principal vanishing point of Piero della Francesca's *Flagellation*.

▷ **Exercise 7.** Use tracing paper to find the vanishing point of Paolo Uccello's *Night Hunt* (1460). The conveniently placed logs in the foreground help delineate the converging parallel lines of the perspective.

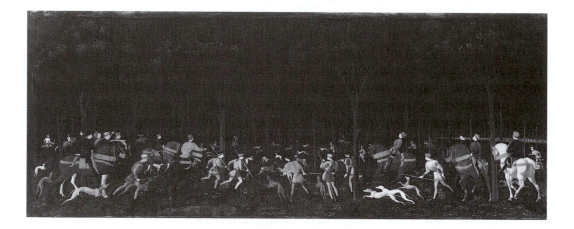

Next we figure out how to draw perspective views of simple three-dimensional objects, like boxes. A rectangular box has three sets of parallel sides: the top and bottom, the left and right, and the front and back. Depending on the orientation of the box to the viewer's eye, this will require one, two, or three vanishing points. We discuss only one- and two-point perspectives.

One-Point Perspective

Imagine a box placed on the object plane. In a one-point perspective, the box is assumed to be placed with the front face parallel to the picture plane and the bottom parallel to the object plane. To make things easier, we assume that the box sits on the object plane with the front face on the picture plane at the ground line. On the next page on the left is a bird's-eye view of the box. The rectangle $ABCD$ denotes the footprint of the box. Dashed lines are drawn from the eye at E to the corners of the box. These lines intersect the picture plane at points $A' = A$, $B' = B$, C' and D'. On the right is a drawing of the picture plane, showing the ground line, horizon, and the principal vanishing point V:

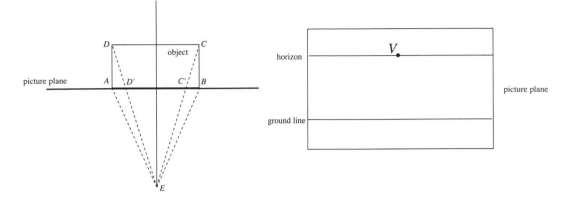

Since the edges AD and BC are perpendicular to the ground line in the object plane, their perspective images will meet at the vanishing point. The dotted lines are dropped from the points A, B, C', and D' in the bird's-eye view above to the *rabattement* of the picture plane. Since points A and B lie on the ground line, they do not move during the perspective transformation. Thus the points A'' and B'' lie directly beneath A and B on the ground line in the picture plane. The point C'' must lie on the line $B''V$, which is the perspective image of the edge BC and lies directly beneath C', the image of C in the bird's-eye view of the perspective. Similarly, D'' lies on $A''V$ directly beneath D':

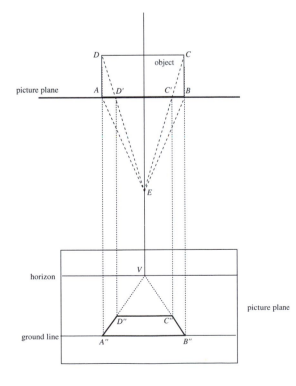

To get a three-dimensional image, measure the height of the box. Since we have assumed that the front face of the box lies on the picture plane, the height can be directly transferred to the picture plane. The side edges of the top are also parallel and perpendicular to the picture plane, and so their images must intersect at V. We draw the front face of the box and connect the upper front corners with the vanishing point V in the illustration on the left. On the right, we have completed the perspective view of the box, realizing that the top rear corners must lie directly above C'' and D'', the images of the rear corners:

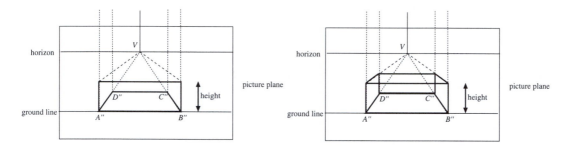

▷ **Exercise 8.** Draw a one-point perspective of the following box. A bird's-eye view is shown. An indication of the (apparent) height of the box is also given. [Note: If one face of the box were on the picture plane, one can transfer the actual height to the drawing. In this case, assume that we have already scaled the height of the front face correctly.]

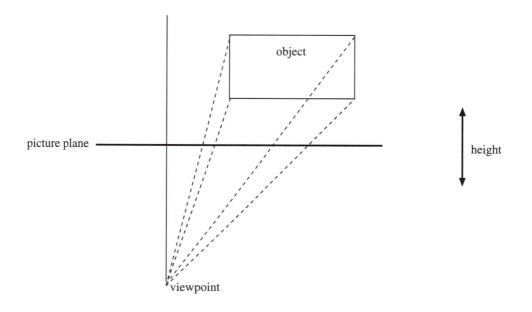

Two-Point Perspective

In a two-point perspective, the box is assumed to be sitting on or parallel to the object plane, but with none of the sides parallel to the picture plane. Below is a bird's-eye view of the footprint of such a box with corners $ABCD$. The eye is at E, and the points A', B', C', and D' mark the intersection of the lines of sight with the picture plane. To find the vanishing points, consider the sight lines from the eye at E running parallel to the sides of the box. These lines will be perpendicular, since the sides of the box are perpendicular, and will intersect the picture plane at points V_1 and V_2:

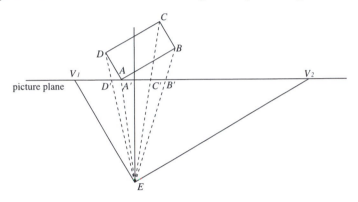

In the rabattement, drop the points V_1 and V_2 to the horizon on the picture plane, and drop the point $A' = A$ to the ground line. Since EV_1 is parallel to AD, the image of AD runs from A'' to V_1. Similarly, the image of AB runs from A'' to V_2. Drop the points D and B to D'' and B'' on these lines directly below D' and B'. Since BC is parallel to AD, $B''C''$ meets $A''D''$ at V_1, and similarly, $D''C''$ intersects $A''B''$ at V_2. This places the point C'' directly below C':

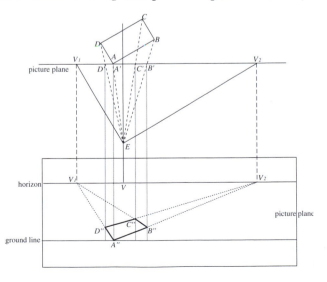

We have now drawn the perspective footprint of the box. To complete the three-dimensional image, we build the box up using the height at A. Since we have chosen A to be on the ground line, the height is transferred directly without distortion. The top front edges run parallel to AD and AB, and so their images run from the top front corner to the vanishing points V_1 and V_2:

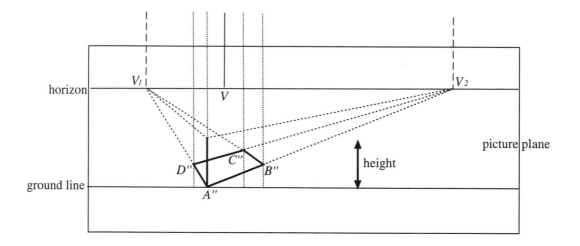

We complete the image by drawing the top edges of the box:

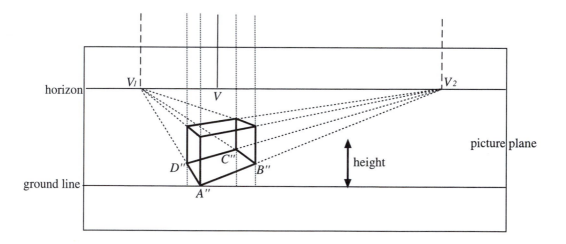

▶ **Exercise 9.** Draw two-point perspective views of the box on the next page using the three different positions of the picture plane shown. Note that the three positions for the picture plane intersect the box at three of the vertices. Transfer the height to the picture plane at these points.

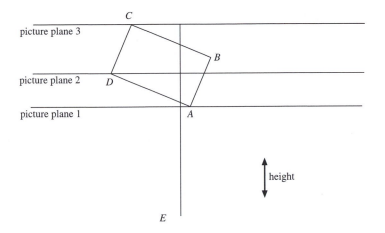

▶ **Exercise 10.** What is the effect of moving the picture plane?

▷ **Exercise 11.** Draw two-point perspective views of the box below using the three different positions of the eye shown.

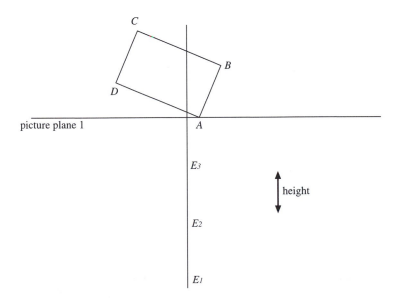

▷ **Exercise 12.** What is the effect of moving the viewpoint or eye?

▷ **Exercise 13.** Draw one-point perspective views of the box on the next page using the three different relative positions of the ground line and horizon shown. Note that since the box is positioned with its front face on the picture plane, I have drawn the view of the front face, so you need only draw the other faces. In the first picture the box is below the horizon, in the second it is centered on the horizon, and in the last it is floating above the horizon.

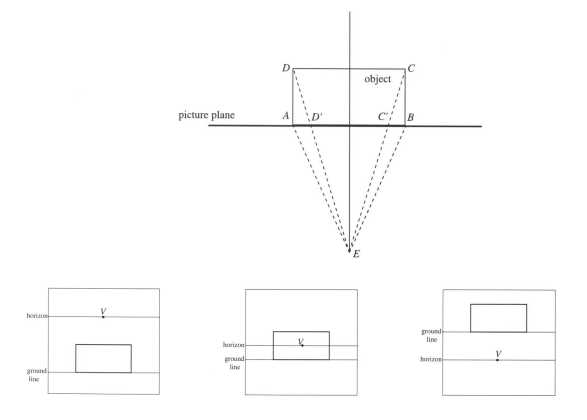

▷ **Exercise 14.** What is the effect of moving the groundline?

SUGGESTED READINGS

Franco and Stefano Borsi, *Paolo Uccello*, Harry N. Abrams, New York, 1994.

J.V. Field, *The Invention of Infinity*, Oxford University Press, New York, 1997.

Robert W. Gill, *Basic Perspective*, Thames and Hudson, New York, 1974.

Martin Kemp (ed.), *Leonardo on Painting*, Yale University Press, New Haven, 1989.

Martin Kemp, *The Science of Art*, Yale University Press, New Haven, 1990.

William J. Mitchell, *The Reconfigured Eye*, MIT Press, Cambridge, 1992.

Dan Pedoe, *Geometry and the Visual Arts*, Dover, New York, 1983 (a republication of *Geometry and the Liberal Arts*, St. Martin's Press, 1976).

M.H. Pirenne, *Optics, Painting and Photography*, Cambridge University Press, New York, 1970.

C.R. Wylie, *Introduction to Projective Geometry*, McGraw-Hill, New York, 1970.

10. Drawing Three Dimensions in Two

◆ 10.2. OPTICAL ILLUSIONS

Note: Many of the exercises in this section require drawings. They can be done by hand using a compass and straightedge, but they may be easier to do if you have access to a computer with *Geometer's Sketchpad* or a similar software package.

"I wouldn't have believed it unless I had seen it with my own eyes." Statements like this are evidence of the faith we put in our power of visual observation. We accept most of what we see without question. However, optical illusions are very common in our society. Most are designed to help give people a more accurate understanding of the world around them. Some are specifically designed to give a false impression. The optical illusions we discuss in this section are mainly those that intend to confuse the brain into making false or unjustified conclusions.

Any two-dimensional representation of a three-dimensional object must make use of optical illusions. The perspective drawings of the previous section are meant to give the impression of three-dimensional space. For example, the figures showing the object and picture planes are drawn so you visualize two planes intersecting in a line even though this cannot happen on a flat piece of paper. Most paintings also intend to represent space on a flat surface. Hence, the artist uses perspective, which if executed properly, enhances the viewer's interpretation of the scene.

Some optical illusions are the result of the way the optic nerves transfer information to the brain. The children's magic trick of a mysterious "third finger" is based on this type of misinterpreted information between the eye and the brain.

Hold your index fingers about two inches in front of your eyes pointing toward each other four inches apart. While staring straight ahead, slowly bring your fingers together. You should see a third finger with a finger tip on each end between two normal fingers.

If you look with each eye separately, you can see where the problem occurs. The left eye thinks the point where your fingers touch is way to the right—probably in front of the right eye because the fingers are so close to your face. Similarly, the right eye sends the message that the fingers meet in front of the left eye. When your brain puts these two concepts together, it creates a third finger by overlapping the pictures sent from the two eyes.

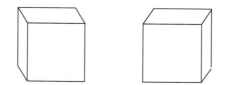

Stereopticon slides, 3D movies, and magic-eye pictures, which were so popular in the mid-1990s, are also examples of taking advantage of the way in which the brain interprets visual cues to intentionally create an illusion of depth. Each eye is shown a slightly different flat picture (or even an exact copy slightly displaced) mimicking the different views the eyes would see when looking at a truly three-dimensional object. Below is an exaggerated example of two views of a cube, from the viewpoint of left and right eyes.

In the old 3D movies, two slightly differently colored images were projected on the same screen, but glasses with different colored lenses filtered the images so each eye saw only one view. (If you watched without the glasses, the 3D parts looked fuzzy as if the image had been smeared when printed.) The magic eye pictures use roughly the same theory, but the strategy is more complex.

In this section, we will concentrate on the more geometric optical illusions. Such illusions have been an important object of study by many psychologists, who have created and studied illusions since the late nineteenth century. While they found many ways to classify these illusions, we agree with J.O. Robinson that "classification is a taxonomic exercise and does not itself provide explanations." Categories have included illusions of extent, of direction, of contour, of perspective, of length and distance, of size and area, and of angle, direction, straightness, and curvature. The interested reader is referred to the book by Robinson.

The figures below form the classic Müller-Lyer illusion, first exhibited in 1889:

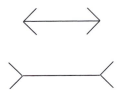

▷ **Exercise 1.** In which figure does the horizontal line look longer? Measure the horizontal segments to convince yourself they are actually the same lengths.

▶ **Exercise 2.** In the Müller-Lyer illusion above, the angle between one piece of the "arrowhead" and the horizontal line is 45°. Draw several similar pairs of figures but change this angle, trying several very small angles to get very pointed arrows and large angles giving almost flat arrows. Which angle gives the most pronounced illusion?

▷ **Exercise 3.** Draw several more pairs of figures changing the length of the arrowheads. Do shorter or longer fins affect the illusion? Try to explain your responses in complete sentences.

Some similar pairs of figures are shown below using different ways to effect the same type of Müller-Lyer illusion. These are due to, from right to left, Müller-Lyer, Delboeuf (1892), Jastrow (1891), and Brentano (1892).

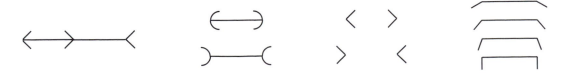

▷ **Exercise 4.** Add lines between some of the dots in the picture below to make the distances between the dots on the middle row appear different at different places along the row. All lines must connect two dots. No half lines are allowed.

The next figure illustrates a Ponzo illusion, discovered in 1912. Converging lines surround two or more identical figures.

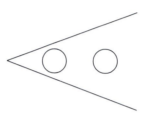

▷ **Exercise 5.** Convince yourself that the two circles are actually the same size.

▶ **Exercise 6.** (a) Construct a Ponzo illusion by keeping the angle between the converging lines the same but replacing the two circles of the illustration above with two short vertical lines.
(b) Construct a Ponzo illusion by replacing the two circles with three short vertical lines.
(c) Construct a Ponzo illusion by replacing the two circles with three circles.
(d) Do you find a difference in the effect of the illusion from changing the inside figures or increasing the number of figures included?

▷ **Exercise 7.** (a) Construct a Ponzo illusions using two circles as in the illustration above but changing the diameter of the circles.
(b) Construct a Ponzo illusion using two circles as in the illustration above but changing the angle between the converging lines.
(c) What is the effect of making these changes?

Because people react differently to optical illusions, you may want to show your drawings from Exercises 6 and 7 to a few friends to compare their reactions. Following is an illusion consisting of two sets of Titchener circles. These figures demonstrate T. Obonai's general statement that figures seem larger when adjacent to smaller figures and they seem smaller when adjacent to larger figures:

▷ **Exercise 8.** Convince yourself that the two center circles are actually the same size.

▶ **Exercise 9.** If all seven circles in a set of Titchener circles are the same size, each of the six outer circles is tangent to (touches but does not overlap) the inner circle. Figure out how to construct such a set of circles using only a compass and straightedge. Explain your steps clearly, and use complete sentences.

▷ **Exercise 10.** Explore the effect of increasing and decreasing both the size of the surrounding circles and the distance of the surrounding circles from the center circle.

The Müller-Lyer, Ponzo, and Titchener illusions are variations on the theme of size and related properties. Two of the properties people seem most confident in determining is whether lines are straight and when lines are parallel. However, the background can fool the eye even on these basic concepts. Consider the picture below of Zöllner's illusion of 1860. In this illusion, the vertical lines look slanted.

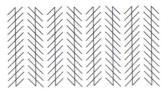

▶ **Exercise 11.** Show that the lines in Zöllner's illusion are actually parallel.

▷ **Exercise 12.** Add background lines to the following to make the nonparallel lines look parallel.

In the next variations of Zöllner's illusion, due to Orbison in 1939, the lines seem to bend: The sides of the square seem to bend inward, while the two parallel lines seem to bow out.

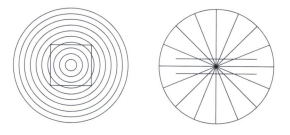

▷ **Exercise 13.** In this exercise we make use of the ideas from the figures above to try to reverse the illusions.
  (a) Draw a square and a field of circles so that it appears that the sides of the square bow out.
  (b) Draw a pair of parallel lines and a field of other lines to make the parallel lines bend inward.

Architects have long been aware of these illusions even if they did not bother to classify them. Builders introduce "optical refinements" to counteract some perspective effects. The Parthenon is an excellent example of this from classical times. The columns appear to be straight and parallel. However, tall straight columns tend to appear to have a bit of a waist in the middle. To counteract that effect, the 34-feet-tall columns of the Parthenon swell a bit: by about three-quarters of an inch. The central column is perpendicular to the ground, but the distance between the other columns gradually decreases, and the outer columns are slanted toward the center, so that the end columns have a tilt of three inches from the perpendicular. The architrave, the long beam stretching over the columns, would appear to sag if it were straight by a variation on Zöllner's illusion, so it actually bows up from the horizontal at the center. In fact, "hardly a single true straight line is to be found in the building."

Another type of illusion takes advantage of our need to make sense out of chaos. Consider the figure below. While the symbol used is neither an H nor an A, most people will see the symbols and read this as "THE CAT" in order to make the phrase mean something.

Another example of seeing something that is not there is given below. There appears to be a faint but illusory circle at the center.

A related phenomenon is called the Hermann Grid illusion. If you stare at the illustration below for a while, gray spots will appear at the intersections of the white bands.

Piet Mondrian made use of this effect in the painting *Composition with Gray Lines*, producing the impression of motion as your eyes move around the painting.

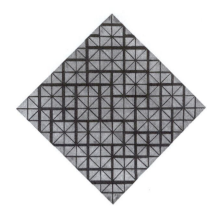

Many other optical illusions are the result of cultural standards or assumptions. From childhood we are taught to interpret certain visual cues standardly used in illustrations. We thus learn to see depth and dimensionality as represented in pictures drawn in the Western tradition. Consider the figures below. The figure below on the left is literally three line segments in a plane, but most people in our society will see two lines—one crossing over the other. The second figure was an exercise in the section on frieze patterns in an early draft of this text, but was removed when several generations of students persisted in claiming that the figure has horizontal symmetry because they saw interlocking triangular links in space instead of line segments in a plane.

▷ **Exercise 14.** Using the "crossing lines" idea from the illustrations above, make a nonsymmetrical sketch that appears to be a symmetrical object.

Necker's cube is another example of seeing something that is not there with the added complication that even the illusion is uncertain. The illusion consists of two overlapping squares connected by lines at the corners. It can be seen as a cube with the dot as the upper right front corner, or as the upper right rear corner. With a bit of effort, you can make it pop back and forth between these two interpretations.

▷ **Exercise 15.** (a) Shade the Necker cube so that the dot appears as the upper right front corner.
(b) Shade the Necker cube so that the dot appears as the upper right rear corner.

A variation on the Necker cube can be drawn as a regular hexagon with a line segment from the center point to each vertex. Coloring or shading the faces shows the possible different views of the "cube."

▶ **Exercise 16.** Below is a section of a popular quilting pattern called Tumbling Blocks. Count the cubes. Now rotate the picture 180° and count them again (or look for another perspective of the cubes to recount). You should get six cubes counting one way and seven counting the other. Try to explain how this could happen.

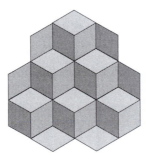

▷ **Exercise 17.** Can you create a similar illusion with fewer cubes? With more cubes? Draw the picture or explain why it cannot be done.

Culturally based illusions have also been used by architects. Generally, they are used to correct false impressions but occasionally to mislead. The next figure is of three bodies that are identical in size. However, the lines behind the figures seem to approach a vanishing point, creating the illusion that the body on the right is much larger than the one on the left. We interpret the body on the right as farther away, and anything farther away that looks the same size must be bigger. Compare this figure with the Ponzo illusions discussed above.

Adelbert Ames designed a room to confuse the onlooker from the Western tradition. Such a viewer expects rooms to be rectangular. Ames changes the angles of the room but adds other subtle changes so the room still appears rectangular. One corner is actually much farther away from the viewer so a person standing in the near corner looks like a giant compared to the person in the far corner. Below is a drawing of identical twins in an Ames room. The Ames illusion only works from the particular viewpoint for which it is designed, but from that viewpoint, the resulting view is astounding. The illusion does not work nearly as well on people who grew up in cultures where nonrectangular rooms are common.

▷ **Exercise 18.** Make a model (about a foot in each dimension) of an Ames room to see whether the illusion holds on a small scale. Note that neither the roof nor the floor should be horizontal. In the perspective picture, the actual Ames room is shown by solid lines, and the perceived rectangular room by dashed lines. Dotted lines from the designated viewpoint align the corners of the perceived room with the corners of the Ames room.

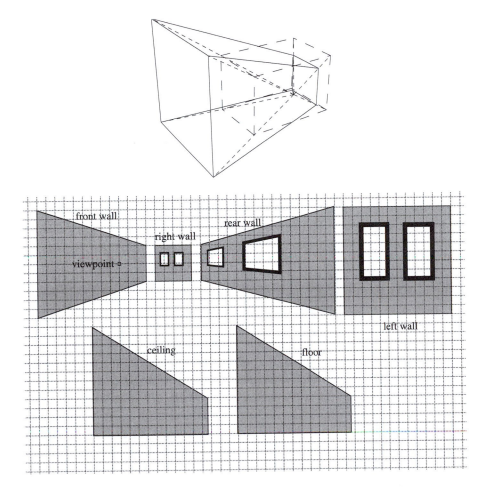

The Italian Renaissance architect Francesco Borromini also designed to confuse, at least once. He understood perspective well enough to manipulate it. By reducing the height of the columns in a corridor of the Palazzo Spada in Rome and by making the sides converge, Borromini creates the impression that the corridor is much longer than it actually is, which has the odd effect of making a person walking towards you through the colonnade appear to grow. Following is a bird's eye view of the diameters and spacing of the columns of such a colonnade: The lighter circles are the perceived columns and the darker ones the actual columns.

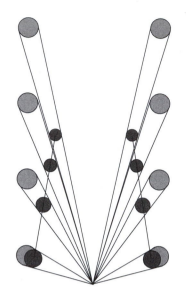

▷ **Exercise 19.** How would you design a corridor to look shorter than it actually is? Try to be specific and answer in complete sentences with appropriate drawings.

While architects can actually create objects to produce optical illusions as in the examples above of the Parthenon and the Palazzo Spada, many two-dimensional sketches give the impression of being constructible objects but cannot really be built. The confusion caused by these "impossible objects" is the result of cultural conditioning that signals how a flat representation should be interpreted for a three dimensional object. Most such impossible drawings are the result of mismatching pieces of possible drawings.

Consider the three nested L shapes below on the left. By themselves, they look as though they lie in a plane, but if we add carefully chosen lines, they take on a three-dimensional aspect. They can become either part of a frame lying down or a frame sitting up on edge, depending on the lines we add. The notion that there are two perpendicular surfaces in the resulting illustrations is the result of accepting cultural conventions on perspective drawing.

If we put four of the L-shaped corners above together following accepted convention, we get a perfectly constructed frame, as on the left on the next page. If we choose mismatching corners, they still fit together but the object they imply is not constructible. The different ends of the lines connecting the corners represent different parts of an object. In the impossible drawing, the middle line on the left is the outside edge of the frame. This line is connected to the middle line on the bottom which is the inside edge of the frame.

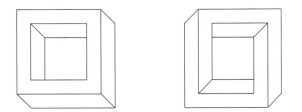

▷ **Exercise 20.** Try to identify where the following figures go wrong. Try to write for an "uninformed" audience.

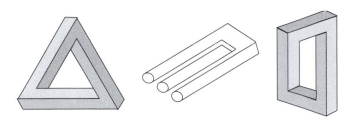

▷ **Exercise 21.** The impossible figure below on the left is called the tribar. A model can be constructed as shown on the right. Each of the bars is at right angles to the others, but from one particular point of view this looks like the picture on the left. Build a model as shown.

▷ **Exercise 22.** Create your own drawing of an "impossible object," different from those shown above.

Escher made use of these impossible figures in several works, such as *Waterfall* of 1961, which uses the tribar.

SUGGESTED READINGS

Hadley Cantril (ed.), *The Morning Notes of Adelbert Ames, Jr.*, Rutgers University Press, New Brunswick, 1960.

Bruno Ernst, *The Magic Mirror of M.C. Escher*, Taschen, New York, 1978.

Bruno Ernst, *Optical Illusions*, Taschen, New York, 1992.

Judith Gries, "Creating an Allée," *Fine Gardening* 54, March–April 1997.

Rosalind Krauss, *The Optical Unconscious*, MIT Press, Cambridge, 1993.

Fred Leeman, *Hidden Images*, Harry N. Abrams, New York, 1975.

M. Luckiesh, *Visual Illusions*, Dover, New York, 1965.

J.O. Robinson, *The Psychology of Visual Illusion*, Hitchinson & Co., London, 1972.

Nigel Rodgers, *Incredible Optical Illusions*, Barnes & Noble, New York, 1998.

Marvin Trachtenberg, *Architecture: From Prehistory to Post-Modern*, Harry N. Abrams, New York, 1986.

11. Shape

◆ 11.1. NONEUCLIDEAN GEOMETRY

SUPPLIES

clay, Play-Doh™, or Silly Putty™
lots of paper equilateral triangles: 1 or 2 inches on a side
a ball, the size of a tennis ball or bigger
rubber bands

▷ **Exercise 1.** Form a disc out of clay or whatever flexible substance you want to use. Find a way to increase the circumference of the disc without changing the diameter. You cannot tear the disc or overlap it onto itself.

▶ **Exercise 2.** Form a disc out of clay or whatever. Find a way to decrease the circumference without changing the diameter. You cannot tear the disc or overlap it onto itself.

▷ **Exercise 3.** In the plane, exactly six equilateral triangles fit together at each point. Use equilateral triangles to construct a model of a space where only five triangles meet at each point. Continue the model as far as you can. That is, once you have 5 triangles meeting at the center of one disc, move to the edge of that disc and add triangles until there is the appropriate number at that vertex as well. Then move further out on your surface, making sure that 5 triangles meet at each vertex.

▷ **Exercise 4.** Use paper equilateral triangles to construct a model of a space where seven triangles meet at each point. Continue the model as far as you can (in 5 or 10 minutes anyway). That is, once you have 7 triangles meeting at the center of one disc, move to the edge of that disc and add triangles until there is the appropriate number at that vertex as well. Then move further out on your surface, making sure that 7 triangles meet at each vertex. At each edge, be sure that only two triangles meet.

Of course, none of the objects constructed in Exercises 1–4 are flat, since we are purposefully constructing noneuclidean objects. You should notice that the objects of Exercises 2 and 3 have similarities, as do the objects of Exercises 1 and 4. In Exercises 3 and 4 you constructed *polyhedral objects*, while the objects of Exercises 1 and 2 are smooth, but there are still basic similarities of shape. The objects of Exercises 2 and 3 have *elliptic geometry*, while the objects of Exercises 1 and 4 have *hyperbolic geometry*. This nomenclature comes from the Greek: elliptic comes from ελλειψις, which means coming short or deficient, while hyperbolic comes from υπερβολη, which means excessive.

▷ **Exercise 5.** Relate the Greek roots of the words "elliptic" and "hyperbolic" to your findings of Exercises 1–4.

▷ **Exercise 6.** Which type of geometry does the surface of a ball or sphere have?

A circle is defined to be the set of all points equidistant from a given center. In flat euclidean space you have gotten used to the ideas that area $= \pi r^2$ and circumference $= 2\pi r$, where r is the radius of the circle. In Exercises 1 and 2 you showed that while the idea of a circle still makes sense, the relationships between circumference and radius of these formulae need not hold if you do not restrict yourself to flat space.

▷ **Exercise 7.** What can you say about the area of a circle in elliptic space? In hyperbolic space?

▷ **Exercise 8.** In euclidean space you have not only circles but also lines. One easy and intuitive way to define a line is to say that it is the shortest distance between two points. Can you think of other ways to describe a line?

An axiom of euclidean space (in fact, Euclid's first postulate) is that given two distinct points, there is a unique line that contains both points. That is, given any two points, you are guaranteed there is a line that goes through both and that there is only one such line. Now that you have two new types of (nonflat) space, let us consider the concept of a line in each.

Elliptic Geometry

Remember that your "space" is the surface of whichever object we are dealing with. That means you cannot cut through the center or jump off the surface. For the following group of exercises use your ball as a model of elliptic space.

▶ **Exercise 9.** What does a line on a sphere look like? Use your ball as a model for the surface and rubber bands to find shortest distances. The stretchiness of the band should make it find the shortest path between two points. Extend the line (rubber band) past the points to generalize what a line on a sphere should be. (Note: In order to stay on the surface of the sphere, the "lines" may not look like the lines with which you are familiar.)

▶ **Exercise 10.** Is there a unique line through any two points on the sphere? The answer is NO. Find an example to prove this.

Now that you have some idea of what lines on a sphere look like, think about parallel lines. What are some of the things you know about parallel lines in euclidean space?

1. They are always the same distance apart.
2. If you draw a line that crosses two parallel lines, the alternate interior angles are congruent.

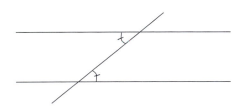

3. They never intersect.
4. Given a line and a point not on that line, there is exactly one line through the point that is parallel to the original line.

▷ **Exercise 11.** Latitude lines on a globe are always the same distance apart. Are latitude lines examples of parallel lines on a sphere? Why or why not?

▷ **Exercise 12.** Can you find a pair of lines on a sphere that never intersect? Find an example or explain why one can't exist.

▷ **Exercise 13.** What does this mean about parallel lines on a sphere?

Hyperbolic Space

Now your "space" is the floppy hyperbolic surface you created in Exercise 4 by taping together seven triangles at each corner point. You still cannot cut through or jump off the surface.

▷ **Exercise 14.** What does a line in hyperbolic space look like? Use your paper model for the surface and rubber bands to find shortest distances. It may be easier to cut the band into a strip to work on this exercise. Still, the stretchiness should make it find the shortest path between two points. (Note: Again, the "lines" may not look like lines to you.)

▶ **Exercise 15.** Is there a unique line through any two points on the floppy surface? The answer is YES. Convince yourself of this by experimenting.

▷ **Exercise 16.** Can you find two lines in hyperbolic space that never intersect? Find an example or explain why one cannot exist.

▶ **Exercise 17.** Given a line and a point not on the line, can you find a line through the point that never intersects the original line?

▷ **Exercise 18.** For the original line and point in Exercise 17, can you find a second line through the point that never intersects the original line?

▷ **Exercise 19.** What does this mean about parallel lines in hyperbolic space?

Noneuclidean Polygons

If the rules for lines and circles change, what rules for other shapes change? What is a triangle? A square? A pentagon? And what can you say about such objects in your new spaces? The next group of exercises investigates these ideas. One of the first things students learn about triangles is that the sum of the measures of the interior angles is 180 degrees.

▶ **Exercise 20.** Find a triangle on the surface of the sphere whose angles add up to 270°.

▷ **Exercise 21.** Can you find a triangle on the sphere whose angles add up to 360°? More than 360°? Less than 180°?

▷ **Exercise 22.** For your triangles of Exercise 21, is the area equal to, greater than, or less than one-half of the base times the height? Explain your answer.

The standard definition of a square is a four-sided figure with four equal angles and four equal sides. The fact that each of the angles is a right angle follows from the fact that the angles in any quadrilateral in euclidean space must add up to 360°. In noneuclidean geometry that will no longer be true, so we require only that a square have four equal angles and four equal edges.

▷ **Exercise 23.** Can you find a square on the surface of the sphere? Can you find a regular pentagon? Other regular polygons? What can be said about their vertex angles? (Contrast with the answers in Section 1.2.)

▶ **Exercise 24.** Draw a triangle on your floppy hyperbolic space, making sure to draw a big one that spans several of the small equilateral triangles from which your floppy plane was built. What can you say about the sum of the angles of this triangle?

▷ **Exercise 25.** Can you find a triangle in hyperbolic space whose angles add up to more than 360°? Less than 180°?

▷ **Exercise 26.** For a triangle in hyperbolic space, is the area equal to, greater than, or less than one-half of the base times the height? Explain your answer.

▷ **Exercise 27.** Can you find a square in hyperbolic space? A regular pentagon? Other regular polygons? What about their vertex angles?

The Poincaré Model of Hyperbolic Space

There are many ways of representing elliptic space on a flat piece of paper, which we will discuss in the next section on mapping projections. There are also several ways to represent hyperbolic space, and we will here show only one: the Poincaré disc model. In this model, the entire infinite hyperbolic plane is identified with the points in a disk inside a circle. Thus, points on the boundary circle must be considered as infinitely far away. These points are called *ideal points*, or points at infinity. The distance from the center of the disc to the outer circle must also be considered as infinite, so if you imagine walking from the center outwards, you must imagine that your legs get shorter and shorter, so the boundary circle is unreachable.

Therefore, any point in hyperbolic space is represented by a point in the disc. Lines are represented in two ways. The simplest lines are represented by diameters of the circle. Other lines that do not pass through the center are represented by arcs of circles that meet the boundary circle at right angles.

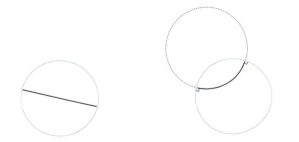

A hyperbolic triangle will be defined by the intersection of three lines on the model.

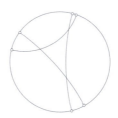

Tilings can be formed in hyperbolic space and will be represented in the disc model as below. Since the way distance is measured is not uniform over the whole disc, note that each of the figures must be considered to be congruent.

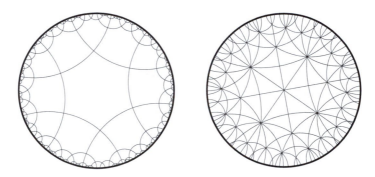

This model of hyperbolic space inspired Escher to produce a series of woodcuts, among which is *Circle Limit IV*.

SOFTWARE

1. The Poincaré disc model of hyperbolic space can be programmed in the interactive *Geometer's Sketchpad*. This is available from Key Curriculum Press.
2. *KaleidoTile*, by Jeffrey Weeks, of the Geometry Center, University of Minnesota, is freeware, for Macintosh only. It will draw selected euclidean, elliptic, and hyperbolic tilings. It can be found at ftp://geom.umn.edu/pub/software/KaleidoTile.
3. *Hyperbolic MacDraw*, by Jeffrey Weeks, of the Geometry Center, University of Minnesota, is freeware, for Macintosh only. This is a drawing program for hyperbolic space, and one of the options is the Poincaré disc model. It can be downloaded from ftp://geom.umn.edu/pub/software/GeometryGames.

SUGGESTED READINGS

H.S.M. Coxeter, *Introduction to Geometry*, Wiley & Sons, New York, 1969.

David W. Henderson, *Experiencing Geometry on Plane and Sphere*, Prentice Hall, Upper Saddle River, 1996.

Silvio Levy, "Automatic Generation of Hyperbolic Tilings," *Leonardo* 25(3), 1992.

Richard Trudeau, *The Noneuclidean Revolution*, Birkhauser, Boston, 1987.

Edward C. Wallace and Stephen F. West, *Roads to Geometry*, Prentice Hall, Englewood Cliffs, 1992.

Jeffrey R. Weeks, *The Shape of Space*, Marcel Dekker, New York, 1985.

11. Shape

◆ 11.2. MAP PROJECTIONS

SUPPLIES

a styrofoam ball (4 to 6 inches in diameter)
string
pin
clear stiffish (0.2 to 0.4 mils) film, such as used for transparencies
colored markers
an orange

▷ **Exercise 1.** Without talking to anyone, take about 10 minutes and sketch a map of your campus.

▷ **Exercise 2.** Switch your map with someone else in the class, preferably with someone you do not know well and who isn't going to look over your shoulder while you critique his or her map. Consider that person's map and write a short paragraph about what is important to him or her. Note especially things that you left off of your map.

▷ **Exercise 3.** Without talking to anyone, take about 10 minutes and sketch a map of the world.

▷ **Exercise 4.** Discuss the following questions about your world map:
a. What is in the center of your map?
b. Which direction is up?
c. Which feature of the world is biggest?
d. Which feature has the most detail?
e. Is anything labeled? If so, what and why?

Any map is biased by the culture that creates it. Most students in the United States put either the U.S. or the Atlantic Ocean in the center of the map. The maps with which we have grown up were generated by a Eurocentric culture. North America is drawn better and often bigger than the rest of the continents. Labels show with which parts of the world you are most familiar.

The maps with which we are most familiar describe north as up. There is no definite "up" in space. My older sister Anne at age 50 still refuses to read maps (a considerable inconvenience on road trips), claiming to have been traumatized by her third grade teacher: She had spent a lot of time and trouble drawing an assigned map of South America, only to have it ridiculed as being upside down.

While we have progressed far beyond the point of having to write "Heere lyes Monsters" five miles from the city limits, you should have seen that most of the maps drawn by the class get rather vague as one moves farther from home. Distances and size get distorted in one's perception. If there is anyone from Texas in the class, note how his map shows Texas dead center and twice the size of anything else (if there is anything else at all shown on the map). All of the maps drawn by your class were probably based loosely on the Mercator projection, the most likely to have been hanging in your elementary school classroom.

I grew up in a family that was not particularly intellectual—not illiterate, but we did not discuss global issues around the supper table. I still remember the distinct shock of surprise that I felt when my first-grade teacher mentioned, in an unwarrantably off-hand manner, that of course the earth was round. I am assuming that you are also aware of this fact. The rest of this chapter is devoted to the mathematical problem of mapping a round earth onto flat paper and a comparison of some of the mapping projections used by cartographers.

Map Projections

Okay, we are all agreed that the earth is round, or more properly, spherical, or even more properly, an oblate spheroid. The perfect map would be a scale model of the earth, precisely replicating every detail of the topography and roadways, but this would be difficult to carry around. First we discuss how to lay out a coordinate system for the earth. From our discussion of non-euclidean geometry in Section 11.1, we know that the shortest distance between 2 points on the surface of a sphere is along a *great circle*: a circle, like the equator, around the fattest section of the sphere. Circles of *longitude* are other great circles, each passing through both poles. Traditionally, these are indicated by degrees of the 360° equatorial circle, with the 0° longitudinal line (the *prime meridian*) passing through the Greenwich Observatory in England, and moving eastward and westward from there (which says a lot about who was making the maps). Located 180° around the earth from the prime meridian is the International Date Line.

▶ **Exercise 5.** At the instant when the sun is directly over the point where the prime meridian and the equator intersect, give the longitude where it is 1 o'clock. What time is it at the International Date Line? What time is it in Washington, DC, whose latitude is 38°54′N and longitude is 77°2′W? You may approximate the longitude of DC using 75°W.

▷ **Exercise 6.** Mark a point on your ball and label it the North Pole. Using your string and the fact that any two great circles through the North Pole will also intersect at the South Pole, mark and label the South Pole. Draw longitudinal lines at 0°, 90°E, 180°, and 90°W.

The longitude of a position measures how far east and west one is from the prime meridian, in spherical terms. *Latitude* measures north and south. The equator marks the points at 0° latitude, and latitude is measured north of that point in degrees towards the North Pole at 90°N, and southward to the South Pole at 90°S.

▷ **Exercise 7.** The equator is the great circle equidistant from both poles. The easiest way to mark it is to stick your pin into the ball at the North Pole. Measure with your string and a ruler the distance between the poles. Tie a bit of string to the pin at the North Pole and a loop for a marker so that the length is half the distance between the poles. Draw the equator and label it. Draw the circles of latitude at 30°N, 60°N, 30°S, and 60°S. Draw a triangle on your ball and label its vertices *A*, *B*, and *C*.

▷ **Exercise 8.** Take a piece of paper (or aluminum foil works well) about 4 inches on a side and try to smooth it onto your ball. What happens?

▷ **Exercise 9.** Peel an orange (trying to keep the peel in one piece) and try to flatten the orange peel onto a piece of paper. What happens?

▷ **Exercise 10.** Explain the connection between Exercises 8 and 9 and mapping the earth.

All projections of a round ball onto a sheet of paper must change something, or the map would not lie flat. Various features of the earth can be distorted in this mapping process. The most important of these are:

1. Distortion of area: This will alter perceived sizes of countries.

2. Distortion of lines and length: This will alter perceived distance between cities.

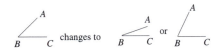

3. Distortion of angle measure: This will alter perceived compass bearings, which affect navigation.

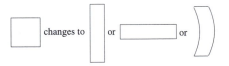

4. Distortion of shape:

5. Distortion of a point into a line:

Most mapping projections distort several of these factors. The impossibility of mapping the sphere in a continuous manner to the plane, without tearing or puncturing the sphere, is a result in the mathematical subfield of *topology*, and is beyond the scope of this text. However, we have included a brief introduction to topology in Chapter 13. Long before a mathematically rigorous proof of the impossibility of a perfect map was derived, it was accepted as intuitively true, as witnessed by the poem *The Definition of Love*, verses 4 through 7, written in 1681 by Andrew Marvell. The term *planisphere* refers to a planar projection of the sphere, or a map of the world, most commonly referring to what is called the orthographic projection:

> For Fate with jealous eye does see
> Two perfect loves, nor lets them close:
> Their union would her ruin be,
> And her tyrannic power depose.
>
> And therefore her decrees of steel
> Us as the distant Poles have placed,
> (Though Love's whole world on us doth wheel)
> Not by themselves to be embraced,
>
> Unless the giddy heaven fall,
> And earth some new convulsion tear;
> And, us to join, the world should all
> Be cramped into a planisphere.
>
> As lines (so loves) oblique may well
> Themselves in every angle greet:
> But ours so truly parallel,
> Though infinite, can never meet.

▷ **Exercise 11.** Take your ball and a clean sheet of paper. Find a way to map your ball on your sheet of paper. Transfer the key markings of your ball to your map. Explain in words how you made your map. Answer the following questions about your map.
  a. Does a line (great circle) on the ball map to a line on the paper? (Consider the equator and longitudinal great circles as well as random great circles.)
  b. Does a circle on the ball map to a circle on the paper? (Consider latitudinal circles and smaller circles centered on the equator.)
  c. Does your map preserve area?
  d. Does your map preserve length?
  e. Does your map preserve angle measure?
  f. Does your map preserve shape?
  g. Does your map stretch any points?

There are several strategies that people traditionally have used in making maps. You may have used one of these in Exercise 11 or you may have done something totally different.

▶ **Exercise 12.** *Cylindrical projection*: Wrap a sheet of transparency film around the ball along the equator. These points will transfer exactly. For points above and below the equator, you have several choices illustrated on the next page. Draw each map, transferring the longitude and latitude lines and any other features you have marked on your ball.

A. Equal-area cylindrical projection: Project points straight out to the transparency film parallel to the equator.

B. Equirectangular cylindrical projection: Unroll the ball onto the paper so the distance of a point from the equator will be correct on the map. Use your string to measure how far a point is from the equator and to transfer this distance to your map.

C. Central cylindrical projection: Project points by imagining a line from the center of the ball through the surface at the point and onto the map.

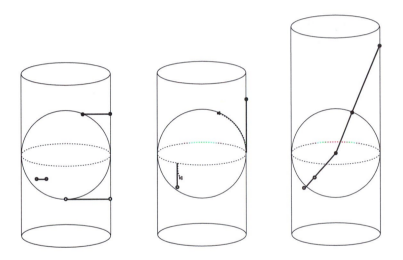

▷ **Exercise 13.** Answer the following questions for each of the cylindrical projections A, B, and C.

a. Does a line (great circle) on the ball map to a line on the paper? (Consider the equator and longitudinal great circles as well as random great circles.)

b. Does a circle on the ball map to a circle on the paper? (Consider latitudinal circles and smaller circles centered on the equator.)

c. Does your map preserve area?

d. Does your map preserve length?

e. Does your map preserve angle measure?

f. Does your map preserve shape?

g. Does your map stretch any points?

▷ **Exercise 14.** *Orthographic projection*: Set the transparency film on the ball at the North Pole. Map the North Pole to the point of contact. Every other point gets mapped to the point on the transparency film directly over it. Note that the southern hemisphere cannot be seen, so this map does not include it.

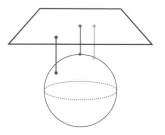

▷ **Exercise 15.** Answer the following questions for the orthographic projection.
a. How does the equator map onto the paper? The longitudinal great circles?
b. How do the latitudinal circles map onto the paper?
c. Does your map preserve area?
d. Does your map preserve length?
e. Does your map preserve angle measure?
f. Does your map preserve shape?
g. Does the map stretch any points?

▷ **Exercise 16.** *Stereographic projection*: Set the transparency film on the ball at the North Pole. Map the North Pole to the point of contact. Every other point gets mapped by imagining a line from the South Pole through the point on the ball and then to the point on the map where the line intersects the transparency film.

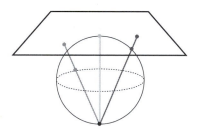

▷ **Exercise 17.** Answer the following questions for the stereographic projection.
a. How does the equator map onto the paper? The longitudinal great circles?
b. How do the latitudinal circles map onto the paper?
c. Does your map preserve area?
d. Does your map preserve length?
e. Does your map preserve angle measure?
f. Does your map preserve shape?

g. What happens to the South Pole? Does your map stretch any points?

▷ **Exercise 18.** *Gnomonic projection*: The last projection is similar to the one of Exercise 16, but imagine the paper cutting through the equator of the ball. Points in the northern hemisphere will end up mapped to points inside the circle of the equator and points in the southern hemisphere outside the equatorial circle.

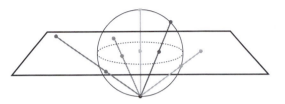

▷ **Exercise 19.** Answer the following questions for the gnomonic projection.
   a. How does the equator map onto the paper? The longitudinal great circles?
   b. How do the latitudinal circles map onto the paper?
   c. Does your map preserve area?
   d. Does your map preserve length?
   e. Does your map preserve angle measure?
   f. Does your map preserve shape?
   g. Does your map stretch any points?

   Here is an argument that no map projection can preserve distance between all points on the sphere. Consider five points on the ball: the North Pole, the South Pole, the point X where the prime meridian intersects the equator, the point Y on the equator at 90° longitude, and Z on the equator at 180° longitude. We know that the distances from Y to X is the same as from Y to Z, and the distance from Y to the North Pole is also the same, as is the distance from Y to the South Pole. Thus $\triangle XYN$, $\triangle ZYN$, $\triangle XYS$, and $\triangle ZYS$ are all congruent equilateral spherical triangles.

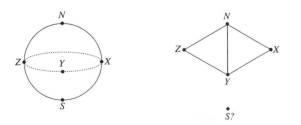

If we had a distance-preserving map, then on the map $\triangle XYN$, $\triangle ZYN$, $\triangle XYS$, and $\triangle ZYS$ would still be congruent equilateral triangles. Hence, we must have the picture on the previous page, but where can we put S? The only way to get the distances correct (Y to S equal to X to S equal to Z to S) is to put S exactly where N is, but we know that the North and South Poles can't be the same. Consequently, there is no way to represent a round world on flat paper without distorting distance.

Common Mapping Projections

The cylindrical projections you drew in Exercise 12 all have the longitudinal lines drawn as equally spaced vertical lines, but the spacing of the latitudinal lines varied widely: from bunched together at the poles to equally spaced to diverging at the poles. The Mercator projection of 1569 is a variation of these cylindrical projections where the spacing of latitudes is adjusted to preserve angles. Such angle-preserving projections are called *conformal*.

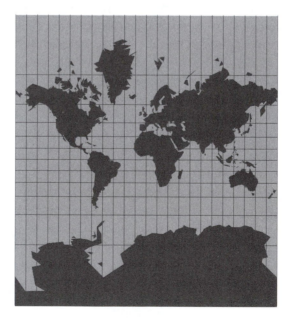

Gerardus Mercator, a Flemish mapmaker, invented this projection for its usefulness to navigators: an angle measure or compass bearing on the surface of the sea is the same as that shown on the Mercator map. Thus, to steer from one port to another, all the captain had to do was to draw a line on the map between the two ports, measure the angle, and use that for his compass heading. Such lines of constant bearing are called *rhumb lines* or *loxodromes*.

However, the Mercator projection greatly distorts length and stretches the poles into lines, so the sailor, while he will know he is heading in the right direction and will get to port eventually, needs a specially designed ruler to figure out the true distance in order to know when he will get there. Further, a constant compass bearing will not give the shortest path between two ports. The shortest distance between two points on the sphere is an arc from a great circle. Below are shown the rhumb line path and the great circle path from New York to Cape Town on a Mercator map. The rhumb line appears straight on the map, but the great circle is actually shorter.

Rhumb lines or loxodromes form equiangular or logarithmic spirals on the surface of the earth: the line must cross each longitudinal line at the same angle. Escher made use of the spiral form of the rhumb lines in his etching *Sphere Spirals*.

Related to the various cylindrical projections are the *pseudocylindrical projections*: Like the cylindrical projections, latitude is represented by parallel straight lines, and the longitudinal lines are equally spaced, but may not be straight. These were commonly developed as attempts to correct the extreme distortion of the cylindrical projections at the poles. One example is the *sinusoidal projection*, first used in 1570, which preserves area and scale along the latitudes and the central meridian, but distorts other lengths:

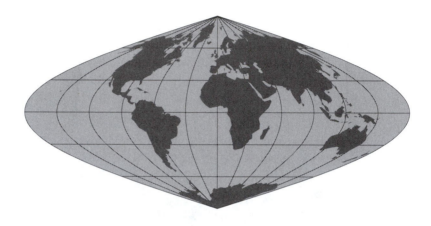

Another pseudocylindrical equal-area projection is the *Mollweide projection* of 1805:

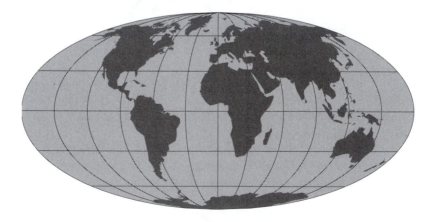

Combining the sinusoidal and Mollweide projections and slitting the result open, one gets the interrupted *Goode homolosine projection* of 1925. You can see the bumps where the two projections are spliced together:

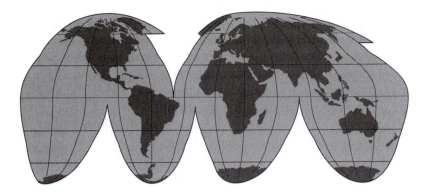

The idea of interrupting or slitting the projection has been used commonly since the Renaissance. These maps usually have much less distortion, but many interruptions. Below is a world map with 12 gores, or sections, from 1507. Cut out and taped together, it makes a quite reasonable approximation of a round ball:

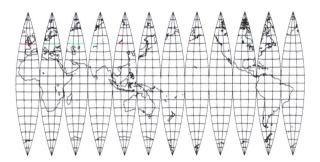

Dating back to at least Dürer in 1538 is the idea of projecting the globe onto a polyhedron. The most famous of these is Buckminster Fuller's Dymaxion Air-Ocean World Map of 1946, in the form of a cuboctahedron (later revised to use an icosahedron). Claims that this map has less distortion than any other known map cannot be countenanced:

Since all flat maps of a spherical surface must be distorted, the major decision to make in choosing a map is what type of distortion it should have. The Mercator projection is angle-preserving for navigational purposes. It is also one of the most familiar projections. The sinusoidal and Mollweide projections preserve area and some distances, though they are not rectangular.

SOFTWARE

1. *Mathematica* includes a package *WorldPlot*, which includes a number of the common projections.

FULLER PROJECTION
Dymaxion™ Air-Ocean World

SUGGESTED READINGS

Wellman Chamberlin, *The Round Earth on Flat Paper*, National Geographic Society, Washington, DC, 1950.

Charles H. Deetz and Oscar S. Adams, *Elements of Map Projection*, U.S. Government Printing Office, Washington, DC, 1938.

David Greenhood, *Mapping*, University of Chicago Press, Chicago, 1964.

Mark Monmonier, *Drawing the Line*, Henry Holt, New York, 1995.

John M. Novak, "WorldPlot," *The Mathematica Journal* 3, 1993.

Frederick Pearson, II, *Map Projections: Theory and Applications*, CRC Press, Boca Raton, 1990.

John P. Snyder, *Flattening the Earth*, University of Chicago Press, Chicago, 1993.

John P. Snyder, "Delighting in Distortions," *Mercator's World* 1, 1996.

John P. Snyder and Philip M. Voxland, *An Album of Map Projections*, U.S. Geological Survey, Washington, DC, 1994.

Denis Wood, *The Power of Maps*, Guilford Press, New York, 1992.

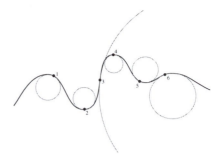

11. Shape

◆ 11.3. CURVATURE OF CURVES

SUPPLIES
    ruler
    circle template
    Slinky™
    knife
    vegetables, such as potatoes, cucumbers, cabbage, and kale

In this section, we want to figure out a way to mathematically describe the shape of things. In a calculus class, one of the first things you would learn to calculate is the slope of the tangent line. We are not assuming that you know any calculus, but people were drawing tangent lines long before calculus was invented, so we won't let that stop us. The tangent line to a curve at a point is the line that best approximates the curve near that point. The slope of this tangent line describes how steep the curve is. A positive slope indicates that the line goes up as you travel from left to right, and a negative slope that the line goes down as you travel from left to right. A slope of $\frac{a}{b}$ means that for every b units traveled to the right, the line goes up a units if a is positive, or down a units if a is negative. In the illustration below, imagine that the curve represents a road you are riding along:

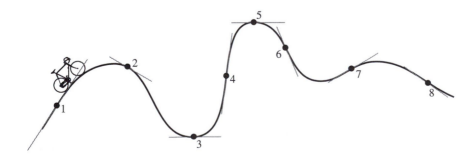

At point #1, the road slants uphill, with a slope of $+2 = \frac{2}{1}$. At point #2, the road goes downhill, with slope -1; at point #3, the slope is 0, denoting a horizontal tangent line; at point #4, the slope is a very large positive number. At point #5, the slope is again 0; point #6 has moderate negative slope; point #7 has a small positive slope; and point #8 has small negative slope. The larger the slope, ignoring sign, the steeper the road, and the smaller the slope, the nearer the road is to horizontal. At points with negative slope, you can coast. At points with positive slope you need to exert some effort, with more effort required for larger slopes. At points with zero slope, the bike will remain at rest. At points where the road is completely vertical, and therefore not rideable, the slope is infinite.

▷ **Exercise 1.** For the curve below, draw in the tangent line at each of the marked points. Guess at the slope, paying particular attention to the sign, but otherwise describing the slope as "large positive," "moderate positive," or "small positive," etc.

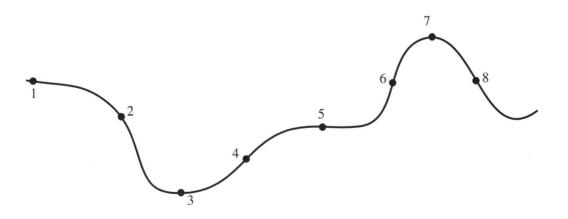

While tangent lines describe how steep or shallow a curve is, there is another quantity that describes more or less the turning radius of a curve, or how curvy the curve is. Again, for precise calculations, we would need calculus, but we can understand the general geometric idea without it. To measure the curvature of a curve, we approximate the curve by tangent, or osculating, circles (osculate is from the Latin word for kiss), rather than by tangent lines. Below is a selection of circles of varying radii:

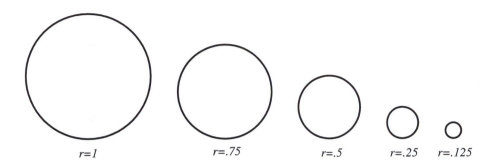

The radius of the circle determines how tight the turning radius of a circle is: The larger the radius, the less you need to turn the steering wheel of a car and thus the less curvy the circle is. We define the *curvature* of a circle with radius r by

$$\kappa = \frac{1}{r}.$$

Thus the circles above have curvatures, from left to right, of $\kappa = 1$, $\kappa = \frac{4}{3}$, $\kappa = 2$, $\kappa = 4$, and $\kappa = 8$. The larger the curvature, the smaller the radius of the circle and the tighter the curve, and the smaller the curvature, the larger the radius and the shallower the curve.

Below is illustrated a curve with osculating circles drawn at certain points:

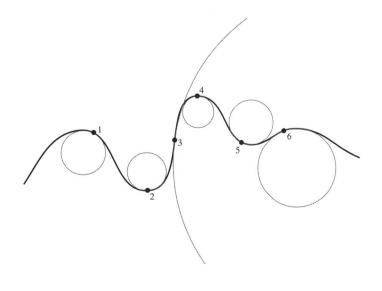

You can see that the smaller circles, with larger curvature, occur where the curve makes a tight turn. The larger circles, one of which is so large that we drew only an arc of it, occur where the curve is almost straight. The circles are chosen to be the best possible fit to the curve near the point.

There is a relationship between the tangent line and the osculating circle: The circle will also be tangent to the tangent line at the point on the curve. The ray perpendicular to the tangent line, called the *normal vector* at the point, will point toward the center of the circle. The normal vector is always at right angles to the curve, but this allows us to give a plus or minus sign to the curvature: Let the curvature be positive when the normal arrow pointing toward the center of the osculating circle points up from the curve and negative if it points down from the curve. It should be noted that this assignment of positive and negative values is somewhat arbitrary, but consistency is important. If we take the circle at point #1 to have radius 1, then the curvature at point #1 is $\kappa(P_1) = -1$, at point #2, $\kappa(P_2) = +\frac{3}{2}$, while $\kappa(P_3) = -\frac{1}{10}$, $\kappa(P_4) = -2$, $\kappa(P_5) = +1$, and $\kappa(P_6) = -\frac{1}{2}$.

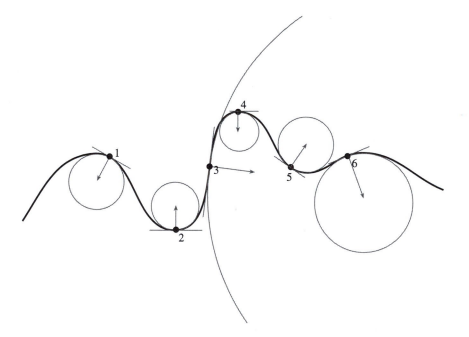

For Exercises 2 through 7, use a circle template to draw the osculating circle at each marked point and then calculate the curvature at each point. If you cannot fit a circle to the curve at some point, explain why not and what this means about the curvature at that point.

▷ **Exercise 2.** An ellipse:

▶ **Exercise 3.** A catenary:

▷ **Exercise 4.** A sine curve:

▷ **Exercise 5.** A logarithmic spiral:

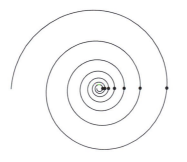

▷ **Exercise 6.** A lemniscate:

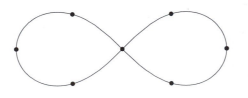

▶ **Exercise 7.** A limaçon:

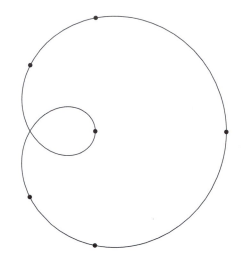

While so far we have concentrated on curves drawn on the plane, the definition of curvature in terms of the osculating circle can be extended to curves in space, and this will be used to measure the way the surface of a three-dimensional object curves.

▷ **Exercise 8.** Figure out the curvature of a Slinky™. Does this depend on whether the Slinky™ is at rest or stretched?

There are two definitions of the curvature of the surface of a three-dimensional object: One, called the *mean curvature*, is defined in terms of curves on the surface and will be discussed here, and the other, called the *Gaussian or sectional curvature*, will be discussed in more detail in the next section. Just as curves can be approximated by tangent lines, so can surfaces be approximated by tangent planes:

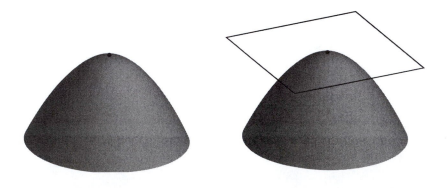

Any plane perpendicular to the tangent plane slices the surface along a curve, called a *normal section*. We can measure the curvature of this normal section using our circle template and determine its curvature $\kappa = \frac{1}{r}$, where r is the radius of the circle of curvature or osculating circle.

Of course, there are infinitely many ways to slice the surface through the given point, even if we do insist that the slice be perpendicular to the tangent plane. But one of these must have the greatest possible curvature and one the least. Let κ_1 be the greatest curvature of all the normal sections at the point and κ_2 the least possible curvature. These are called the *principal curvatures* of the surface at the given point. The *mean curvature H* of the surface is defined by the average of the principal curvatures:

$$H = \frac{\kappa_1 + \kappa_2}{2}.$$

Below are some bowl shapes. The one on the left has both principal curvatures negative, since the centers of the circles of curvature lie below the surface. This shape has mean curvature -1. The bowl on the right has positive mean curvature, since the centers of the principal circles of curvature lie above the surface. These signs are somewhat arbitrary, but should be applied consistently.

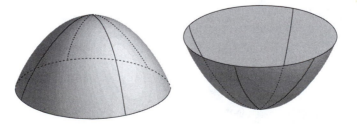

At point *P*, the cylinder below has one negative principal curvature and one zero. Thus the mean curvature is negative:

In the following hyperbolic surface, the principal normal sections are shown. In the case where the radii of the circles of curvature are the same, then since one must be positive and the other negative, the mean curvature will be zero.

▷ **Exercise 9.** Pick a point on your potato. Slice the potato perpendicular to the surface in the direction of greatest curvature. Use your circle template and the flat face of the cut potato to find one of the principal curvatures at that point. Put the potato back together and repeat to find the other principal curvature at the same point. Calculate the mean curvature at this point.

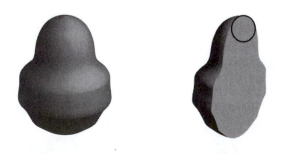

▷ **Exercise 10.** Find the point on your potato with the smallest mean curvature and the point with the largest mean curvature.

▷ **Exercise 11.** Repeat Exercises 9 and 10 for a much smaller potato.

▷ **Exercise 12.** Repeat Exercises 9 and 10 for a cucumber.

▷ **Exercise 13.** Repeat Exercises 9 and 10 for a cabbage leaf taken from the outer leaves.

▷ **Exercise 14.** Repeat Exercises 9 and 10 for a cabbage leaf taken from the inner leaves.

▷ **Exercise 15.** Repeat Exercises 9 and 10 for a kale leaf.

The *Gaussian, or total curvature, K* can also be defined in terms of the principal curvatures κ_1 and κ_2 by

$$K = \kappa_1 \kappa_2.$$

While the mean curvature H depended on the somewhat arbitrary choice of signs, the Gaussian curvature will be positive when the centers of both principal circles of curvature lie on the same side of the surface, and negative if they do not. Thus for both of the bowl shapes illustrated previously, $K > 0$. For the cylinder, $K = 0$, while for the hyperbolic surface $K < 0$.

Both types of curvatures are important, and measure subtly different properties. The mean curvature describes precisely the conditions with which soap films naturally comply, as we will see in Section 11.5. The Gaussian curvature, on the other hand, describes the type of geometry a space has at each point. Also, the mean curvature is an *extrinsic property*: one that depends on how the shape sits in the surrounding three-dimensional universe. Gaussian curvature is *intrinsic*: It depends only on the shape itself. To see this, take a flat piece of paper: the principal curvatures are both zero, so $H = 0$ and $K = 0$. Bend the piece of paper into a cylinder as shown below. Now one of the principal curvatures is still zero, $\kappa_1 = 0$, but the other is the reciprocal of the radius of the cylinder, so $\kappa_2 = \frac{1}{r}$. Thus $K = 0 \cdot \frac{1}{r} = 0$, but $H = \frac{\frac{1}{r} + 0}{2} = \frac{1}{2r}$.

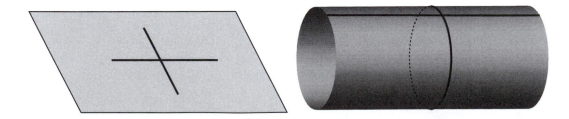

▷ **Exercise 16.** Find the Gaussian curvature for the point on the potato from Exercise 9.

▷ **Exercise 17.** Find the Gaussian curvature for the point on the potato from Exercise 11.

▷ **Exercise 18.** Find the Gaussian curvature for the point on the cucumber from Exercise 12.

▷ **Exercise 19.** Find the Gaussian curvature for the point on the cabbage leaf from Exercise 13.

▷ **Exercise 20.** Find the Gaussian curvature for the point on the cabbage leaf from Exercise 14.

▷ **Exercise 21.** Find the Gaussian curvature for the point on the kale leaf from Exercise 15.

▷ **Exercise 22.** Find the mean and Gaussian curvatures of a sphere of radius r.

SUGGESTED READINGS

James Casey, *Exploring Curvature*, Friedrich Vieweg & Sohn, Braunschweig, 1996.

Alfred Gray, *Modern Differential Geometry of Curves and Surfaces*, CRC Press, Boca Raton, 1993.

Stefan Hildebrandt and Anthony Tromba, *The Parsimonious Universe*, Springer-Verlag, New York, 1996.

Jeffrey R. Weeks, *The Shape of Space*, Marcel Dekker, New York, 1985.

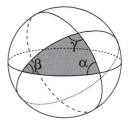

11. Shape

♦ **11.4. CURVATURE OF SURFACES**

SUPPLIES
    ball
    knife
    a selection of vegetables, such as potatoes, eggplant, onions, cucumbers,
        cabbage, and kale
    protractor
    pin
    a bit of string
    map or plan measure (optional)

In Chapter 7: Polyhedra, we discussed the euler characteristic, $\chi = v - e + f$, which turned out to be 2 for all of the regular and semiregular polyhedra discussed in Sections 7.1, 7.2, and 7.3. Another quantity associated with every polyhedron is the angular deficit. Recall that when six equilateral triangles meet at a vertex, then the sum of the face angles at that vertex will be $6 \cdot 60° = 360°$. Therefore, this configuration of triangles will lie flat on the plane. On the other hand, the tetrahedron has three equilateral triangles meeting at each vertex, for a vertex angle sum of $3 \cdot 60° = 180°$. If you cut a small region surrounding a vertex of the tetrahedron, you will have a small triangular pyramid that will not lie flat. If you slit it open along one edge up to the vertex in question and lay it out flat, it forms a gap of 180°. This measures how far the tetrahedron comes from lying flat at this vertex.

The quantity $\delta = 360° - $ (*vertex angle sum*) is called the *angular deficit at the vertex*, and measures how far from being flat the vertex configuration is. Thus the tiling by triangles has a deficit of $\delta = 0°$, while the tetrahedron has a deficit of $\delta = 180°$ at each vertex. The icosahedron, on the other hand, has 5 equilateral triangles meeting at each vertex, giving a deficit of $\delta = 360° - 5(60°) = 60°$. The smaller the angular deficit is, the flatter the shape is at the vertex, while the larger the angular deficit is, the pointier the shape is at the vertex.

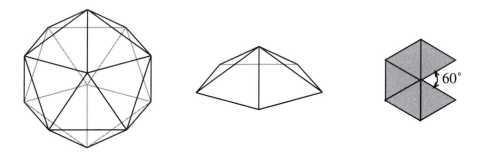

Since a regular or semiregular polyhedron has the same vertex configuration at each vertex, it will have the same vertex angle deficits at each of its vertices. The *total angular deficit*, Δ, of a polyhedron is the sum of the vertex angle deficits for each vertex.

▷ **Exercise 1.** Fill in the table below.

Angular Deficits

Polyhedron	δ	# vertices	Δ
Tetrahedron	180°	4	720°
Cube			
Octahedron			
Dodecahedron			
Icosahedron			

For irregular polyhedra, the number and type of faces meeting at each vertex may not be the same, so to compute the total angular deficit Δ, you must compute δ for each vertex and add them. In other words, if a polyhedron has v vertices, we can label them $V_1, V_2, V_3, \ldots, V_v$, and the total angular deficit of the polyhedron is

$$\Delta = \delta(V_1) + \delta(V_2) + \cdots + \delta(V_\nu)$$
$$= [360° - (\text{sum of the angles meeting at } V_1)]$$
$$+ [360° - (\text{sum of the angles meeting at } V_2)] + \cdots$$
$$+ [360° - (\text{sum of the angles meeting at } V_\nu)].$$

▶ **Exercise 2.** Find the vertex angular deficits of each vertex of the pyramid below, which has a square base and four equilateral triangles for sides, and find the total angular deficit Δ.

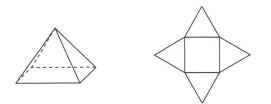

▷ **Exercise 3.** Find the vertex angular deficits of each vertex of the regular-pentagonal prism below, and find the total angular deficit.

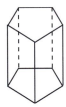

▷ **Exercise 4.** Find the vertex angular deficits of each vertex of the pentagonal antiprism below, and find the total angular deficit.

▷ **Exercise 5.** Using the computations above, guess what the total angular deficit of any convex polyhedron is.

The result of Exercise 5 seems surprising, considering how much the angular deficits of each vertex can vary. This seeming oddity is similar to that experienced in Chapter 7, where we showed that the *Euler characteristic*, $\chi = v - e + f$, is also constant for all of the convex polyhedra, since for each of these shapes $\chi = 2$. There is a connection between these results, which we explore next.

Descartes's Formula

Consider a possibly irregular but convex polyhedron with v vertices, e edges, and f faces. Label the vertices V_1, V_2, V_3, . . . , V_v with vertex angular deficits $\delta(V_1)$, $\delta(V_2)$, $\delta(V_3)$, . . . , $\delta(V_v)$. Label the faces $F_1, F_2, \ldots, F_f$ and let $N(F_j)$ be the number of vertices (and edges) of face F_j. From Section 1.2, we know that the sum of the angles of the vertices of a single face with N vertices (or edges) is $180° \, (N - 2)$. *Descartes's formula* relates the total angular deficit and the Euler characteristic:

$$
\begin{aligned}
\Delta &= \delta(V_1) + \delta(V_2) + \cdots + \delta(V_v) \\
&= [360° - (\text{sum of the angles meeting at } V_1)] \\
&\quad + [360° - (\text{sum of the angles meeting at } V_2)] + \cdots \\
&\quad + [360° - (\text{sum of the angles meeting at } V_v)] \\
&= 360°v - [(\text{sum of the angles meeting at } V_1) \\
&\qquad\qquad + (\text{sum of the angles meeting at } V_2) + \cdots \\
&\qquad\qquad + (\text{sum of the angles meeting at } V_v)] \\
&= 360°v - [\text{sum of all the angles of all the polygonal faces}] \\
&= 360°v - [(\text{sum of the angles of the vertices of face } F_1) \\
&\qquad\qquad + (\text{sum of the angles of the vertices of face } F_2) + \cdots \\
&\qquad\qquad + (\text{sum of the angles of the vertices of face } F_f)] \\
&= 360°v - [180°(N(F_1) - 2) + 180°(N(F_2) - 2) + \cdots \\
&\qquad\qquad + 180°(N(F_f) - 2)] \\
&= 360°v - 180°[N(F_1) + N(F_2) + \cdots + (N(F_f)] \\
&\quad + [180° \cdot 2 + 180° \cdot 2 + \cdots + 180° \cdot 2] \\
&= 360°v - 180°[(\text{total number of edges for all the faces before} \\
&\qquad\qquad\qquad \text{assembling the polyhedron}] + 360°f
\end{aligned}
$$

Note that the total number of edges for all the faces is $2e$, since these edges are glued together in pairs to form the polyhedron, so

$$
\begin{aligned}
\Delta &= 360°v - 180° \cdot 2e + 360°f \\
&= 360°(v - e + f) \\
\Delta &= 360°\chi.
\end{aligned}
$$

Thus for any convex polyhedron, we have $\Delta = 360° \cdot 2 = 720°$, as predicted in Exercise 5.

Angle Sum for Spherical Polygons and Area

The standard formula for the surface area of a sphere is $A = 4\pi r^2$, where r is the radius of the sphere. We wish to use this formula to find the area of a triangle drawn on the surface of a sphere. Such triangles are formed on the surface of the sphere by arcs from great circles: circles passing around the fattest part of the sphere. Since you are probably much more comfortable measuring angles in degrees than in radians, let us convert the area formula into degrees. Since $180° = \pi$ radians, we know that $1° = \frac{\pi}{180°}$, and so the formula for the area of a sphere can be restated as

$$A = 720°\left(\frac{\pi}{180°}r^2\right).$$

In the following exercises, be careful to leave the quantity $(\frac{\pi}{180°}r^2)$ intact, as this will make the numerical patterns we are searching for more obvious.

▶ **Exercise 6.** Consider the three-sided region formed on the surface of a sphere by the equator and two longitudinal circles, meeting at the north pole at an angle of 90°. This spherical triangle thus has three angles, each measuring 90°. Find the sum of the angles and the area of this spherical triangle (in terms of the radius r).

▷ **Exercise 7.** Consider the spherical triangle formed on the surface of a sphere by the equator and two longitudinal circles, meeting at the north pole at an angle of 30°. Find the sum of the angles and the area of this spherical triangle.

▷ **Exercise 8.** Consider the spherical triangle formed on the surface of a sphere by the equator and two longitudinal circles, meeting at the north pole at an angle of 120°. Find the sum of the angles and the area of this spherical triangle.

▷ **Exercise 9.** Guess a formula for the area of a spherical triangle in terms of the angle sum and r, the radius of the sphere. [Hint: Consider the angle sum of the triangle minus 180°.]

▶ **Exercise 10.** Use Exercise 9 to find the area of a spherical triangle with angles 45°, 60°, 90°.

▷ **Exercise 11.** Use Exercise 9 to find the area of a spherical triangle with angles 61°, 62°, 63°.

If we consider an angle α drawn on the surface of a sphere, extending the sides, a wedge-shaped region (and its double on the back side of the ball) is formed. This wedge-shaped spherical region together with its double on the back of the sphere is called a *lune* with angle α.

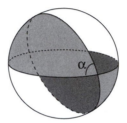

Since the lune forms $\frac{2\alpha}{360°} = \frac{\alpha}{180°}$ of the whole sphere, the area of a lune with angle α, denoted by α-lune for $0 \leq \alpha \leq 180°$, is

$$A(\alpha\text{-lune}) = \frac{\alpha}{180°} 720° \left(\frac{\pi}{180°} r^2 \right) = 4\alpha \left(\frac{\pi}{180°} r^2 \right).$$

We can use this formula to find the area of any spherical triangle with angles α, β, and γ.

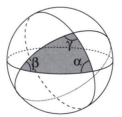

▷ **Exercise 12.** On your ball, draw the triangle with angles α, β, and γ as above and shade the lunes corresponding to α, β, and γ with three different colors.

From the shadings on the ball of Exercise 12, you can see that the three lunes cover the entire surface of the sphere, but both the original triangle and its double on the back side of the sphere are shaded by each of the three lunes. Thus,

$$A(\alpha\text{-lune}) + A(\beta\text{-lune}) + A(\gamma\text{-lune}) = A(\text{sphere}) + 4A(\text{triangle}).$$

Therefore,

$$4\alpha\left(\frac{\pi}{180°}r^2\right) + 4\beta\left(\frac{\pi}{180°}r^2\right) + 4\gamma\left(\frac{\pi}{180°}r^2\right) = 720°\left(\frac{\pi}{180°}r^2\right) + 4A(\text{triangle}),$$

$$(\alpha + \beta + \gamma)\left(\frac{\pi}{180°}r^2\right) = 180°\left(\frac{\pi}{180°}r^2\right) + A(\text{triangle}),$$

$$A(\text{triangle}) = (\alpha + \beta + \gamma - 180°)\left(\frac{\pi}{180°}r^2\right)$$

▷ **Exercise 13.** Using the fact that any polygon with *n* sides can be cut up into (*n* − 2) triangles as in Section 1.2, find a general formula for the area of a spherical polygon with *n* sides in terms of the angle sum.

▶ **Exercise 14.** If a triangle on a sphere of unknown radius is measured and it is found that its area is 120 square feet and its angles measure 90°, 60°, and 60°, find the radius of the sphere.

The Gauss-Bonnet Formula for the Sphere

Again, there is an unexpected relationship between the area of a sphere and the Euler characteristic, called the *Gauss-Bonnet formula*. Divide a sphere of radius *r* into *f* spherical polygons $F_1, F_2, F_3, \ldots, F_f$, and let N_1 denote the number of edges in face F_1, N_2 the number of edges in face F_2, etc. Let *v* be the number of vertices and *e* the number of edges for the spherical polyhedron. In the picture below, the sphere is divided into 8 faces, each with 3 sides, so $f = 8$, $e = 12$, $v = 6$, and $N_i = 3$ for each of the eight faces.

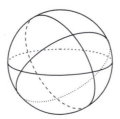

$$A(\text{sphere}) = A(F_1) + A(F_2) + A(F_3) + \cdots + A(F_f)$$

$$= [(\text{sum of the angles of } F_1) - (N_1 - 2)180°]\left(\frac{\pi}{180°}r^2\right)$$

$$+ [(\text{sum of the angles of } F_2) - (N_2 - 2)180°]\left(\frac{\pi}{180°}r^2\right)$$

$$+ \cdots + [(\text{sum of the angles of } F_f)$$

$$- (N_f - 2)180°]\left(\frac{\pi}{180°}r^2\right)$$

$$= \left(\frac{\pi}{180°}r^2\right)[(\text{sum of the angles of } F_1)$$

$$+ (\text{sum of the angles of } F_2) + \cdots$$
$$+ (\text{sum of the angles of } F_f)]$$

$$- \left(\frac{\pi}{180°}r^2\right)[N_1 + N_2 + \cdots + N_f]180°$$

$$+ \left(\frac{\pi}{180°}r^2\right)[2 + 2 + \cdots + 2]180°$$

$$= \left(\frac{\pi}{180°}r^2\right)[\text{sum of all of the angles of all the polygons}]$$

$$- 180°\left(\frac{\pi}{180°}r^2\right)[\text{sum of all of the edges of all of the polygons}]$$

$$+ 180°\left(\frac{\pi}{180°}r^2\right)[\text{sum of one 2 for each face}].$$

Note that the total number of edges for all the faces is $2e$, since these edges are glued together in pairs to form the polyhedron, so

$$A(S) = \left(\frac{\pi}{180°}r^2\right)[\text{sum of one } 360° \text{ for each vertex}]$$

$$- 180°\left(\frac{\pi}{180°}r^2\right)[2e] + 180°\left(\frac{\pi}{180°}r^2\right)[2f]$$

$$= \left(\frac{\pi}{180°}r^2\right)360°v - \left(\frac{\pi}{180°}r^2\right)360°e + \left(\frac{\pi}{180°}r^2\right)360°f$$

$$= 360°\left(\frac{\pi}{180°}r^2\right)[v - e + f]$$

$$= 2\pi r^2\chi(S).$$

Since S is a sphere and we know that $\chi(S) = 2$, this gives an alternative proof of the area of a sphere: $A(S) = 2\pi r^2 \chi(S) = 720°(\frac{\pi}{180°}r^2)$. Recalling from the previous section that the Gaussian curvature of a sphere of radius r is $K = \frac{1}{r^2}$, we can restate this result as the Gauss-Bonnet formula for the sphere:

$$KA = 2\pi\chi.$$

An analogous development takes place for the area of a triangle on the floppy hyperbolic plane built in Section 11.1:

$$A(\text{triangle}) = (180° - \text{sum of the angles in the triangle})k$$

for a constant k that depends on the particular hyperbolic plane and acts somewhat like the radius for a sphere.

▷ **Exercise 15.** Explain why this means that there is a maximum value for the area of any triangle on a given hyperbolic plane.

A space has constant Gaussian curvature K if the curvature is the same at every point in the space. The sphere is one example of a space with constant positive Gaussian curvature. For any such space, the Gauss-Bonnet theorem says that

$$KA = 2\pi\chi.$$

Angle Deficit and Gaussian Curvature

The idea that led us to Descartes's formula and to the Gauss-Bonnet theorem for the sphere is to measure the relative flatness and pointiness of the shapes under study, by means of measuring the angle deficit. This central idea can be extended to many shapes. For example, if you make a cone from a sheet of paper and slit it open, you can measure the resulting angle and compute the angular deficit $\delta = (360° - \text{angle})$ to get an idea of how far from flat the cone is.

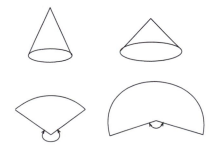

▷ **Exercise 16.** Build three different cones, ranging from extremely pointy to quite flat, and measure their angular deficits at the vertex.

One can also apply this concept to hyperbolic spaces. For example, in Exercise 4 of Section 11.1 we constructed a space by taping seven equilateral triangles together at a vertex. This space will have angular deficit $\delta = 360° - 7(60°) = -60°$, indicating that you have 60° too much material to lie flat. A vertex will have a negative angular deficit (or *excess*) if the space is hyperbolic at that vertex. A zero angular deficit means that the space lies flat at that vertex. A positive angular deficit means that the space is elliptic at the vertex.

▷ **Exercise 17.** Compare the sign (positive, zero, or negative) of the angular deficit with the sign of the Gaussian curvature at a typical elliptic point, a flat point, and a hyperbolic point.

The idea of angular deficit can be extended to lumpy things whose curvature varies from one point to the next. Since lumps don't usually slit open nicely, we have to attack the lump slightly differently. Take a potato and choose a more or less disc-like region of radius one on it. Peel a strip going all the way around the potato following the boundary of the region as pictured below:

Lay the strip out flat on a piece of paper and draw lines along both cut ends of the strip of peel. The angle between these lines can be measured with a protractor. This angle measures how pointy or flat the potato is for that region, and is an approximation for the Gaussian curvature, K, discussed in the previous section. A piece of peel surrounding a region with $K = 0$ is illustrated below on the left, $K = \pi$ in the center, and $K = 2\pi$ on the right:

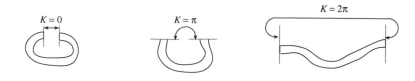

▷ **Exercise 18.** Measure the curvature for one region for each of your vegetables.

▷ **Exercise 19.** Cut a slice from an onion about halfway between the equator and the end, and so that the slice separates into rings. Measure the curvature for three of the layers or rings, choosing an outer layer, a middle layer, and one from close to the center. What do you notice?

▷ **Exercise 20.** Find regions of one of your vegetables with approximate curvature:

(a) 0

(b) $\frac{\pi}{2}$

(c) π

(d) 2π

(e) $-\frac{\pi}{2}$

▶ **Exercise 21.** Describe a typical vegetable with:
(a) positive curvature
(b) zero curvature
(c) negative curvature

Here is an alternative way of measuring the curvature of lumpy things, if you can lay hands on a map or plan measure, which is a little device with a wheel that one runs along a road map with a dial which reads how far the wheel has traveled. It is a sort of miniature hand-held odometer. Tie one end of the bit of string to a pin or nail and tie a pen to the other end so that the distance from the center of the pin to the pen point is 1 inch. Stick the pin well into a potato. The string and pen allow you to draw a *geodesic circle*: the collection of points that are one inch from the pin measured over the surface of the potato. Run the map measure around this curve to calculate L, the length of the geodesic circle. If you don't have a map measure, then you can lay a bit of string along the curve drawn with the pen, carefully following all the bumps and bends. You can then mark the beginning and the end of the curve on the string, remove it, and stretch the string out straight to find the length of the curve. The Gaussian curvature of the region inside the geodesic circle is approximately

$$K = 2\pi - L.$$

For a circle of radius 1 drawn on a flat piece of paper, obviously this technique should give you $K = 0$, since the circumference of a circle is $L = 2\pi r$ and $r = 1$. For elliptic regions, the circle will be a bit smaller than it would if it were flat, and for a hyperbolic region a bit larger.

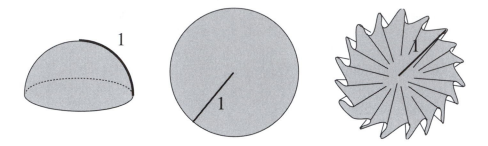

If a radius of 1 inch is not convenient, the formula above can be adapted to give the approximation $K = 2\pi - \frac{L}{r}$, where the geodesic circle has radius r, i.e., L is the length of the curve given by all points r units from your chosen center.

▷ **Exercise 22.** Use a map measure to measure the Gaussian curvature of several regions on a potato.

In order to proceed with the study of curvature, one needs to know calculus. For those who do, the formula that measures the Gaussian curvature at a point, and which we have approximated above, is:

$$K(p) = \lim_{r \to 0} \frac{2\pi - \frac{L}{r}}{\frac{\pi r^2}{3}}.$$

For spaces of nonconstant curvature, where the curvature varies from point to point, the Gauss-Bonnet Theorem states that:

$$\int_M K dA = 2\pi\chi,$$

where M is the two-dimensional surface. This formula can be generalized to n-dimensional spaces called closed manifolds, p2ovided that n is even.

SUPPLIES

1. A map measure made by Silva is available at some hiking and orienteering suppliers. A similar plan measure by Alvin can be found in architects' supply stores.

SUGGESTED READINGS

Ethan Bloch, *A First Course in Geometric Topology and Differential Geometry*, Birkhäuser, Boston, 1997.

James Casey, *Exploring Curvature*, Friedrich Vieweg & Sohn, Braunschweig, 1996.

Peter Doyle, Jane Gilman, and William Thurston, *Geometry and the Imagination*, reprint, Geometry Center, University of Minnesota, 1991.

Jeffrey R. Weeks, *The Shape of Space*, Marcel Dekker, New York, 1985.

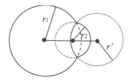 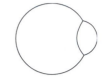

11. Shape

◆ **11.5. SOAP BUBBLES**

SUPPLIES
   wire (I used 14 gauge copper wire)
   wire cutters
   solder
   soldering iron
   bucket or deep bowl
   dishwashing soap
   glycerin
   plate
   straw

Soap bubbles are fun to blow, beautiful to contemplate, and involve some surprising and interesting mathematics. But before we can investigate these, we must make some preparations. First mix a good quantity of soap solution by stirring together about 12 cups of water, a cup of dishwashing soap, and a quarter-cup of glycerin (available at most drugstores), or similar proportions to make enough bubble mixture to have 2 or 3 inches in the bucket. The glycerin is used to slightly prolong the lifetime of the bubbles. Stir the mixture gently until mixed and try not to make it foamy: If there is too much foam on the surface, you will ruin the bubbles we will make. The best conditions for blowing bubbles are damp cool days, out of the direct sun. Fortunately, here in New York state, we are blessed with an extraordinary number of days that satisfy precisely these conditions.

   The standard bubble wand is very good for blowing the standard spherical bubbles, but we also want to make some unusual frameworks for soap films. Solder together bits of wire to make the following bubble frames: First a simple circle about three inches in diameter; then two parallel circles, three inches in diameter and an inch apart; third, a larger circle twisted into the boundary of a Möbius band as shown; and last, a knotted loop of wire.

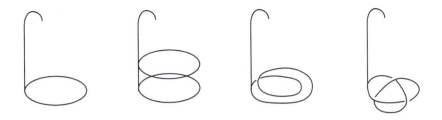

Now go ahead and blow lots of bubbles. Get it out of your system before we proceed.

Dip the frame consisting of two parallel circles into the soap solution. On withdrawing it from the bucket, you should get one of three shapes: two separate disc-like regions spanning the two circles shown below on the left, a sort of curved cylinder (called a *catenoid*) spanning between the two circles shown below in the middle, or a compromise between the two consisting of a single disc suspended between the two circular frames and connected to them by bands of soap film shown on the right.

▷ **Exercise 1.** Dip the two-circle frame and try to create all three forms. Blowing gently on the first form will sometimes make it transform into the second or third, but sometimes this will merely pop your bubble. Touching the central disc of the third form with a soapy finger will often make it transform into the second form.

▶ **Exercise 2.** Estimate the angle made in the third form between the central disc and the walls.

▷ **Exercise 3.** Estimate which of the three forms has the least area.

▶ **Exercise 4.** Dip the Möbius frame into the soap solution. You should be able to get three different forms for the soap film eventually. Sketch these.

▷ **Exercise 5.** Estimate which of the Möbius forms has the least area.

▷ **Exercise 6.** Dip the knot frame into the soap solution. How many different forms of the soap film can you get? Sketch these.

▷ **Exercise 7.** Estimate which of the knot forms has the least area.

The study of the mathematics of soap bubbles and films was initiated by the Belgian physicist Joseph Plateau, in spite of his being blind. He went blind as a result of staring at the sun without eye protection for almost half a minute, while doing an experiment when he was in his twenties. He was aided in his later investigations of soap films by his family and assistants. He published the results of his research in 1873, and since then many mathematicians and physicists have studied problems related to soap films. Plateau formulated a number of hypotheses for the shape of soap films, several of which are called *Plateau's problem*.

To return to our exploration of actual soap films, build a new dipping frame consisting of two intersecting circles:

▷ **Exercise 8.** Dip this frame into the soap solution. Sketch the form that you get. Estimate the angle made by the intersection of the soap film with itself.

▷ **Exercise 9.** Pour a small amount of bubble solution onto a plate and blow through a straw to build bubble clusters. Sketch a cluster of bubbles, paying particular attention to the way they intersect.

One of Plateau's observations is that double bubbles always intersect at an angle of 120°. If the bubbles are precisely the same size, then the wall separating them will be straight:

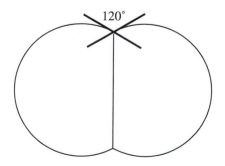

For a double bubble formed by two bubbles of unequal size, the wall separating them will be curved in toward the larger bubble. A spherical soap bubble, like a balloon, is forced into its shape by the air contained inside, which is at a higher pressure than the outside atmosphere. A law of physics, called Laplace's law, states that the pressure on a film is given by the surface tension of the film times the mean curvature of the film, or $P = tH$, where P is the pressure inside the bubble, H the mean curvature of the soap bubble, and t is the surface tension, a constant depending only on the soap solution. For spherical bubbles, the curvature is $H = \frac{1}{r}$ where r is the radius of the sphere. We can use this fact to calculate precisely how two bubbles must meet.

The wall between a double bubble is an arc from a circle whose center lies on the line through the centers of the two bubbles:

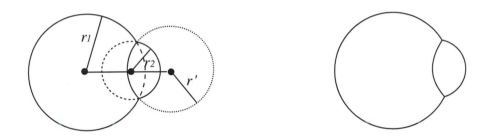

If we have two spherical bubbles of radius r_1 and r_2 with $r_1 \geq r_2$, which merge to form a double bubble, then the pressure exerted on the wall between the two from the first bubble is $P_1 = \frac{t}{r_1}$, and the pressure on the wall from the second bubble is $P_2 = \frac{t}{r_2}$. Thus, the total pressure on the wall is $P = \frac{t}{r_2} - \frac{t}{r_1} = \frac{t}{r'}$, where r' is the radius of the circle whose arc forms the wall separating the double bubble. Therefore $\frac{1}{r'} = \frac{1}{r_2} - \frac{1}{r_1}$.

Furthermore, any point where the two bubbles and the wall between them meet must be in equilibrium. In the picture below, the two bubbles are centered at A and B, while C marks the center of the circle forming the arc for the intermediate wall. Radial lines are drawn from A, B, and C to the point D, and tangent lines at D to each of the three circles are shown as dashed. The tension at D will be balanced only if these dashed lines form angles of 120°:

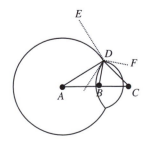

▷ **Exercise 10.** Using the fact that $\angle EDF = 120°$ and the fact that for a circle a radial line and the tangent must be at right angles, show that $\angle ADB = \angle BDC = 60°$.

Using this exercise, we can construct the wall for any double bubble. Start with the first bubble centered at A and a line through A on which the centers will lie. Choose a point D on the circumference of the circle, which will be one of the points where the bubbles intersect. Then $r_1 = AD$ is a radius for the first bubble. Draw an angle $\angle ADB = 60°$ choosing B on the original line through A for the bubble centers. B will be the center of the second bubble and $r_2 = BD$ its radius. Draw an angle $\angle BDC = 60°$ so that C lies on the line for the bubble centers as shown. Point C will be the center of the curved wall between the two bubbles, and $r' = CD$ a radius of this circle:

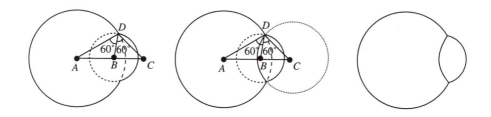

▶ **Exercise 11.** Draw a double bubble and the intermediate wall if one bubble has radius 2 inches and the other 1 inch.

▷ **Exercise 12.** Draw a double bubble and the intermediate wall if one bubble has radius 1.5 inches and the other 1 inch.

Any time three bubble walls come together, an angle of 120° will be formed. Next, build some frames in the shapes of some of the regular polyhedra:

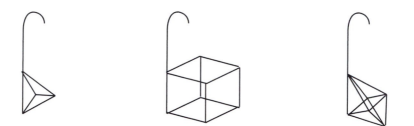

Dip the tetrahedral frame into the bubble solution once and you usually get the following form:

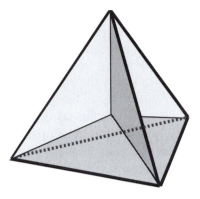

You can see and admire the perfect 120° angle formed wherever three walls meet. But what of the angle at the center, where the four lines meet? This is another angle observed by Plateau in his experiments, and the only other angle ever formed by soap films. This angle is exactly 109°28′16″. This might seem an oddly precise number, but a little trigonometry shows that the cosine of this angle is $-\frac{1}{3}$ and is formed at the center of a tetrahedron. While noticed by Plateau, it was not until more than 100 years later that a satisfactory mathematical explanation was given by Jean Taylor.

If you dip the tetrahedral frame in again fairly slowly, you will sometimes get the same figure with an added interior bubble:

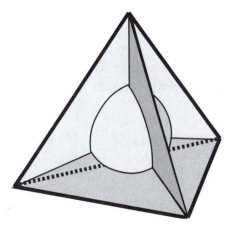

▷ **Exercise 13.** Dip the cubic frame into the bubble solution and sketch the result. Dip it again until you get a central bubble and sketch that result. On both sketches note the locations of the 120° angles and the 109°28′16″ angles.

▷ **Exercise 14.** Dip the octahedral frame into the bubble solution and sketch the result. Dip it again until you get a central bubble and sketch that result. On both sketches note the locations of the 120° angles and the 109°28′16″ angles.

The most basic bubble shape, the sphere, is well known as the surface with the least area for a given volume. Shapes that minimize surface area are called *minimal surfaces*. While soap films effortlessly assume the shape of minimal surfaces, the mathematics of showing that any frame that is topologically a circle (though it may be knotted) spans a minimal surface is quite elegant, but difficult.

Another example of a minimal surface is the catenoid, shown above as one of the shapes formed by the frame consisting of two parallel circles. The sides of the catenoid form a curve called a catenary: the curve made by a piece of chain whose ends are suspended from two posts (*catena* is Latin for chain).

▶ **Exercise 15.** The catenoid has surface area a bit less than the cylinder spanning the same two rings, which has surface area $2\pi rh$, where r is the radius of the rings and h the distance between them. For the following problems, approximate t(e surface area of the catenoid by the surface area of the cylinder.
(a) When (depending on r and h) is the area of the two discs less than the area of the cylinder?
(b) When is the area of the cylinder with a central disc less than the area of two discs?

Another minimal surface is found by building a helical frame and dipping it into the bubble solution. This is called the *helicoid*:

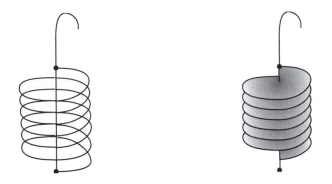

We have found that a frame can make more than one shape, but of these one will have minimal area. One of Plateau's questions was whether a frame can span more than one minimal surface. That the minimal surface for a frame need not be unique is shown by the following picture, where a single frame has three different, but all area-minimizing, spanning surfaces.

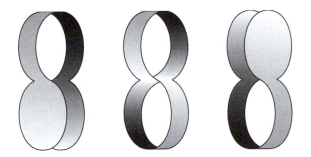

Unlike a spherical bubble or the double bubble discussed above, films such as the catenoid and the helicoid have the property that the pressure on both sides of the soap film is the same. If the pressure is to be the same on both sides of the film, then Laplace's law implies that the mean curvature must be the same on both sides; thus we must have

$$H = \frac{\kappa_1 + \kappa_2}{2} = 0$$

for any minimal surface. Mathematicians call any surface with zero mean curvature a minimal surface, whether it minimizes area or not.

▷ **Exercise 16.** Explain why the catenoid has mean curvature $H = 0$.

▶ **Exercise 17.** Explain why the helicoid has mean curvature $H = 0$.

The flat plane, the catenoid, and the helicoid were the first known minimal surfaces. In 1835, Scherk found a number of infinite minimal surfaces, and the list was extended by Schwarz in 1865. These periodic minimal surfaces are related to the soap films formed on the cubical and tetrahedral frames. The minimal films formed in the following exercises can be put together to form infinite structures, just as we built the infinite polyhedra of Section 7.6.

▷ **Exercise 18.** Take the tetrahedral frame and dip it in the bubble solution. With a wet finger carefully pop sections of the soap film until you get a single sheet of soap film, connected to the frame only along the black edges in the picture below.

▷ **Exercise 19.** Take the cubic frame and dip it in the bubble solution. With a wet finger carefully pop sections of the soap film until you get a single sheet of soap film, connected to the frame only along the black edges in the picture below.

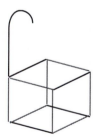

More infinite periodic minimal surfaces were found by Schoen and Meeks in the 1970s and early 1980s, but it was thought that there were no more minimal surfaces of finite topological type until the mathematical world was startled by the discovery, by Celso Costa, a Brazilian graduate student, of a new minimal surface, the *Costa surface*, topologically equivalent to a torus with three punctures. At the 9th annual Breckenridge International Snow Sculpture Championships, held in Colorado in January 1999, a team consisting of Stan Wagon, Helaman Ferguson, Dan Schwalbe, and Tamas Nemeth, sponsored by Wolfram Research, sculpted a Costa surface of snow:

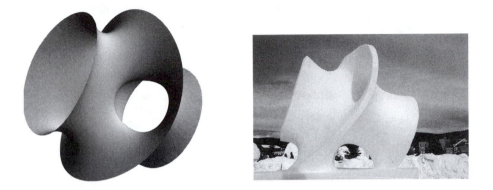

Following this discovery, David Hoffman and William Meeks went on to produce infinite families of new minimal surfaces, such as the one below:

SUGGESTED READINGS

Frederick J. Almgren, *Plateau's Problem*, W.A. Benjamin, New York, 1966.

Frederick J. Almgren and Jean E. Taylor, "The Geometry of Soap Films and Soap Bubbles," *Scientific American* 235, July 1976.

C.V. Boys, *Soap Bubbles: Their colors and the forces which mold them*, reprint of 1905 edition, Dover, New York, 1959.

Richard Courant and Herbert Robbins, *What Is Mathematics?*, Oxford University Press, London, 1941.

Alfred Gray, *Modern Differential Geometry of Curves and Surfaces*, CRC Press, Boca Raton, 1993.

Stefan Hildebrandt and Anthony Tromba, *The Parsimonious Universe*, Springer-Verlag, New York, 1996.

David Hoffman, "The Computer-aided Discovery of New Embedded Minimal Surfaces," *Mathematical Intelligencer* 9(3), 1987.

Cyril Isenberg, *The Science of Soap Films and Soap Bubbles*, Dover, New York, 1992.

Robert Osserman, *A Survey of Minimal Surfaces*, Dover, New York, 1986.

Peter S. Stevens, *Patterns in Nature*, Little, Brown and Co., Boston, 1974.

C.L. Strong, "The Amateur Scientist: How to blow soap bubbles that last for months and even years," *Scientific American* 220, May 1969.

D'Arcy Thompson, *On Growth and Form*, abridged edition, Cambridge University Press, New York, 1961.

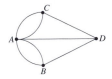

12. Graph Theory

◆ 12.1. GRAPHS

Graph theory has almost nothing to do with all the graphs that you plotted in high school algebra. To see what graph theory is about, first consider these three problems. Get together with two or three other people and worry about these, and then discuss your suggestions and ideas about the solutions with the rest of the class before proceeding to the rest of this section.

The Königsberg Bridge Problem

The origination of graph theory traces back to Leonhard Euler in 1736 (as quoted in Biggs et al., 1986):

> The problem, which I am told is widely known, is as follows: in Königsberg in Prussia, there is an island A, called the *Kneiphof*; the river which surrounds it is divided into two branches . . . and these branches are crossed by seven bridges. . . . Concerning these bridges, it was asked whether anyone could arrange a route in such a way that he could cross over each bridge once and only once. I was told that some people asserted that this was impossible, while others were in doubt; but nobody would actually assert that it could be done.

Can such a route be arranged? Why or why not?

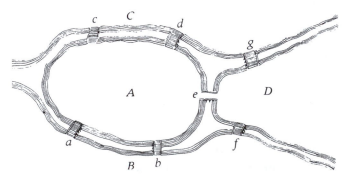

The Handshake Problem

A party consists of exactly three couples. Ben and Alice are one of the couples. Several of the people shake hands with the following sensible conditions:

1. No two people shake hands with each other more than once.
2. No one shakes the hand of the person with whom they came.

After the handshaking is completed, Alice asks everyone how many hands he or she shook. She gets a different number from each person. How many hands did Ben shake?

The Utility Company Problem

Three houses, owned by the Adamses, the Bakers, and the Carrolls, have been built. Each, naturally enough, wishes to be connected to three utility companies: electric, gas, and water. Suppose that all utility lines must run on the surface of the earth, and no two can intersect. How can this be done?

After thinking about these problems, discuss your findings with your friends. You may want to join with others to get six people so that you can all shake hands. Note that if someone shakes your hand, it is assumed that you have also shaken that person's hand, since it is quite rude to let your hand go all limp like a dead fish.

▷ **Exercise 1.** Can you find a path crossing all seven of the Königsberg bridges once? Explain your answer.

▷ **Exercise 2.** How many hands did Ben shake? Explain your answer.

▷ **Exercise 3.** Can you connect each of the three houses to all three utility companies? Explain your answer.

One way to visualize solutions to this type of problem is to draw a diagram called a graph representing the situation.

Definition: A *graph* is a finite non-empty set *V* of *vertices* (or points) along with a set *E* of *edges* between pairs of vertices. A *loop* is an edge from a vertex to itself.

The graph representing the Königsberg bridge problem is sketched on the next page. The land masses are the vertices and the bridges are the edges. Call the north shore C, the south shore B, the island A, and the peninsula D.

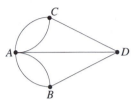

In this example, the set of vertices is $V = \{A, B, C, D\}$. There are seven edges, representing the seven bridges. The edge from vertex A to vertex D could be represented by $\{A, D\}$. Edges do not have a preferred direction, so the edge from A to D is the same as the edge from D to A; i.e., $\{A, D\} = \{D, A\}$. A loop would be denoted by $\{A, A\}$ for example, though the graph above contains no loops. But this notation doesn't distinguish the two different edges between A and C. One could also label all the edges as a, b, c, etc., but this notation doesn't tell you which vertices are connected by each edge. There does not seem to be a perfect notation for all situations.

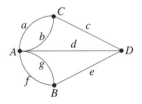

What does the graph above tell you about whether people can cross each of the bridges exactly once before returning home?

Definition: A *path* in a graph is a string of edges where the ending point of one edge in the path is the starting point of the next edge in the path, and where each edge can be used at most once. A *circuit* is a path that begins and ends at the same vertex. An *Euler path* is a path that includes every edge (so that each edge is used exactly once), and an *Euler circuit* is a circuit that includes every edge.

In the example above, one path is the sequence of edges *abde*, and *acd* and *ab* are circuits. In the Königsberg bridge problem, you were looking for an Euler circuit.

Graphs can be used to represent many situations. For example, if we let the students in a class be the set of vertices, we can draw an edge from person *A* to person *B* if they live in the same building. Each vertex will have a loop, since obviously you live in your building.

▶ **Exercise 4.** Suppose there is a section of this course where everybody dropped out except Barry, Larry, Terry, Sherri, Mary, and Carrie. Barry, Terry, and Sherri live in one building, and the others live in another building. Draw a graph representing the situation.

▷ **Exercise 5.** Let Barry, Larry, Terry, Sherri, Mary, and Carrie be represented by vertices. Draw an edge between two vertices if the people are of the same sex. (Is Terry short for Terence or Teresa?)

In both of these exercises, the graph should have broken the set of vertices into two subsets, representing the two living groups in Exercise 4 and the two sexes in Exercise 5.

Definition: A graph is called *connected* if there is a path between each pair of vertices. Otherwise, it is called *disconnected*.

The graphs of Exercises 4 and 5 are disconnected. The subsets into which the graph collects the vertices are called the *components* of the graph.

▷ **Exercise 6.** Create a graph using the students of your class as the vertices and make up your own relationship for the edges. State your criterion for when an edge exists between two students. Watch for loops. You may not use the relationships of Exercises 4 and 5. Is your graph connected or disconnected? If it is disconnected, list the vertices in each component.

▶ **Exercise 7.** Consider the graphs below. Determine whether there is an Euler circuit for each graph.

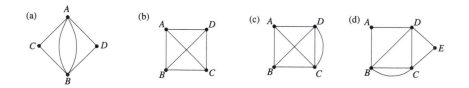

▷ **Exercise 8.** State a general rule for when there cannot be an Euler circuit. You may want to make up some other graphs to test your rule.

▶ **Exercise 9.** Even if there is not an Euler circuit, there can still be an Euler path. Determine which of the following graphs have an Euler path.

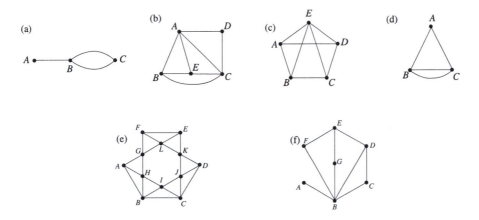

▷ **Exercise 10.** State a general rule for when there cannot be an Euler path. You may want to make up some other graphs to test your rule.

The rules for determining that there is **not** an Euler path or circuit are based on what is called the degree of the vertices.

Definition: The *degree* of a vertex is the number of edges meeting at that vertex. Note that a loop adds two to the degree of a vertex, since it touches the vertex twice.

▷ **Exercise 11.** Draw a graph representing the handshake problem. Let the people be the vertices and draw an edge between two people if they shook hands. What do you know about the degrees of the vertices other than that of Alice? The graph should help if you had trouble with Exercise 2.

▷ **Exercise 12.** Restate (if necessary) your general rules from Exercises 8 and 10 in terms of degrees.

▷ **Exercise 13.** Based on your rules for when an Euler path or circuit will not exist, make a conjecture for when Euler paths and circuits **will** exist. Test your theory on the following graphs and adjust it as necessary.

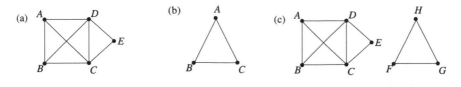

▷ **Exercise 14.** Explain why the sum of the degrees of all the vertices in a graph is twice the number of edges.

▶ **Exercise 15.** Explain why the number of vertices with odd degree must be even.

A *complete graph* is a graph such that every pair of vertices is joined by exactly one edge. The complete graph on n vertices is denoted by K_n. Below are drawings of K_n for $n = 1, 2, 3, 4, 5$.

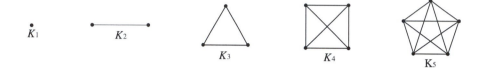

▷ **Exercise 16.** Draw K_6 and K_7.

A related construction gives us the *complete bipartite graphs*. In a bipartite graph the vertices are divided into two groups, satisfying the following two conditions: that no vertex can be connected by an edge to any vertex in the same group (including itself), and that each vertex in the first group must be connected by exactly one edge to each vertex in the second group. The complete bipartite graphs are denoted by $K_{n,m}$ where there are n vertices in the first group and m in the second. Thus, $K_{2,4}$ would be represented by the following picture:

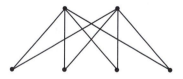

▶ **Exercise 17.** Draw $K_{3,2}$ and $K_{3,4}$.

▷ **Exercise 18.** How many edges will $K_{n,m}$ have?

It turns out that this is precisely the idea we need to describe the Utility Problem. Below, the utility companies are represented by E, G, and W, and the houses by A, B, and C, giving the graph $K_{3,3}$:

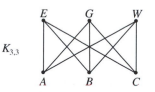

▶ **Exercise 19.** Can you redraw the edges of $K_{3,3}$, bending as necessary, so that they do not intersect?

If a graph can be drawn on the plane so that edges intersect only at the vertices, then the graph is called *planar*. Thus, the Utility Problem can be solved precisely if $K_{3,3}$ is planar. The graph K_4 as drawn above has two edges crossing without a vertex, but it can be redrawn by bending one edge so that these edges do not intersect. Thus, K_4 is planar.

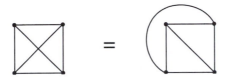

▷ **Exercise 20.** Show that $K_{2,4}$ and $K_{2,5}$ are planar.

▷ **Exercise 21.** Try to redraw K_5 so that the edges do not intersect.

To see that $K_{3,3}$ is not planar, we can rearrange the vertices to make a hexagonal circuit. In the drawing on the left, House A has electric and gas service, House B has gas and water, and House C has water and electricity. Draw another edge, giving House A water. To connect House B to the electric company, we clearly cannot cross through the hexagon without intersecting House A's water line, but we can go around the hexagon. But then House C cannot be connected to the gas company without running the gas line either across the hexagon, in which case it will intersect House A's water line, or around the outside, when it will intersect House B's electric line. Thus, $K_{3,3}$ is not planar.

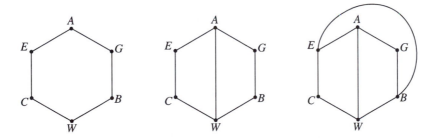

▷ **Exercise 22.** Explain why K_5 is not planar.

It turns out that $K_{3,3}$ and K_5 essentially define nonplanarity for graphs. It can be shown that a graph is planar if and only if it does not contain $K_{3,3}$ or K_5 as a subgraph.

Definition: A *digraph* (directed graph) is a graph with a direction assigned to each edge. Thus, each edge is represented by an ordered pair of vertices.

On the drawing of a digraph, directed edges are represented by arrows, so that if the arrow points from vertex *B* to vertex *A*, then there is a directed edge from *B* to *A*. Many relationships are not reciprocal. Just because you like Sam, there is no guarantee that Sam likes you. If you are constructing a graph representing affection, there would be a directed edge from you to Sam, but there may not be a directed edge from Sam to you. Directed edges can be thought of as one-way streets.

Digraphs show more information than a regular graph. One example of the use of digraphs is called the **House Swap Problem**. From a collection of four houses, each owner prefers some house other than the one currently owned. The owners rank each of the four houses (including their own) from most desirable to least desirable. Their responses are listed in the table below:

House Swap Problem

Owners ↓ preferences →	First	Second	Third	Fourth
A	C	A	D	B
B	A	D	B	C
C	B	C	D	A
D	A	C	D	B

▷ **Exercise 23.** Using the owners as vertices, construct a digraph showing first choice. Since owner *A* thinks that owner *C* has the best house, draw an edge with an arrow pointing from *A* to *C*. Do this for each of the owners. Since there are four owners, your graph should have four directed edges.

Theory guarantees that there will be at least one circuit in such a graph. Since every person has a first choice, every vertex has an edge leaving it. That means you are never stuck at a vertex with no way off. Since there are a finite number of vertices (people) in the graph, you must revisit someone. No matter where you start, if you can always keep moving along edges and there are only a fixed number of vertices, you will eventually return to a vertex. (Note that it may not be the vertex on which you originally started.) Your circuit begins and ends at the revisited vertex.

▶ **Exercise 24.** Find the circuit in the digraph of Exercise 23. Assign the houses according to that circuit, and remove the people who have changed houses from your list. Draw a new graph showing the remaining owners' preferences. Since someone's first choice may no longer be available, use the highest-ranked house on their preference list that has not been sold. Again, there will be a circuit. This process can be repeated until everyone has been assigned a house, though it may work out that not everyone moves.

Note that this is a *selfish algorithm*. Even if person C could guarantee that everyone else gets his or her first choice by agreeing to accept her second choice, she will still want her first choice. Also note what anyone whose last name starts with a letter near the end of the alphabet already knows: Person A is not inherently better than person F. Being first on the list is not a reason to automatically get your first choice.

▷ **Exercise 25.** From a collection of six houses, the owner preferences are listed in the table below. Determine who should live in each house. [This problem is due to Bill Lucas, of the Claremont Colleges.]

House Swap Problem

Owners ↓ preferences →	First	Second	Third	Fourth	Fifth	Sixth
A	F	D	E	B	A	C
B	D	F	C	E	A	B
C	E	A	B	C	F	D
D	F	E	C	B	A	D
E	C	A	E	D	B	F
F	E	C	F	D	B	A

SUGGESTED READINGS

N.L. Biggs, E.K. Lloyd, and R.J. Wilson, *Graph Theory: 1736–1936*, Clarendon Press, Oxford, 1986.
Gary Chartrand, *Introductory Graph Theory*, Dover, New York, 1977.

Helen Christensen, *Mathematical Modeling for the Marketplace: Applying Graph Theory in Liberal Arts and Social and Management Sciences,* Kendall/Hunt, Dubuque, 1988.

F.S. Roberts, *Discrete Mathematical Models, with Applications to Social, Biological, and Environmental Problems*, Prentice Hall, Upper Saddle River, 1976.

Richard J. Trudeau, *Introduction to Graph Theory*, Dover, New York, 1993.

12. Graph Theory

◆ 12.2. TREES

SUPPLIES
   two sheets of Plexiglas™ (or glass)
   thumbtacks
   glue
   bucket or deep bowl
   dishwashing soap
   glycerin

In any graph, what matters is the vertices (or nodes) and the connections between them. Thus, all of the pictures below represent the same graph:

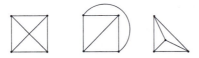

   All of the examples we have considered so far have *loops* or *circuits*: paths on the graph that begin and end at the same vertex. Consider the following examples of graphs without loops or circuits. Note that the last two are actually different representations of the same graph:

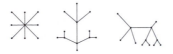

Definition: A *tree* is a graph with no loops or circuits.

   Count the number of edges and vertices in the examples above, letting v denote the number of vertices and e the number of edges.

Definition: *For any graph* Γ, $\chi(\Gamma) = v - e$ *is the Euler characteristic of* Γ.

The actual numbers of edges and vertices do not have much to do with anything, but for both of the trees we have one more vertex than edges. Could this be coincidence? Intuitively, a tree is built by starting with a line segment with one edge and two vertices, so $\chi = v - e = 1$, and adding branches at one of the already existing vertices of the tree. Adding a branch in this manner adds one edge and one vertex. Thus, $\chi = v - e$ should still be 1.

Theorem: Let T be a tree. Then $\chi(T) = 1$.

All trees have the same Euler characteristic. This, of course, does not mean that all trees are the same, but it does tell something about the shape. The nice thing about trees is that there is only one way to get from one vertex to another. In a graph, the presence of loops or circuits provides more than one way to connect two vertices.

▶ **Exercise 1.** Compute χ, the Euler characteristic of the graphs below:

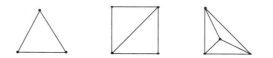

The Euler characteristic for a graph is related to the number of loops or circuits in the graph. One way to count circuits is to remove edges, as few as possible, from the graph until a tree is left.

▷ **Exercise 2.** For the graphs of Exercise 1, count the number of circuits for each graph. Figure out a formula for χ, the Euler characteristic of a graph, in terms of the number of loops and circuits.

The Euler characteristic does not give a precise description of a graph but does give valuable information about the shape.

In graph theory, the particular drawing of a graph or tree doesn't matter, but only the relationship among the vertices and edges. For example, the drawings below are all representatives of the same tree:

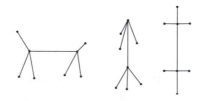

▶ **Exercise 3.** Decide which of the drawings below represent the same trees:

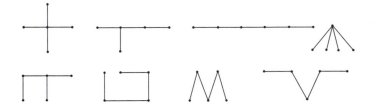

▷ **Exercise 4.** Below are the numbers of combinatorially different trees for each number of vertices. Draw these trees.

Numbers of Vertices and Trees

Number of vertices	1	2	3	4	5	6	7	8
Number of trees	1	1	1	2	3	6	11	23

In Section 12.1, we were looking for Euler circuits in order to solve problems such as the Königsberg bridge problem. Such Euler circuits are ways to cover every edge of a graph while returning to the starting point. What good are they? Ask your local postal carrier or trash collector. If roads are edges and intersections are vertices, you would like to travel every road exactly once before returning to the main office. Notice that if the post office allows mailboxes on both sides of the road, each road must be traveled twice (two edges), so every vertex must have even degree. This means that there would always be an Euler circuit for any map. (Of course, if you have two mail trucks, things get more complicated. . . .)

Now we are looking at trees, which are in some sense the opposite of Euler circuits. They have no circuits of any kind. So again we ask, what good are they? One of the most common applications of trees is in the area of statistics. If you are trying to make a sequence of decisions and determine the probability of each outcome, it helps to draw a tree. (Should I go to class, go to breakfast, or sleep in? If I go to class, I have an 80% chance of passing the exam on Friday. If I go to breakfast, there is a 50% chance I will see someone who went to class. Then I can get the notes and maybe pass. If I sleep in, there is a 5% chance the professor will ask only things I knew before I came to college . . . . What to do? This problem can be expressed with a tree diagram.) However, we prefer to find examples where the geometry is more obvious. In an earlier paragraph, we talked about edges of a graph being roads in a town. The lengths of the roads did not matter, only the way they met at the intersections. But suppose you are the one building the roads. You would like to build as few as possible and to make them as short as possible. Trees can help you build as few as possible. Steiner trees can help minimize the lengths. We start with a ridiculously simple example, but that makes it easier to see what is really important.

Suppose there are three towns—one at each corner of an equilateral triangle. Your job is to build roads to connect the towns.

▷ **Exercise 5.** Why is it a waste of material to connect each town to the other as in the picture below?

Notice that any circuit will give you more than one way to get from one point to another. Hence, we are looking for trees. When we wanted to get rid of a circuit in Exercise 2, we simply deleted an edge from a graph. If we do this to the graph above, we get the new road plan below:

This is great for the people living in town B, but it is not so great for those who live in A and work in C. They have to travel twice as far to get to work as they did under the original plan! Consequently, every town board is going to swamp you (the builder) with reasons why they should be the central town in your system. Is there some way to build roads that is more efficient in terms of both material used and personal convenience?

▷ **Exercise 6.** Using a software package such as Geometer's Sketchpad, put three points at the corners of an equilateral triangle. Put a fourth point somewhere inside the triangle as shown on the following page. Form a line segment connecting each of the first three points to the fourth point and calculate the sum of the lengths of the three line segments. Move your fourth point around until the total length is minimized. What do you notice?

A

B C

▶ **Exercise 7.** Repeat the previous exercise using several nonequilateral but acute triangles. What does the "road structure" look like around the central point in each exercise?

One simple way of arriving at the solutions to problems like these is to use the area-minimizing tendencies of soap films. We make use of the scheme suggested by Kappraff using two sheets of Plexiglas™ (or glass) and thumbtacks. A small dab of glue will help hold the thumbtacks in place temporarily. The slight flexibility of Plexiglas™ helps when you use more than three thumbtacks. If you do use glass, be very careful when holding the edges. Mix up some soap and glycerin solution as in Section 11.5. Martin Gardner suggests using Plexiglas™, which is safer than glass, but he also drills holes and makes more permanent structures.

▷ **Exercise 8.** Using two sheets of glass or Plexiglas™, three thumb tacks, and the soap solution, place the three tacks on one sheet at the corners of an equilateral triangle and cover with the second sheet. Dip the whole apparatus into the soap solution. Sketch the resulting soap film. Compare your sketch with the figure that minimized total length in Exercise 6.

▷ **Exercise 9.** Using either the thumbtacks and soap solution or the software, experiment with several obtuse triangles. [Some triangles will have soap film only between the thumbtack vertices. In others the soap film will form edges to a new vertex.] What can you say about the angles of the triangles that need a point other than one of the triangle vertices to minimize the "road distance" between the towns? How does this compare to the information you found in Section 11.5 when considering the angles between double bubbles?

For towns arranged in a triangle, you will need to add at most one more point to find a distance-minimizing tree. If you try to add a fifth vertex, you will either end up adding circuits to your plan (which we know are a waste of material), or you will add a vertex in the middle of an edge where it adds nothing new, as seen in the figures below:

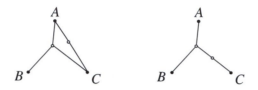

Suppose you have four towns instead of three. First, we will assume that the towns are at the corners of a square.

▶ **Exercise 10.** Which of the road plans below will use less material? Can you think of a better plan?

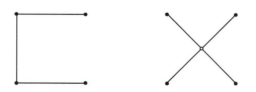

▷ **Exercise 11.** Build a model using the glass sheets and four tacks. You should see a soap film with two additional intersections inside the square. (If you don't have access to the soap solution, use Geometer's Sketchpad or similar software to put two vertices inside a square and move them around until you have minimized the total distance as in Exercise 6.) Explain why a second interior vertex is useful inside a square when it was a waste in the triangular setup.

▷ **Exercise 12.** How many planes meet at each interior vertex of the minimal configuration for the square model? What are the angles separating these planes? Make a conjecture about the angles around any vertex added to minimize distance between fixed points.

What we have called interior or additional vertices are commonly called *Steiner vertices* after Jakob Steiner, a Swiss geometer in the nineteenth century who first studied the problem of minimizing the lengths of trees connecting a finite number of points. While it is fairly easy to study what happens for a small number of points, the difficulty grows exponentially as you add more points. We will stop at four.

▷ **Exercise 13.** Model the situation created by having four towns at random points in the plane. Discuss the Steiner trees you find to minimize the lengths of the trees connecting your towns. What angles between your towns give one or two Steiner vertices? Recall the acute versus certain obtuse triangular setups above. Can you find more than one Steiner tree for some setups? Here are some layouts to try:

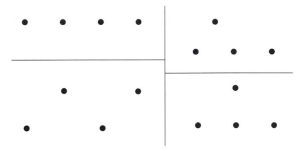

▷ **Exercise 14.** Consider the town placement below. Two Steiner trees can span this setup, but one is shorter than the other. See whether you can form the longer tree using the soap solution. You should be able to transform the longer tree to the shorter, more efficient one by blowing gently parallel to the branch connecting the two Steiner vertices.

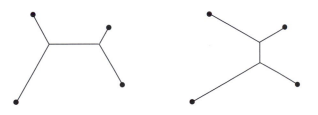

SOFTWARE

1. *Geometer's Sketchpad* can be used to model Steiner trees. It is available from Key Curriculum Press.

SUGGESTED READINGS

N.L. Biggs, E.K. Lloyd, and R.J. Wilson, *Graph Theory: 1736–1936*, Clarendon Press, Oxford, 1986.

Gary Chartrand, *Introductory Graph Theory*, Dover, New York, 1977.

Martin Gardner, "Mathematical Games: Casting a net on a checkerboard and other puzzles of the forest," *Scientific American* 254, June 1986.

Jay Kappraff, *Connections: The Geometric Bridge Between Art and Science*, McGraw-Hill, New York, 1991.

Richard J. Trudeau, *Introduction to Graph Theory*, Dover, New York, 1993.

12. Graph Theory

◆ 12.3. MAZES

SUPPLIES
graph paper

The simplest and by far the most common mazes are based on trees. Here is one simple way to draw such a maze: Outline a grid (squares are easiest, though you can use triangles or hexagons); then walk through, knocking out walls, being careful never to return to a cell where you have already been (and so avoiding forming loops or circuits). You can start by carving out the solution, and then fill in the rest of the maze, as demonstrated below:

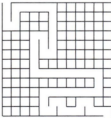 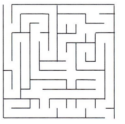

▷ **Exercise 1.** Trace all the paths in the maze above and show that a tree is formed.

▷ **Exercise 2.** Draw a different maze using a 10 by 10 grid of squares.

A *unicursal maze*, or *labyrinth*, is a maze with only one path:

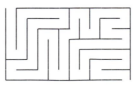

▷ **Exercise 3.** Draw a different labyrinth.

One of the earliest known mazes, known as the Cretan maze, is easily constructed: Begin with a cross and a dot in each of the four sectors. Connect each end of the cross to the nearest dot counterclockwise of the endpoint, after looping around one of the other dots, being careful never to cross paths already drawn. The first arc loops clockwise then counterclockwise, the second counterclockwise then clockwise, the third clockwise then counterclockwise, etc.

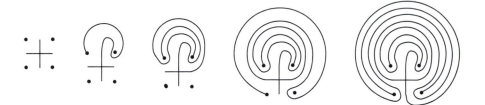

Another way of analyzing the Cretan maze above is by means of the following diagram, taken from Lietzmann. Connect the ends marked "1" etc., remembering always to keep the entrance between the two ends marked "8" open.

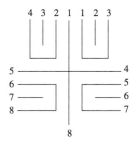

▶ **Exercise 4.** Figure out how to number the following diagram to make a Cretan maze. Draw the maze.

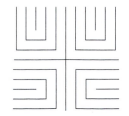

▷ **Exercise 5.** Draw a sixfold Cretan maze. Start with the drawing below:

One way to solve a maze that is based on a tree is to fill in all the dead-ends: Shade them back to the last branching point. Repeat until only the solution is left.

This method is maze-centered: It is useful if you have a picture of a maze, but not very good if you are actually lost in the maze.

▷ **Exercise 6.** Solve the maze below by shading the dead-ends.

▷ **Exercise 7.** Solve the maze below by shading the dead-ends.

Another way to solve a maze focuses on the person in the maze: Place your hand on the right-hand wall as you enter the maze (or the left, it does not matter as long as you're consistent), and follow along, never removing your hand. This rule will often lead you into a dead-end, and then back out. While this algorithm works for many mazes, it may not find the shortest path through the maze. Here is the maze of Exercise 6 solved using the right-hand rule. The dotted line shows the path taken through the maze.

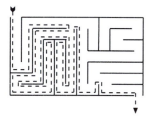

▶ **Exercise 8.** Solve the maze below by the right-hand rule.

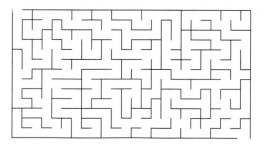

▷ **Exercise 9.** Solve the maze below by the right-hand rule.

Mazes whose paths form a tree, and which can consequently be solved by the right-hand rule, are called *simply connected*. If a maze is not simply connected, the right-hand rule may not work. For example, here is the simply connected maze of Exercise 6 on the left, the set of all paths outlined by a dotted tree in the center, and finally this tree formed by the paths redrawn in simplified form:

▷ **Exercise 10.** Draw the tree formed by the paths of the maze of Exercise 8. Simplify as much as possible. Mark the entrance and the exit on the tree.

Hedge mazes are common in England. The object of such mazes is usually to get to a central courtyard. One of the most famous is the one at Hampton Court.

▶ **Exercise 11.** Below is a schematic drawing of the maze at Hampton court. Draw the graph formed by the paths of the maze.

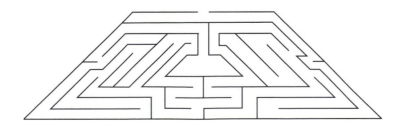

▷ **Exercise 12.** Use the right-hand rule to solve the Hampton Court maze. Then, using a differently colored pencil, apply the left-hand rule. Can you find a connection between these patterns and the graph you found in Exercise 11?

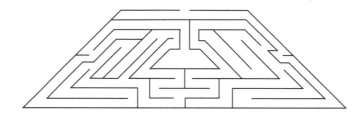

In a maze whose paths form a graph with loops or circuits, the right-hand rule will sometimes fail, leading you around in circles and never reaching the goal.

▷ **Exercise 13.** Show that the right-hand rule does not solve the maze below. Note that in this maze the object is to get to the mark ⊕.

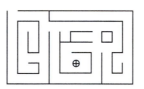

If a maze is not simply connected, so that the paths do not form a tree, you may need to use a different maze solution algorithm. There are a number of such algorithms, but we will only explain one of the most common: Tremaux's algorithm. This rule focuses on the person in the maze, but also requires that you leave markers at certain branch points of the maze, so fill your pockets with rocks before you set off. Or leave chalk marks on the paving. Bread crumbs are not advisable.

Tremaux's Algorithm

Remember that the paths of any maze form the edges of a graph, with vertices at the junctions where several paths meet and at the ends of the dead-ends. For example, on the next page are a maze, the graph of paths, and the same graph simplified. The entrance is marked with a star, and the goal of the maze with a ⊕. The small dots are the vertices of the graph:

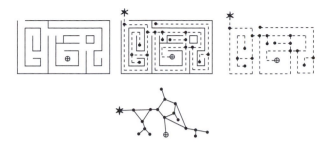

Tremaux's Algorithm

1. Every time you enter a junction, leave a rock indicating the path you just came from.
2. Every time you leave a junction, leave a rock indicating the path you are entering.
3. When you come to a junction you have never visited before, choose any path (except the one you came in on).
4. If you come to a dead end, return to the last vertex. You don't have to leave a rock at a dead end.
5. If you are on a path that you have not been on before and reach a vertex where you have already left a rock, double back to the last vertex. (You will leave two rocks on the path where you come and go, so you know not to go that way again.)
6. If you are on a path that you have been on before and reach a vertex where you've already left a rock, choose a new path if possible. If there are no untraveled paths at that vertex, choose the one that you have only traveled once.
7. If you have followed the rules above, you should never travel on any path more than twice.
8. After you have reached the goal of the maze, the paths marked exactly once will give you a direct path back to the entrance.

To see why Tremaux's algorithm works, note that it is clear that as long as you do not get stuck in a dead-end or start walking in circles around a circuit in the graph, you will eventually reach the goal. Thus, we must check that Tremaux's algorithm will work for those cases.

Dead-Ends

Consider the dead-end on the next page. The first picture shows the person approaching the vertex. In the second picture, the person leaves the junction, leaving two rocks: one for the path she came in on and one for the present path. On discovering that it was a dead-end, she returns to the last vertex visited, following Rule 4 and leaving another rock at the entrance to the dead-end path. Since that path now has two rocks, it has been thoroughly explored and should not be traveled again. In the last picture, Rule 6 dictates that the only remaining untraveled path must be chosen.

Loops and Circuits

Explore the results of Tremaux's algorithm on loops and circuits in the next group of exercises.

▶ **Exercise 14.** Work out Tremaux's algorithm for the following graph with a loop. Consider all three choices of which path to take the first time you leave the branch point.

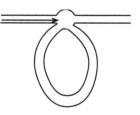

▷ **Exercise 15.** Work out Tremaux's algorithm for the following graph with a circuit. Consider all possible choices.

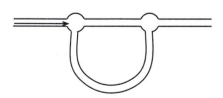

▷ **Exercise 16.** Work out the next move according to Tremaux's algorithm for the following scenarios. Note that some of these scenarios have more than one option. In these cases, you may give any one allowable move. The dots represent rocks left earlier, and the × represents your current position.

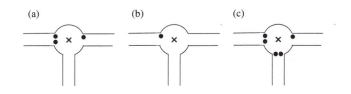

▷ **Exercise 17.** Below is a schematic of the maze at Chevening, England, built in the 1820's. Solve the maze using Tremaux's algorithm.

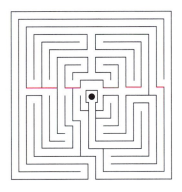

▷ **Exercise 18.** Draw the graph for the Chevening maze.

▷ **Exercise 19.** Below is the Philadelphia maze by puzzle maker Henry Dudeney. Solve the maze.

SUGGESTED READINGS

W.W. Rouse Ball and H.S.M. Coxeter, *Mathematical Recreations and Essays*, Dover, New York, 1987.

H.E. Dudeney, *Amusements in Mathematics*, Dover, New York, 1970.

W. Lietzmann, *Visual Topology*, American Elsevier, New York, 1965.

W.H. Matthews, *Mazes and Labyrinths*, Dover, New York, 1970.

Jearl Walker, "The Amateur Scientist: Methods for going through a maze without becoming lost or confused," *Scientific American* 259, December 1988.

Jearl Walker, "The Amateur Scientist: Mirrors make a maze so bewildering that the explorer must rely on a map," *Scientific American* 254, June 1986.

13. Topology

◆ 13.1. DIMENSION

SUPPLIES
string
paper
pennies

What is topology? You may think it has something to do with maps, because of the word topography, but that is because both words are derived from the Greek τοποσ, which means space. Topology is a relatively new field of mathematics and is related to geometry. In both of these subjects one studies the shape of things. We can begin to explore the difference between geometry and topology with an excerpt from Jerome K. Jerome's *Three Men in a Boat*:

> . . . when George drew out a tin of pineapple from the bottom of the hamper
> . . . we felt that life was worth living after all . . . there was no tin-opener to be
> found . . . I took the tin off myself and hammered at it till I was worn out and
> sick at heart, whereupon Harris took it in hand. We beat it out flat; we beat it
> back square; we battered it into every form known to geometry—but we could
> not make a hole in it. Then George went at it, and knocked it into a shape, so
> strange, so weird, so unearthly in its wild hideousness, that he got frightened.

In geometry, one characterizes, for example, a can of pineapple by its height, radius, surface area, and volume. In topology, one tries to identify the more subtle property that makes it impossible to get the pineapple out of the tin, no matter what shape it is battered into, as long as one does not puncture the can.

We wish to study the elusive properties not detected by geometry. Since length and angle measure are thoroughly covered in geometry, we ignore these factors. Any two line segments, even of different lengths, are considered to be topologically equal, since by stretching, one could be turned into the other. All angles are equal, since one could be bent to form the other. Thus, a square is the same topological shape as a rectangle or any other quadrilateral. By straightening one of the angles, a rectangle is seen to be the same as a triangle. One is also allowed to flex or bend lines into squiggles or curves. Thus, one considers the examples in each row of the illustration below to be the same topologically, since they differ only in the geometric properties of length, angle measure, and curvature.

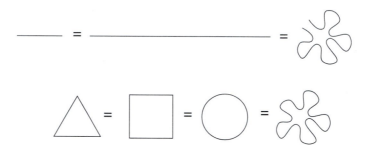

In topology, we concentrate on how a line segment and a circle differ. It seems obvious that the line and circle differ in some way not explicitly described by length or curvature or any other classical geometric property. Two objects are equal topologically if there is a deformation from one to the other. The bending and stretching that we allowed in the pictures above are examples of such deformations. Thus, topology is sometimes called "rubber-sheet geometry" since one pretends that everything is formed of extremely flexible rubber. A mathematical cliché is that a topologist cannot tell a doughnut from a coffee cup.

Two shapes are equal geometrically if they are congruent. Congruent figures will have the same geometric properties: length of corresponding sides, angle measure, area, volume, perimeter, and curvature. But two congruent triangles are not truly equal, since they consist of different points. Congruence says nothing about the positioning of the figures, or their color, or smell, or what they taste like. Congruence carries information only about the geometric properties. Different fields in mathematics usually require different ideas of equality, defined for the objects in which one is currently interested and the properties one wishes to preserve. We wish to define and study a notion of topological equality that will determine when objects are the same topologically, though they may differ geometrically or in other ways.

Two objects are equal topologically (the technical term is *homeomorphic*) if one can be deformed continuously into the other, and, furthermore, if the action of this deformation can be reversed continuously. An example of an operation that is not continuous is cutting a figure in two. The inverse of cutting is pasting, which reverses the cutting operation, and thus is also not a homeomorphism.

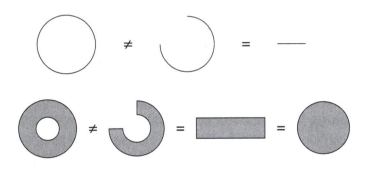

Instead of finding a way to tell whether two shapes are the same topologically, it is often easier to tell how they are different. We search for *topological invariants*: properties or quantities that will stay the same no matter how we stretch or deform a figure. In several sections of this book, particularly in the discussion of *Flatland* in Section 6.1, we have referred to the dimension in which some action is taking place. However, we have not defined the concept of dimension, and the dimension of an object does turn out to be a topological invariant. Most people have an intuitive notion of dimension, but these ideas are often misleading. For example, most people will agree that a straight line is one-dimensional:

But suppose the line is bent? Is it still one-dimensional, or should it be considered as two-dimensional? It is sitting in two-dimensional space, but what is the dimension of the curve itself?

What if there is a knot in the line, as pictured below? Now it exists in three-dimensional space. The illustration cheats, but we have learned to read such diagrams, with the convention that the gaps in the curve represent places where the curve passes under itself. Use your string to make a copy of the knot illustrated and contemplate that for a bit. In spite of its existence in three-dimensional space, in many ways it is still just a line (or a bit of string).

Mathematicians distinguish between the dimension of an object and the dimension of the space in which it sits. Yet even with this distinction, there are many ways to define the dimension of an object, which may give contradictory answers. Alexandre Dranishnikov found a space that has finite "cohomological dimension" but infinite "covering dimension." Mathematicians even study spaces with fractional dimension: $2\frac{1}{2}$-dimensional spaces may seem odd to you, but quite natural to chaos theorists. In this section, we will discuss the basic concept of dimension and try to find two definitions or ways of thinking about dimension: covering dimension and embedding dimension.

In addition to the problem of trying to define dimension, there is the problem of comprehending more than three dimensions, since we seem to live in three-dimensional space.

Let us go back to the straight line. If you travel along the line, you can go only forward and backward. We think of backward as the reverse (or in some sense the negative) of forward, so there is really only one choice of direction, up to plus or minus sign. This leads to the idea of a line being a one-dimensional object. Even if the line is bent or knotted, you still only have one way to move if you are to stay on the line. Traveling on a line is like traveling by monorail or ski lift: you can go only forward or backward. There are only two possible destinations: the beginning of the line and the end.

Next, consider a plane or a flat sheet of paper. You can imagine a line on your plane, and on that line you can move forward and backward. However, you can also turn perpendicular to the line and move left or right without leaving the plane. If you combine these two directions, you can move diagonally, in curves or any way you want to move in your plane. This is one way to convince yourself that a plane is two-dimensional. Travel on a plane is like traveling by foot: you can go anywhere you like on the surface.

To move perpendicular to both your original line in the plane (forward and backward) and the line perpendicular to it (which lets you go left and right), you have to move up or down off the plane, which takes you into three-dimensional space. Three perpendicular directions, three dimensions. Travel in three dimensions is like traveling by spaceship. You can go up, down, sideways, forward, backwards; wherever you like until you run into a solid object or run out of fuel.

People sometimes comment that time is our fourth dimension. We can see the validity of this notion by looking at it in lower dimensions.

▶ **Exercise 1.** Plot a point on a piece of paper. Now, think of slowly moving the point across the paper one inch to the right. What shape do you get if you sketch on the paper all the positions the point was in during the move? What is the dimension of this shape?

▷ **Exercise 2.** Take your shape from the previous exercise. Move it one inch down the paper. What shape do you get if you sketch all the positions it was in during the move? What is the dimension of this new shape?

▷ **Exercise 3.** Take your shape from the previous exercise. Since there is not another perpendicular direction on the paper, think of moving your object one inch up off the paper. What shape would you get if you could sketch all the positions your object was in during the move? What is the dimension of this shape?

The idea is to copy your original object in a new direction and then connect each point of the original to its double without crossing any part of either copy. Try to imagine the fourth dimension by repeating the process, though you cannot avoid overlapping unless you let time be your new direction.

Covering Dimension

One of the ways mathematicians have formalized the idea of dimension is with the definition of covering dimension. The next few exercises will help you explore this idea.

You will need string, paper, pennies (and your imagination when we get to higher levels). The idea of *covering dimension* involves covering up the object under study by spots. Pennies make convenient and inexpensive spots, so we'll use them. (Another great source of spots is the office hole punch but these are harder to keep in place.) Note that while our spots are a fixed size (pennies in this case), the definition should work for arbitrarily small spots.

Otherwise, you can "cover up" any problem points by using a big enough (or small enough) spot.

▶ **Exercise 4.** Draw a finite collection of points spaced well apart. Now cover your collection of points with spots (i.e., the pennies) so that you have the fewest possible overlaps. What is the thickest intersection or overlapping of pennies you have?

▷ **Exercise 5.** Lay out a section of string with as many bends as you want but do not let it cross over itself. Cover your string with spots so that you have the fewest possible overlaps. (Note that putting two pennies edge to edge will always leave a tiny gap between them. The pennies must overlap to truly cover the string.) What is the thickest intersection of pennies you have?

▶ **Exercise 6.** Cut out any shape from a flat sheet of paper. Cover your shape with spots so that you have the fewest possible overlaps. What is the thickest intersection of pennies you have?

▷ **Exercise 7.** Think about three-dimensional objects. You will need to imagine little spherical balls instead of pennies to do the covering. How many balls will you need to overlap to fill up three-dimensional objects?

▷ **Exercise 8.** Conjecture a definition of covering dimension.

Remember our first example of this section:

When we first bent the line, it was still a line but it sat in two-dimensional space.

▷ **Exercise 9.** Using your definition of covering dimension, convince yourself that the bent line is actually a one-dimensional object.

Exercise 9 leads us to think that the covering dimension of an object is a topological invariant: a property that stays the same no matter how the line is distorted, as long as we don't tear it. Of course, we have shown this only for the simplest of objects, a line, and the simplest kind of deformation, but that is enough to show the general idea. Covering dimension is an example of an *intrinsic property*: something that remains the same no matter how you distort the object and move it around in the surrounding space.

Embedding Dimension

Next, let us discuss the space surrounding our objects. The standard one-dimensional Euclidean space is an infinite straight line, usually pictured as the x-axis. The standard two-dimensional Euclidean space is the plane, and the standard three-dimensional space is the familiar space we have lived in all of our lives. When we bent the line segment above, we had to take it out of one-dimensional space into a two-dimensional plane represented by the sheet of flat paper forming the page. The bent line is said to be *embedded* in the plane. An embedding of an object is a way of sticking the object into the surrounding space, without gluing any of the points together. The *embedding dimension* of an object is the smallest-dimensional standard space an object can live in. This is an *extrinsic property:* one that depends not only on the object itself but on precisely how it is related to the surrounding space. The bent line segment, while it is drawn in the plane, could be straightened out and put back in a one-dimensional space without hurting it. In contrast, a circle has covering dimension 1 (try covering it with pennies), but embedding dimension 2, since the smallest of the standard spaces it will fit in is the plane. The circle is a one-dimensional object that cannot live in one-dimensional space.

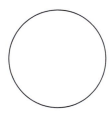

When we knotted the line, it was still a line but it sat in three-dimensional space.

▷ **Exercise 10.** Using your definition of covering dimension, convince yourself that the knotted line is still a one-dimensional object.

Above, we discussed straightening the bent line segment and putting it back into one-dimensional space. In trying to find the embedding dimension of the knotted line segment above, note that we could untie the knot and put it back into one-dimensional space.

▶ **Exercise 11.** Find an object with covering dimension 1 that cannot be fit back into one- or two-dimensional space, so that it will have an embedding dimension of 3. [Note: You cannot use the example that immediately follows this. There are much simpler ones.]

It is possible to find a knotted line that cannot be straightened or untied. The Fox-Artin curve is an example of a line that can be embedded only in three-dimensional space: you cannot get to either end in order to begin untying it!

▷ **Exercise 12.** A point has 0 dimensions. Find a zero-dimensional object that requires one-dimensional space to exist. [Hint: We never said an object had to be connected.]

You can find objects with covering dimension 0 that fit into two- or three-dimensional space, but they can always be squashed down into one-dimensional space without losing any of the points, and so objects such as the one pictured below have embedding dimension 1.

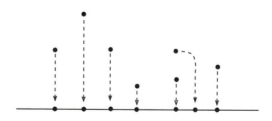

A plane is a two-dimensional object. If we bend it, we can make it sit in three-dimensional space.

▷ **Exercise 13.** Using your definition of covering dimension, convince yourself that the bent plane is actually a two-dimensional object.

Since we could straighten the plane and put it back into two-dimensional space, it has embedding dimension 2.

▷ **Exercise 14.** Find an object with covering dimension 2 that cannot be fit back into two-dimensional space.

Sometimes topologists allow us to cut objects apart as long as we promise to put the cut edges back together just the way we found them. This operation allows changes in the embedding (the way the object sits in space) but not in the object itself. Consider a rubber band as a short two-dimensional cylinder. If you cut the band, twist it twice, and then glue the edges back together, any one living in the space (like A Square in Flatland) wouldn't notice the difference. The embedding has changed, but the object is intrinsically the same.

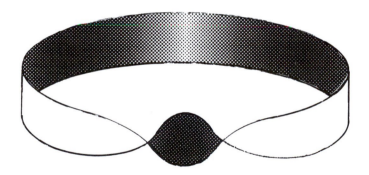

Notice that a zero-dimensional object can require up to 1 dimension to be properly embedded, by Exercise 12. A one-dimensional object can require up to 3 dimensions to be embedded, as in Exercise 11 and the example of the Fox-Artin curve. In Section 13.2 you will see that some two-dimensional objects require more than three dimensions to exist. In fact, some two-dimensional objects require five dimensions to exist, but none requires more than five.

If an object has covering dimension n, you will need at least n-dimensional space to represent it: A line cannot fit into zero-dimensional space without squashing points together, and a plane cannot fit on a line without squashing.

▷ **Exercise 15.** From the above examples, what is the dimension of the space you think you would need to guarantee the fit of an n-dimensional object?

SUGGESTED READINGS

A.N. Dranishnikov, "On a Problem of P.S. Alexandroff," *Matematicheskii Sbornik* 135, 1988.

A.N. Dranishnikov, "On a Problem of P.S. Alexandroff," translated by D. Dimovskii, *Mathematics of the U.S.S.R.-Sbornik* 63, 1989.

James R. Munkres, *Topology: A First Course*, Prentice Hall, Englewood Cliffs, 1975.

C.P. Rourke and B.J. Sanderson, *Introduction to Piece-wise Linear Topology*, Springer-Verlag, New York, 1982.

13. Topology

◆ 13.2. SURFACES

In this section, we wish to discuss *surfaces*: topological objects with covering dimension 2. An object is a surface if every point has a disc-like region (not including the boundary circle) surrounding it. For example, a sphere (to be understood as the skin only of a three-dimensional ball) is a surface. We designate the sphere by $\mathbb{S}$. Each point on the sphere is surrounded by a disc-like region, slightly curved, but to a topologist's eye, a disc none the less.

The surface of the earth forms a sphere, and for much of the history of mankind it was thought to be flat. Pretend you are living on a large-scale surface. If it is very foggy (or if you are quite nearsighted), and you grope around, the surface should feel just like a disk. The curvature would be almost completely undetectable.

On the other hand, a balloon on a string is not considered to be a surface. It has disc-like regions surrounding every point, except where the string is attached. At that point, the surrounding region is a disc with a whisker sticking out:

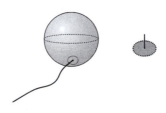

Another surface is the *torus*, denoted $\mathbb{T}$, shaped like a bagel or doughnut (but, again, only the skin: perhaps an inner tube is a better image). Again, if it were sufficiently foggy so that you could see only the region very near you, your immediate neighborhood would feel like a disc.

▷ **Exercise 1.** What topological shape does a doughnut hole have?

One way of joining two surfaces to get a new surface is called the *connected sum* operation. Take two surfaces S and T and cut a small disc from each. This leaves two punctures with circular boundaries. Join the surfaces together by gluing the two circles to get a new surface denoted by $S\#T$. Below is illustrated this process for joining two tori to form a two-holed torus, which is thus $\mathbb{T}\#\mathbb{T}$. To make the process clearer, we have stretched the punctures out to form little necks that join to form a tube connecting the tori.

The result is topologically the same as a torus with two holes:

Similarly we can construct a three-holed torus, or a torus with any number of holes:

▶ **Exercise 2.** Which of the shapes described above do you get when you form the connected sum of two spheres, $\mathbb{S}\#\mathbb{S}$?

▷ **Exercise 3.** Which of the shapes described above do you get when you form the connected sum of a sphere and a torus, $\mathbb{S}\#\mathbb{T}$?

▷ **Exercise 4.** Classify each of the pictures below as an *n*-holed torus, i.e., figure out how many holes it has.

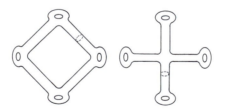

▶ **Exercise 5.** Below is the framework of a cube, built of pipes to form a surface. Classify this space as an *n*-holed torus.

A cylinder is not a surface as we have defined one, since there are points (on the edges) that are not surrounded by disc-like regions. If one were groping around in a fog, one would think one had found the end of the world: a place where the surface ends. The sphere and the torus have no edges. The cylinder is called a *surface with boundary*. Surfaces with boundary have the property that each point either has a disc-like region surrounding it, in which case the point is called an *interior point*, or is on the edge of a region which looks like a disc cut in half, in which case it is called a *boundary point*. The boundary of the cylinder is two circles, one at the top and another at the bottom.

Another surface with boundary can be made by cutting a circular puncture in a sphere. This hole can then be stretched out and the punctured sphere laid out flat to form a disc:

▷ **Exercise 6.** What does a sphere with two punctures form?

▷ **Exercise 7.** How would a topologist view a pair of pants (as a sphere or torus with how many punctures)?

To build a cylinder, take a strip of paper, and tape the edges together, lining up the edges marked with the arrows so that the direction of the arrows match up:

Another surface with boundary can also be made from a strip of paper by taping a pair of edges together, but with the arrows on the edges to be taped pointing in opposite directions. Matching the arrows requires twisting the strip of paper before taping the edges together. This is called a *Möbius band*.

▷ **Exercise 8.** Build a cylinder and a Möbius band as above.

▶ **Exercise 9.** What is the boundary of the Möbius band?

It is easy to imagine a cylinder colored red on the outside and blue on the inside.

▷ **Exercise 10.** Try to color the outside surface of a Möbius band. What happens?

The Möbius band is a *one-sided surface*. Imagine a Flatland inhabitant living embedded in a Möbius band. If he were to travel around the Möbius band, he would come back reversed:

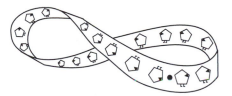

Such one-sided surfaces are called *nonorientable*, since in traveling around such a space one loses all sense of direction or orientation. The two-sided figures, such as the sphere, the torus, and the cylinder, are *orientable*. Every nonorientable surface will contain a Möbius band embedded in it somewhere.

▷ **Exercise 11.** Cut a cylinder in half along the circle that runs around the middle halfway between the upper and lower boundaries. What happens?

▷ **Exercise 12.** Cut a Möbius band along the circle that runs around the middle. What happens?

▶ **Exercise 13.** Cut a Möbius band in thirds around the middle. What happens?

▷ **Exercise 14.** Cut a Möbius band in fourths around the middle. What happens?

Just as a cylinder can be made from a strip of paper with the ends taped together, a torus can be made by taking a tube (or long cylinder) and taping the ends together. Thus one could start with a square of some flexible fabric, tape one pair of opposite edges (marked with a single arrow) together to form a cylinder, then tape the two ends of the cylinder (marked with a double arrow) together to form a torus:

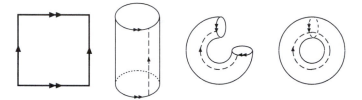

The picture above on the far left is called the *planar diagram* for the torus. (Many video game characters used to live on this diagram. If they went off the left side of the screen, they would magically reappear on the right side.) It provides us with another way to view the torus. You can think of the planar diagram as a set of instructions for building the torus. The edges with their arrows tell us precisely how to assemble the square to make the torus. The planar diagram has the advantage that no points are hidden from view, as they are in the two-dimensional perspective drawing. The disadvantage to the planar diagram is that it doesn't give you much idea what the assembled figure will look like.

Just as the torus was built above by taping the ends of a cylinder together, another surface could be formed by doing the same thing to a Möbius band. In the planar diagram, we will start with a square and mark one pair of edges with arrows pointing in opposite directions to form a Möbius band. The other pair of opposite edges we will mark with parallel arrows. If we glue those edges together first, it will start out just like building a torus. At the next step however, one has trouble matching up the arrows on the ends of the cylindrical tube.

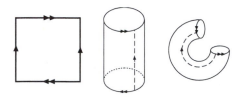

So, once again we appeal to the fourth dimension to pass one end of the tube through its own wall and then tape the ends of the tube together, to form a new surface called the *Klein bottle*, denoted by $\mathbb{K}$, with covering dimension 2 but embedding dimension 4.

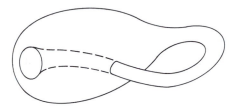

▷ **Exercise 15.** Is the Klein bottle orientable or nonorientable?

▷ **Exercise 16.** What happens when you form the connected sum of a sphere and a Klein bottle, $\mathbb{S} \# \mathbb{K}$?

▷ **Exercise 17.** Which surface do you get when you tape the sides of a square together as in the following diagram:

One last surface without boundary that we have not mentioned is called the *projective plane*, denoted by $\mathbb{P}$. It is obtained by taping together the edges of a square as shown:

▷ **Exercise 18.** Show that the projective plane is nonorientable.

The projective plane is, like the Klein bottle, impossible to construct in three-dimensional space. To assemble the planar diagram, we first pull the edges up. Note that in order the match up the edges, we must form them into a figure eight:

In order to glue the last pair of edges together, we pass through the wall of the surface, appealing to the fourth dimension. The twisted arrangement at the top is called a *crosscap*.

The Rev. Charles Dodgson, better known as his *nom de plum* Lewis Carroll and the author of *Alice in Wonderland*, was a mathematician at Christ Church, Oxford University. In another of his fantasies *Sylvie and Bruno Concluded*, he includes instructions for making a projective plane from three handkerchiefs, which he calls Fortunatus's Purse, in a dialogue between Lady Muriel and Mein Herr:

"You shall first," said Mein Herr, possessing himself of two of the handkerchiefs, spreading one upon the other, and holding them up by two corners, "you shall first join together these upper corners, the right to the right, the left to the left; and the opening between them shall be the *mouth* of the purse."

A very few stitches sufficed to carry out *this* direction. "Now, if I sew the other three edges together," she suggested, "the bag is complete?"

"Not so, Miladi: the *lower* edges shall *first* be joined – ah, not so!" (as she was beginning to sew them together). "Turn one of them over, and join the *right* lower corner of the one to the *left* lower corner of the other, and sew the lower edges together in what you would call *the wrong way*."

"*I* see!" said Lady Muriel, as she deftly executed the order. "And a very twisted, uncomfortable, uncanny-looking bag it makes! But the *moral* is a lovely one. Unlimited wealth can only be attained by doing things in *the wrong way*! And how are we to join up these mysterious – no, I mean *this* mysterious opening?" (twisting the thing round and round with a puzzled air). "Yes, it *is* one opening. I thought it was *two*, at first."

"You have seen the puzzle of the Paper Ring [the Möbius band]?" Mein Herr said, addressing the Earl. "Where you take a slip of paper, and join its ends together, first twisting one, so as to join the *upper* corner of *one* end to the *lower* corner of the *other*?"

"I saw one made, only yesterday," the Earl replied. "Muriel, my child, were you not making one, to amuse those children you had to tea?"

"Yes, I know that Puzzle," said Lady Muriel. "The Ring has only *one* surface, and only *one* edge. It's very mysterious!"

"The *bag* is just like that, isn't it?" I suggested. "Is not the *outer* surface of one side of it continuous with the *inner* surface of the other side?"

"So it is!" she exclaimed. "Only it *isn't* a bag, just yet. How shall we fill up this opening, Mein Herr?"

"Thus!" said the old man impressively, taking the bag from her, and rising to his feet in the excitement of the explanation. "The edge of the opening consists of *four* handkerchief-edges, and you can trace it continuously, round and round the opening: down the right edge of *one* handkerchief, up the left edge of the *other*, and then down the left edge of the *one*, and up the right edge of the *other*!"

"So you can!" Lady Muriel murmured thoughtfully, leaning her head on her hand, and earnestly watching the old man. "And that *proves* it to be only *one* opening!". . .

"Now, this *third* handkerchief," Mein Herr proceeded, "has *also* four edges, which you can trace continuously round and round: all you need do is to join its four edges to the four edges of the opening. The Purse is then complete, and its outer surface—"

"*I* see!" Lady Muriel eagerly interrupted. "Its *outer* surface will be continuous with its *inner* surface! But it will take time. I'll sew it up after tea."

▷ **Exercise 19.** Using three squares of cloth and following Carroll's instructions, try to make a Fortunatus's Purse.

SOFTWARE

1. Jeffrey Weeks has written a shareware program, Torus Chess, for Macintosh computers that allows one to play chess on a torus or a Klein bottle. It is available from www.northnet.org/weeks/TorusGames.

SUGGESTED READINGS

Lewis Carroll, *The Complete Sylvie and Bruno*, Mercury House, San Francisco, 1991.
David W. Farmer and Theodore B. Stanford, *Knots and Surfaces*, AMS, Providence, 1996.

P. A. Firby and C.F. Gardiner, *Surface Topology*, Ellis Horwood Ltd., Chichester, 1982.

Vagn Lundsgaard Hansen, *Geometry in Nature*, A.K. Peters, Wellesley, 1993.

L. Christine Kinsey, *Topology of Surfaces*, Springer-Verlag, New York, 1993.

W. Lietzmann, *Visual Topology*, American Elsevier, New York, 1965.

V. V. Prasolov, *Intuitive Topology*, AMS, Providence, 1995.

Jeffrey R. Weeks, *The Shape of Space*, Marcel Dekker, New York, 1985.

13. Topology

◆ 13.3. MORE ABOUT SURFACES

Just as congruent figures, equal geometrically, will have equal geometric properties, such as equal areas, equal angle measures, and equal curvature, so will topologically equal figures share topological properties. One such topological property is the covering dimension, discussed in Section 13.1. Another example is orientability, as described in Section 13.2. Yet another example of such a property is *connectedness*. Connectedness means exactly what one would think it would: A space is connected if it is all of one piece. The figure eight on the left below is connected, but the two circles on the right are not. Therefore, these figures cannot be the same topologically.

Connectedness and the number of pieces a space has helps identify the critical difference between a line and a circle: removing a single point from the line makes it fall into two pieces, but removing a point from a circle leaves it in one piece.

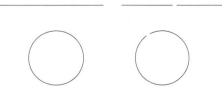

▶ **Exercise 1.** Explain why the following figures are not equal topologically. [Hint: You may have to remove more than one point.]

▷ **Exercise 2.** Classify up to topological equality, i.e., sort into topological types:

ABCDEFGHIJKLMNOPQRSTUVWXYZ0123456789

Yet another topological property is that of enclosing a cavity or hollow space inside, separated by the surface from the exterior space. This is one of the essential differences between a sphere and a disc.

▷ **Exercise 3.** Does the torus contain a cavity?

▷ **Exercise 4.** Does the Klein bottle contain a cavity?

Just as we used the removal of points to determine whether two line figures were the same topologically, we can remove loops or circles drawn on the surface to determine whether two surfaces are the same topologically. Note that in doing this we are removing only the circle itself, not the disc surrounded by the circle.

▶ **Exercise 5.** What happens when you remove a circle from a sphere? Is the result connected or disconnected? Does it matter which circle you choose?

▶ **Exercise 6.** What happens when you remove a circle from a torus? Is the result connected or disconnected? Does it matter which circle you choose?

The *connectivity* of a surface, $C(S)$, is defined to be the number such that one can find $C(S) - 1$ circles on the surface S whose removal would leave S connected, but the deletion of one more (for a total of $C(S)$ circles) makes the surface fall apart. It matters very much which circles you choose. In the illustration below on the left, removing the indicated circle makes the torus fall into two pieces, but the removal of the circle shown on the right does not disconnect the surface.

Thus, if a surface has $C(S) = 3$, then you can find two circles on the surface S such that the removal of these two circles leaves S in one piece. We want this one piece to be such that it can be laid out flat on the plane and it should contain no punctures. Ideally, removing the circles will leave the planar diagram of the surface.

▷ **Exercise 7.** Find $C(\mathbb{T})$, the connectivity of a torus.

▶ **Exercise 8.** Below are illustrated a 2-holed torus and a 3-holed torus. Find the connectivity of each.

▷ **Exercise 9.** Find the connectivity of a torus with n holes.

▷ **Exercise 10.** Find $C(\mathbb{K})$, the connectivity of a Klein bottle.

Since the only thing we know about the projective plane $\mathbb{P}$ is its planar diagram, we must find a way to compute the connectivity from that. Below is the diagram, with the edges marked a and b. To build the projective plane, we would have to glue the two edges marked a together, respecting the direction of the arrows, and repeat for the edges marked b.

We wish to find a circle on $\mathbb{P}$, remove it, and see whether the result is connected or not. Consider the path on $\mathbb{P}$ that starts at vertex P, travels along edge a and then along b. At the end of b, we reach another vertex. This vertex is also the beginning point of edge a, so after assembling the projective plane, this will get glued to point P. In other words, this path forms a circle on $\mathbb{P}$ after assembly. If we remove this path, we would be left with the insides of the square, which is connected.

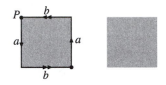

One more circle would disconnect the square, so the projective plane has connectivity $C(\mathbb{P}) = 2$.

In order to define the connectivity of surfaces with boundary, such as the cylinder and the Möbius band, we must allow other types of cuts besides circles. Remember that our objective is to cut the surface open so that it can be laid out flat, without any holes or punctures. For example, if one cuts a cylinder along the middle, it is divided into two smaller cylinders. While this cut has certainly divided the cylinder into two pieces, the pieces do not satisfy the condition of not having any holes.

If we slit the cylinder up the side wall, we get a rectangle, which can easily be laid flat and which has no holes. An additional cut slices the rectangle into two pieces.

The connectivity of the cylinder is 2 since the first cut up the side wall does not disconnect it, but any further cut will. For surfaces with boundary, we use boundary cuts: lines or curves that connect two points on the boundary and that do not intersect themselves. The first cut for a surface with boundary should be a boundary cut; further cuts may be either boundary cuts or circle cuts.

▶ **Exercise 11.** What is the connectivity of a Möbius band?

▷ **Exercise 12.** What is the connectivity of a disc with three punctures?

▷ **Exercise 13.** What is the connectivity of a disc with four punctures?

▷ **Exercise 14.** What is the connectivity of a disc with *n* punctures?

▶ **Exercise 15.** What is the connectivity of a pullover sweater?

▷ **Exercise 16.** What is the connectivity of a torus with one puncture?

This way of cutting a surface open to find its connectivity can also be considered, by the addition of vertices at appropriate points, as introducing edges, vertices, and faces for the surface, as we did for polyhedra in Sections

7.1 and 7.2. For example, consider the cylinder with the boundary cut used above. We introduce vertices where the boundary cut meets the top and bottom circles. Thus we have a structure with 2 vertices, three edges (two forming the top and bottom and one running up the side), and one face. Recall the *Euler characteristic*:

$$\chi(S) = v - e + f$$

where v is the number of vertices, e the number of edges, and f the number of faces. Be careful that all edges begin and end at a vertex (though they are allowed to begin and end at the same vertex) and that all faces are topologically equal to polygons or discs. Thus the Euler characteristic of the cylinder is $\chi(C) = 2 - 3 + 1 = 0$.

▶ **Exercise 17.** Compute the Euler characteristic for the Möbius band.

The Euler characteristic for the torus can be computed either from the planar diagram or from the three-dimensional figure.

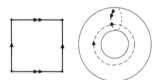

In the planar diagram, it looks as if there are four vertices, but when we glue all the edges together to build the torus, these are all stuck together to make one vertex on the torus. Thus, $v = 1$. It looks as if there are 4 edges, too, but remember that the side edges marked with a single arrow will get glued together and the top and bottom edges, marked with a double arrow, will get glued together to make another edge, so there are really only two edges, and $e = 2$. There is obviously only one face, so $f = 1$. Thus, $\chi(\mathbb{T}) = 1 - 2 + 1 = 0$.

▷ **Exercise 18.** Compute the Euler characteristic for the Klein bottle $\mathbb{K}$.

▷ **Exercise 19.** Compute the Euler characteristic for the sphere $\mathbb{S}$.

▶ **Exercise 20.** Compute the Euler characteristic for the 2-holed torus, $\mathbb{T}\#\mathbb{T}$.

▷ **Exercise 21.** Compute the Euler characteristic for the 3-holed torus, $\mathbb{T}\#\mathbb{T}\#\mathbb{T}$.

▷ **Exercise 22.** Fill in the table below.

Properties of Surfaces

Surface	Orientable	Cavity	Boundary	Connectivity	χ
$\mathbb{S}$	yes	yes	none	1	2
$\mathbb{T}$					
$\mathbb{T}\#\mathbb{T}$					
$\mathbb{T}\#\mathbb{T}\#\mathbb{T}$					
$\mathbb{P}$					
$\mathbb{K}$					
Cylinder					
Möbius band					

The Euler characteristic is an important topological invariant. It would seem to depend on the particular edges, vertices, and faces that the surface is divided up into, but it can be proved (though the proof is too advanced for this book) that any choice of these will give the same Euler characteristic. This seems an astounding result, since obviously it would be easy to alter v, e, and f, but somehow this alternating sum remains the same for the surface. This allows us sometimes to decide when two shapes are not equal topologically: if $\chi(S) \neq \chi(T)$, then S and T cannot be the same topologically. On the other hand, it does not always give conclusive answers. Above, we found that the cylinder and the torus both have Euler characteristic equal to 0. You should have found some other surfaces such that $\chi(S) = \chi(T)$, even though we know from other topological properties such as orientability or boundary that they are not equal.

In Sections 7.1 and 7.2 we found many different polyhedra that all have $\chi = 2$. A look at those polyhedra will show that they are all topologically equivalent to the sphere. All of those convex polyhedra gave different ways of dividing $\mathbb{S}$ into vertices, edges, and faces, but since they were all different structures on the sphere, they all gave the same Euler characteristic.

A surface with boundary S can be considered as obtained from some vertex-edge-face structure on a surface S^* (without boundary) with some of the faces removed, but the vertices and edges left intact. Alternatively, one can consider S^* as the surface S with all the punctures patched by circular discs.

▷ **Exercise 23.** Give a formula for $\chi(S)$ in terms of $\chi(S^*)$ and h, the number of punctures in S.

While no one of the topological properties we have studied is enough to tell the difference between two surfaces, a combination is:

Classification Theorem for Surfaces: Two finite surfaces are topologically equivalent if and only if they have the same number of boundary components, are both orientable or both nonorientable, and have the same Euler characteristic.

One of the greatest unsolved problems in mathematics is the corresponding question for the three-dimensional analogues of surfaces. This question is called the *Poincaré Conjecture*, and so far even the most sophisticated techniques have not been sufficient to distinguish among these. Oddly, enough, spaces of higher dimensions have proved more amenable.

SUGGESTED READINGS

P.A. Firby and C.F. Gardiner, *Surface Topology*, Ellis Horwood Ltd., Chichester, 1982.
D. Hilbert and S. Cohn-Vossen, *Geometry and the Imagination*, Chelsea Publishing Co., New York, 1990.
L. Christine Kinsey, *Topology of Surfaces*, Springer-Verlag, New York, 1993.
W. Lietzmann, *Visual Topology*, American Elsevier, New York, 1965.
Jeffrey R. Weeks, *The Shape of Space*, Marcel Dekker, New York, 1985.

13. Topology

◆ **13.4. MAP COLORING PROBLEMS**

SUPPLIES
    colored pencils (at least 7 different colors)
    bagel

In this section, we discuss an application of the Euler characteristic: map-coloring problems. The object is to find the minimum number of colors needed to color all possible maps (assume that you have been condemned to repeat fourth grade on a limited budget). The first thing to note is that the states can be regarded as polygons, the borders with other states as edges, with vertices placed wherever three or more states meet.

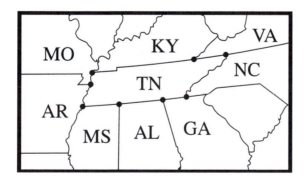

For example, in the map above, Tennessee has eight edges and eight vertices. States that share a border or edge must be colored differently, but one is allowed to color states that meet only at a vertex the same color, as in a checkerboard. We do not allow nonconnected states.

The Four-Color Conjecture. Four colors are enough to color all possible maps drawn on the plane.

The conjecture that only four colors are needed to color any map drawn on the plane was made by Francis Guthrie in 1852 while, oddly enough, coloring a map. The first proof was published in 1879 by Kempe, but this contained an error found in 1890 by Heawood, who salvaged enough to prove that five colors are sufficient to color any map drawn on the plane. The conjecture was finally confirmed in 1976 by Appel and Haken, after a century of false proofs and refinements of techniques. There are, of course, maps on the plane that do not require four colors (imagine flooding all the continents and painting the map solid blue), but the problem we are discussing is how many crayons one must have on hand to be sure of being able to color any map at all and be able to make adjoining countries different colors.

▷ **Exercise 1.** Color the map below with two colors so that countries that share a border are colored differently.

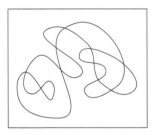

▷ **Exercise 2.** Draw another map (with at least 6 countries) that requires only two colors.

▶ **Exercise 3.** Color the map below with three colors so that countries that share a border are colored differently. Explain why at least three colors are necessary.

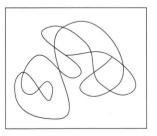

▷ **Exercise 4.** Color the map at the top of the next page with three colors so that countries that share a border are colored differently. Explain why at least three colors are necessary.

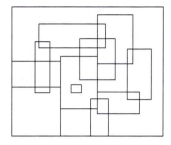

▷ **Exercise 5.** Draw another map (with at least 6 countries) that requires three colors.

▷ **Exercise 6.** Color the map below with four colors so that countries that share a border are colored differently. Explain why at least four colors are necessary.

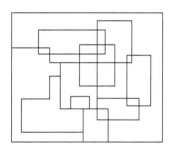

▷ **Exercise 7.** Draw another map (with at least 6 countries) that requires four colors.

The problem of coloring a map on the plane is equivalent to the problem on the sphere. If a map on a globe requires N colors, one can use any of the mapping projections covered in Section 11.2 to obtain a map on the plane with the same coloring scheme. Conversely, if all possible maps on the plane require at most N colors, then given any particular map, we can reverse the projection to get a map on the globe.

We can also consider maps drawn on any of the surfaces we have studied, which may be useful if we are ever called on to draw a map of some as-yet undiscovered toroidal planet. In this section we allow only polygonal countries, and all countries must be connected (so Hawaii must be considered as a separate country).

Let S be a surface with a polygonal map M drawn on S. Let f be the number of faces or countries in the map, with e the number of edges and v the number of vertices. Think about the map as a jigsaw puzzle for a minute.

The jigsaw puzzle has f pieces, each shaped like a polygon. If you think about the puzzle pieces before you assemble the puzzle, each n-sided polygon has n edges and n vertices. When putting the puzzle together, these edges are matched together in pairs. Thus, there are $2e$ edges in the pile of polygons before the puzzle is put together. The number of vertices in the disassembled puzzle depends on how many vertices each face has and also on how many get stuck together in putting the puzzle together. At a point on the map where three countries meet, three separate puzzle piece vertices join to form one map vertex. Since at least three countries meet at each vertex, the number of vertices in the pile of polygons is at least $3v$.

▶ **Exercise 8.** Consider the puzzle below. Compare the number of puzzle pieces and f, the number of polygons in the assembled map. Counting only the interior edges (indicated by heavy lines) and not those making up the outside of the square, both in the puzzle and in the puzzle pieces, compare the number of interior edges in the separated puzzle with e, the number of interior edges in the assembled map. Verify that $2e$ gives the number of interior edges for the separated pieces. Compare the number of interior vertices (indicated by dots) in the separated puzzle with v, the number of interior vertices in the assembled map. Verify that the number of interior vertices for the separated pieces is at least $3v$.

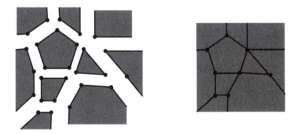

In any polygon, the number of edges is equal to the number of vertices, and the edges of these polygons are matched in pairs in assembling the map, so

$$3v \leq (\text{number of vertices before assembly}) = 2e.$$

Using the Euler characteristic of S we get

$$\chi(S) = v - e + f,$$
$$\chi(S) - f = v - e \leq \frac{2e}{3} - e = -\frac{1}{3}e,$$
$$3(f - \chi(S)) \geq e,$$

$$6(f - \chi(S)) \geq 2e,$$
$$\frac{6(f - \chi(S))}{f} \geq \frac{2e}{f}.$$

Of course, e and f depend on the specific map M drawn on the surface S. The quantity $\frac{2e}{f}$ in the equation above has a particular significance. Remember that $2e$ is the number of edges in the pile of disassembled puzzle pieces, and f the number of pieces in M. Thus, $\frac{2e}{f}$ is the total number of edges divided by the total number of polygons, so it gives the *average number of edges per polygon*. Therefore, from the inequality above, we can get

$(\star)$ $$\frac{2e}{f} \leq 6\left(1 - \frac{\chi(S)}{f}\right).$$

As is the nature of averages, this means that the map M must have at least one country with *no more than* $\frac{2e}{f}$ edges.

▷ **Exercise 9.** Use the inequality $\star$ and the fact that $\chi(\mathbb{S}) = 2$ to show that any map drawn on the sphere must have at least one country with no more than 5 edges and hence no more than 5 neighbors.

▷ **Exercise 10.** Use the inequality $\star$ and the fact that $\chi(\mathbb{T}) = 0$ to show that any map drawn on the torus must have at least one country with no more than 6 neighbors.

We next prove that if $\frac{2e}{f} < N$ for *all possible maps* M that can be drawn on S, then you will never need more than N crayons to color these maps. Note that this is obviously true for maps M where $f < N$, since if there are fewer polygons than colors, one can paint each country a different color.

We are going to use a mathematical technique called *induction*: We show that the fact is true at some starting point, and then show that whenever we can color a map with k countries then we can color a map with $k + 1$ countries. For example, if we had 6 crayons and we know that $\frac{2e}{f} < 6$ for all maps on the sphere, as we learned using the inequality above, then we can obviously color any map on the sphere with 6 countries. The induction argument shows that if we can color 6 countries, then we can color 7 countries, and then if we can color 7 countries, we can color 8, etc.

Thus, we assume the statement that any map with k countries can be colored with our N crayons is true. We must show that any map M with $(k + 1)$ faces can also be colored with these N colors. Since $\frac{2e}{f} = \frac{2e}{(k + 1)} < N$ for the map M, the average number of edges per polygon is less than N, so there is a country in M with fewer than N edges. Designate this particular polygon by P. We can eliminate P by shrinking the polygon to a point by a process called *radial projection*.

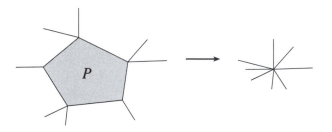

Thus, a new map M' is formed with one fewer country than M, so M' has k faces. Since we are assuming that the result is true for any map with k faces, M' can be colored with only N colors. The coloring on M' gives a coloring on all of M except for the country P. The face P was chosen because it had fewer than N neighbors, so at least one of the N colors was not used in the countries surrounding P. We can color the face P with this unused color to get a coloring of M using only N colors. We summarize what we have proved:

Theorem: If we have N crayons and $6\left(1 - \frac{x(S)}{f}\right) < N$, then we can color all possible maps drawn on the surface of S.

For the sphere, $\chi(S) = 2$, so $\frac{2e}{f} < 6$. Therefore, no more than six colors are needed to color any map on the sphere (and thus on the plane) or the projective plane. That at least four colors are needed to color all maps in the plane is shown by the picture below.

Thus, we need at least four crayons to color an arbitrary map on the sphere, and no more than six. This is called the Six-color Theorem. Heawood proved in 1890 that this could be improved to show that five colors are enough for the sphere, and finally in 1976 the Four-color Theorem was proved: that four colors were enough and that all were needed. Martin Gardner has written a short story about the breakdown of a topologist while visiting, with an anthropologist friend, an island inhabited by 5 primitive tribes each of which shares a border with all of the other tribes.

For the torus or Klein bottle, $\chi(S) = 0$, so $\frac{2e}{f} \leq 6 < 7$, so it follows that seven colors are enough to color any map on the torus or the Klein bottle. However, we must ask whether all of these colors are necessary. Might we make do with fewer? That all seven are necessary is shown by the map below with exactly seven countries, each of which requires a different color. Remember that the sides marked with a single arrow are to be glued together, and the sides marked with a double arrow also get glued together. Thus, the four regions marked "1" will get glued together to form one connected rectangular area on the torus. This gives the Seven-color Theorem for the torus:

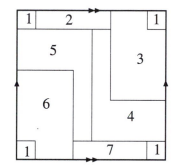

▷ **Exercise 11.** Draw and color the map shown above on the surface of a bagel. The surface of a doughnut is too crumbly. Try to get a bagel, preferably plain or egg, with a reasonably sized hole in it. You will find that markers work better than colored pencils on the surface of a bagel. Show that every country shares a border with each of the other countries, so that no two can be the same color.

▶ **Exercise 12.** Show that the map below gives another coloring on the torus, which also requires seven colors.

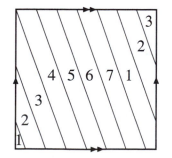

▷ **Exercise 13.** Draw a map on the Klein bottle that requires at least six colors. (There are many such.)

We can answer the question of how many colors are needed for maps on the projective plane similarly: Since $\chi(\mathbb{P}) = 1$, we know that $\frac{2e}{f} < 6$. Thus, no more than six crayons are necessary. A Möbius band can be considered as a projective plane with a hole in it. Thus if we can find a map on the Möbius band that requires 6 colors, then the same map could be drawn on the projective plane once the hole is patched. This is the Six-color Theorem for the Möbius band and projective plane. Similar results can be shown for other surfaces.

▷ **Exercise 14.** Draw the following map on a strip of paper. Tape the ends together with a twist to form a Möbius band. Show that each country is connected and that all six colors are necessary.

	4		5	
1	2	3		1
5		6		

▶ **Exercise 15.** Color the map below with as few colors as possible, noting that one country is not connected and has 2 disjoint pieces, which are already shaded. All of the other countries are connected. How many colors are necessary? Does this contradict the Four-color Theorem?

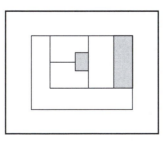

In the April 1 issue of *Scientific American* in 1975, Martin Gardner, in his regular *Mathematical Games* column, wrote a spoof citing major break-throughs in mathematics and science, such as the claim that π can be written as a square root and that Leonardo da Vinci invented the flush toilet. He published a map that claimed to be a counterexample to the Four-color Conjecture. A number of newspapers took his April Fool's column literally and reported these breakthroughs seriously.

▷ **Exercise 16.** Color Gardner's map with four colors.

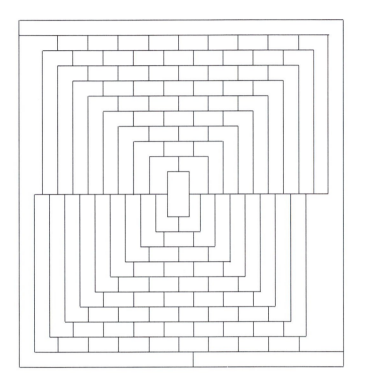

SUGGESTED READINGS

W.W. Rouse Ball and H.S.M. Coxeter, *Mathematical Recreations and Essays*, Dover, New York, 1987.

N.L. Biggs, E.K. Lloyd, and R.J. Wilson, *Graph Theory: 1736–1936*, Clarendon Press, Oxford, 1986.

H.S.M. Coxeter, *Introduction to Geometry*, Wiley & Sons, New York, 1969.

Martin Gardner, "The Island of Five Colors," in C. Fadiman, *Fantasia Mathematica*, Springer-Verlag, New York, 1958.

Martin Gardner, "Mathematical Games: Six sensational discoveries that somehow or another have escaped public attention," *Scientific American* 232, April 1975.

L. Christine Kinsey, *Topology of Surfaces*, Springer-Verlag, New York, 1993.

Hints and Solutions
to Selected Problems

♦ **1.1. MEASUREMENT**

▶ **7.** If a right triangle has legs of lengths 5 and 12, the hypotenuse is $\sqrt{5^2 + 12^2} = \sqrt{169} = 13$ long.

▶ **10. a.** 26 **d.** $30 + 5\pi \approx 45.71$ **f.** $14 + \frac{5\pi}{2} \approx 21.85$

▶ **14. b.** 24 **e.** 25 **h.** $100 + 12.5\pi \approx 139.27$
j. $11 + 6.25\pi \approx 30.63$

▶ **16.** The peak is 25 feet from the attic floor. The rafters must be $1 + 25\sqrt{2} \approx 36.36$ feet. The area of the roof is $2 \cdot 75 \cdot (1 + 25\sqrt{2}) \approx 5453.3$ square feet.

♦ **1.2. POLYGONS**

▶ **1.** The sizes of the tangram pieces are figured out by using the way they fit together and the Pythagorean theorem: the square is 1 on each side; the smallest triangles have legs 1 and hypotenuse $\sqrt{2}$; the parallelogram has sides of lengths 1 and $\sqrt{2}$; the middle-sized triangle has legs $\sqrt{2}$ and hypotenuse 2; the big triangles have legs 2 and hypotenuse $2\sqrt{2}$.

▶ **3.** Many of these have more than one solution, but here is one for each of the first four:

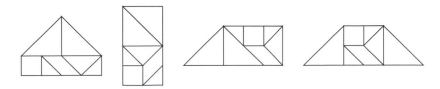

449

▶ **5.** 540°

▶ **6.** 720°

▶ **9.** 24

▶ **12.** interior angle = 45°, vertex angle = 135°

◆ 2.1. BILLIARDS

▶ **2.**

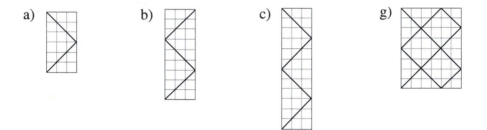

a)   b)   c)   g)

▶ **3.** Answers are not unique.   **a.** 2 by 1   **e.** 3 by 2   **h.** 14 by 6

▶ **4.** Hint: You know that a 6 by 3 table has the same pattern as 4 by 2 and 2 by 1 tables. Consider the smallest tables.

▶ **6.** Hint: Look at the reduced dimensions.

▶ **7. b.** rotation   **e.** vertical reflection   **g.** horizontal reflection

▶ **9.** Hint: How would this table compare with a 10 by 5 table?

▶ **13.** Hint: You can handle integer and fractional lengths. What other kinds of numbers are there?

▶ **14.**   ▶ **15.**   ▶ **17. a,d.**

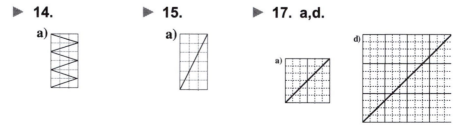

a)   a)   d)

a)

▶ **18. a.** $6\sqrt{2}$ **d.** $12\sqrt{2}$

▶ **22.** Below are pictured the shots where the cue ball hits the west wall first then the east which we can refer to as west-east, followed by east-west, and north-east. There are a total of 12 combinations of 2 walls, but only 8 possible shots. Unfolding trajectories will show why the others are impossible.

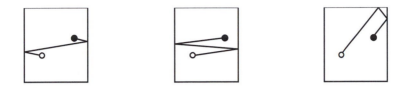

◆ **2.2. CELTIC KNOTS**

▶ **1.** ▶ **4.** ▶ **7.**

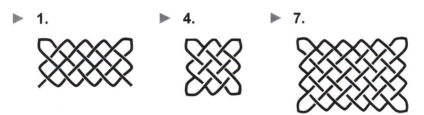

▶ **8.** A celtic knot will have loose ends if either of the dimensions is odd. Only when both dimensions are even are there gray dots in each of the four corners, where the knot can turn around.

Celtic Knots

Reference	Dimensions	Closed	Loose ends
Text	4×6	Yes	
Text	5×6	No	Lower left and lower right
Text	3×7	No	Lower left and upper right
Exercise 1	5×10	No	Lower left and lower right
Exercise 2	3×9	No	Lower left and upper right
Exercise 3	6×9	No	Upper right and lower right
Exercise 4	6×6	Yes	
Exercise 5	6×8	Yes	
Exercise 6	4×12	Yes	
Exercise 7	8×12	Yes	

▶ **11.** Hint: If the knot is closed, both dimensions are even. Consider the numbers for the closed knots in the table from Exercise 8 above. The number of strands will be half of some quantity that depends on the dimensions of the knot. For knots with loose ends, it's a bit more difficult. Use the same quantity as for closed knots and divide by 2 as before, and then add $\frac{1}{2}$.

▶ **12.**   ▶ **15.**

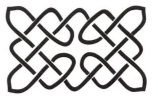

▶ **17.** Here are a few examples. Use these ideas to find a total of 20.

◆ 3.1. RULER AND COMPASS CONSTRUCTIONS

▶ **3.** (1) Draw a circle centered at C and intersecting the line AB at two points, D and E.
(2) Draw circles of equal radius centered at D and E, which meet each other at F below the line AB.
(3) The line CF, meeting AB at G, is perpendicular to AB.

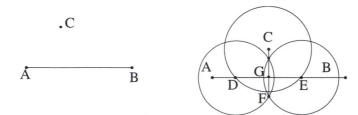

▶ **5.** Given line *AB* and point *C*, draw a line from *C* to an arbitrary point *D* on *AB*. Construct angle ∡*ECD* to be equal to ∡*BDC*, using R & C Construction 5.

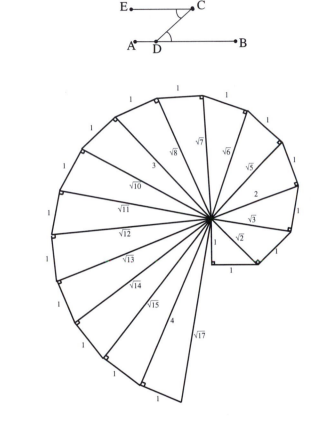

▶ **9.**

▶ **11.** Given the base *AB*, draw circles centered at *A* and *B*, both with radius *AB*. They will intersect at point *C* and △*ABC* is equilateral.

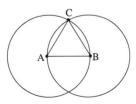

▶ **13.** The easiest way to construct a square is to draw two perpendicular lines *AB* and *CD* as shown. Draw a circle centered where these lines intersect. The four points where the lines *AB* and *CD* intersect the circle are the four corners of a square.

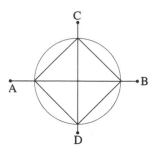

▶ **15.** Hint: Bisect the right angles forming the diagonals of a square. Now draw the octagon.

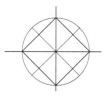

◆ 3.2. THE PENTAGON AND THE GOLDEN RATIO

▶ **3. a.** $\phi^4 = 3\phi + 2$ **b.** $\phi^5 = 5\phi + 3$ **c.** $\phi^6 = 8\phi + 5$

▶ **4. a.** $\phi^{-3} = 2\phi - 3$ **b.** $\phi^{-4} = -3\phi + 5$ **c.** $\phi^{-5} = 5\phi - 8$

▶ **6.** Hint: Since $AB = BD = DE = EA = 1$, and F is the midpoint of AB, $FB = \frac{1}{2}$. Use the Pythagorean Theorem for the right triangle $\triangle FBD$.

▶ **7.** Perform R & C Construction 7, and then set your compass to the length of AC to find point G so that $AG = AC = BG$.

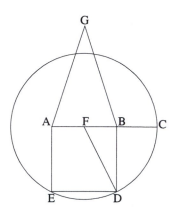

▶ **10.** There is, of course, no one correct answer. The golden rectangle is the fifth one in the top row.

◆ 3.3. THEORETICAL ORIGAMI

▶ **1.** Folding between two points forms a straight line.

▶ **3.** The fold bisects one of the two angles formed by the two lines.

▶ **5.** The fold line is shown as dashed:

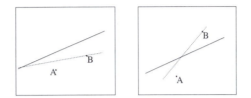

▶ **9.** (1) Let ⦮*ABC* be the given angle.
(2) Using Origami Postulate 4, fold so that line *AB* falls on top of line *CB*. The crease thus formed bisects the angle.

▶ **13.** Let *ABCD* be a square piece of origami paper. Fold it in half as shown below, then fold through *D* so that corner *C* lies on the mid-line. The dashed line from *D* to the point where *C* falls forms a 60° angle and thus one side of the equilateral triangle. Repeat for the other side.

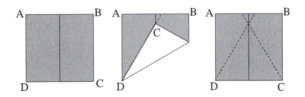

▶ **16.** (b) Hint: Use the Pythagorean Theorem.

▶ **20.** By the Pythagorean Theorem, $x^2 + (\frac{1}{2})^2 = (1 - x)^2 = 1 - 2x + x^2$. Thus, $BE = x = \frac{3}{8}$ and $CE = 1 - x = \frac{5}{8}$. The triangle $\triangle BEM$ has sides $\frac{3}{8}$-$\frac{1}{2}$-$\frac{5}{8}$, in proportion 3-4-5.

◆ 3.4. KNOTS AND STARS

▶ **4.** When you connect every third dot or every fifth dot you get a star with one strand as shown on the next page. These two figures are exactly the same in the end, though the points are connected in different orders. Every fourth dot gives four straight lines intersecting at the center. Every sixth dot gives the same picture as every other dot. Every seventh dot gives a regular octagon.

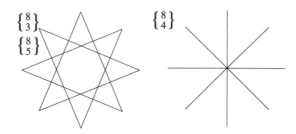

▶ **8.** Also see Exercise 10. Note that $\{^{12}_1\}$ and $\{^{12}_6\}$ are not shown.

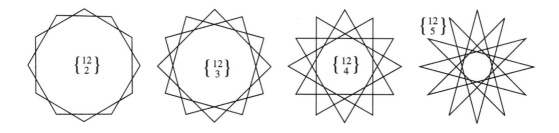

▶ **10.**

Star Polygons

$\{^n_k\}$	Revolutions	Regular/compound	Cycles	Length
$\{^8_2\}$	2	Compound	2	4
$\{^8_3\}$	3	Regular		
$\{^8_4\}$	4	Compound	4	2
$\{^9_2\}$	2	Regular		
$\{^9_3\}$	3	Compound	3	3
$\{^9_4\}$	4	Regular		
$\{^{12}_2\}$	2	Compound	2	6
$\{^{12}_3\}$	3	Compound	3	4
$\{^{12}_4\}$	4	Compound	4	3
$\{^{12}_5\}$	5	Regular		
$\{^{12}_6\}$	6	Compound	6	2
$\{^{15}_2\}$	2	Regular		
$\{^{15}_3\}$	3	Compound	3	5
$\{^{15}_4\}$	4	Regular		
$\{^{15}_5\}$	5	Compound	5	3
$\{^{15}_6\}$	6	Compound	3	5
$\{^{15}_7\}$	7	Regular		

▶ **16.** The vertex angle of the star polygon $\{\frac{8}{3}\}$ is 45°.

▶ **17.** The vertex angle of the star polygon $\{\frac{9}{3}\}$ is 100°, and the vertex angle for $\{\frac{9}{4}\}$ is 20°.

▶ **22.** There are two possibilities, depending on which 9-sided star polygon you choose to model the knot on.

◆ **3.5. LINKAGES**

▶ **1.** *B* draws a circle around *A* with radius *AB*.

▶ **3.** *D* draws a straight line, perpendicular to *AC*.

▶ **5.** Traditional ironing boards use a variable triangle linkage, as shown below:

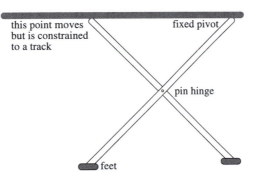

▶ **7.** The same figure scaled to be twice as large.

▶ **11.** The orientation is reversed and *BC* = 3*AB*, so the scaling factor is −3.

▶ **13.** The triangles △*AYB* and △*AXC* below are similar and since *AC* = 4*AB*, the scaling factor is 4.

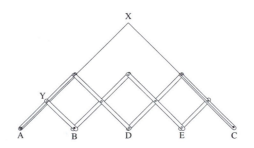

◆ 4.1. REGULAR AND SEMIREGULAR TILINGS

▶ **1.** The equilateral triangle, the square, and the regular hexagon can each tile the plane.

▶ **2.**

Vertex Angles of Regular Polygons

Polygon	Sides	Angle
Triangle	3	60°
Square	4	90°
Pentagon	5	108°
Hexagon	6	120°
Heptagon	7	128.57°
Octagon	8	135°
Nonagon	9	140°
Decagon	10	144°
Dodecagon	12	150°
Pentakaidecagon	15	156°
Octakaidecagon	18	160°
Icosagon	20	162°
Tetrakaicosagon	24	165°

▶ **4.** No, there cannot. The smallest sum of four different vertex angles is $60 + 90 + 108 + 120 = 378° > 360°$.

▶ **7.** 3.3.3.4.4 and 3.3.3.3.6

▶ **9. a.** 5.5.10 and 3.12.12   **b.** 4.6.12 and 4.5.20

▶ **11. b.** 3.12.12, 4.6.12 tile the plane.

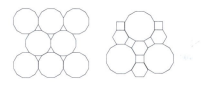

▶ **13. b.** 3.3.6.6 does not tile, but 3.6.3.6 tiles the plane.

▶ **17.** The line connecting the centers of two polygons will intersect the shared edge at the midpoint, forming a 90° angle.

▶ **20.** The dual tiling for 3.12.12 consists of 120°-30°-30° triangles.

▶ **22.** The dual tiling for 3.6.3.6 consists of quadrilaterals with angles 120°-60°-120°-60°.

◆ **4.2. IRREGULAR TILINGS**

▶ **1.** Two copies of any triangle can be put together to form a parallelogram, which then tiles the plane.

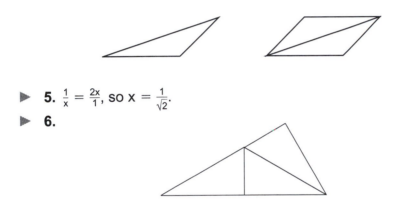

▶ **5.** $\frac{1}{x} = \frac{2x}{1}$, so $x = \frac{1}{\sqrt{2}}$.

▶ **6.**

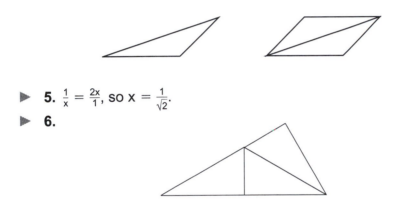

▶ **17.** The dog tiling is formed from a grid of equilateral triangles by side rotation on a pair of equal sides and midpoint rotation on the remaining side.

3 Fold dogs by Kevin Lee

▶ **20.**

▶ **24.**

◆ 4.3. PENROSE TILINGS

▶ **2.**

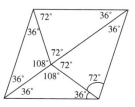

▶ **4.** A periodic tiling can be formed if you do not match up the dots.

▶ **6.** The answer is no. Explain why.

▶ **13.** At most 2 kite heads, and at most 5 kite tails can come together at any vertex.

▶ **16.**

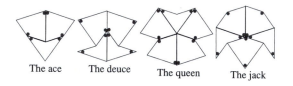

The ace   The deuce   The queen   The jack

◆ 5.1. KALEIDOSCOPES

▶ **6. a.** This figure has no lines of reflection.

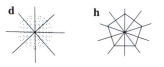

d   h

▶ **8.** The perceived distance between reflections of the dot is twice the distance from the dot to the mirror.

▶ **11.**

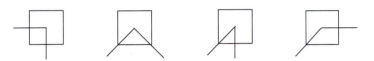

▶ **15.** For 60°, there are 6 lizards: 3 right-handed and 3 left-handed. For 72°, there are 5 lizards, but the number of right- and left-handed lizards varies with the viewing angle. Investigate for 90°, 108°, 120°, and 150°.

▶ **17.** For 60°, there are 6 arrows. For 72°, there are 5 arrows. Investigate for 90°, 108°, 120°, and 150°.

▶ **22.** The Coxeter triangles are 60°-60°-60°, 30°-60°-90°, and 45°-45°-90°.

◆ **5.2. ROSETTE GROUPS: POINT SYMMETRY**

▶ **1.** A 120° clockwise rotation of the triangle is the same as R^2, the 240° counterclockwise rotation.

▶ **4.** $FR = R^2F$, $FR^2 = RF$, and $RFR = F$

▶ **5. b.** $(RF \cdot RF) \cdot (RF \cdot RF) = 1$
 d. $R^{-2}FR^2 = RFR^2 = (RF) \cdot R^2 = R^2F$

▶ **9.** One such figure (there are infinitely many) is:

▶ **12.** One such figure is pictured below, with its symmetry group D_1:

D_1		
	1	F
1	1	F
F	F	1

▶ **19.** The figure on the left has symmetry group D_3, and the one on the right C_3. Give the multiplication table.

◆ 5.3. FRIEZE PATTERNS: LINE SYMMETRY

▶ **1.** This gives a translation of the original motif.

▶ **2.** This gives a rotation of the motif by 180° about a point on the center line.

▶ **6.** glide + reflect in center = translate, reflect perpendicular to center + reflect across center = rotate, rotate + reflect perpendicular to center = reflect across center, etc. Investigate the other combinations and show that there are no new patterns.

▶ **8. a.** $p1m1$ or jump   **e.** $pmg2$ or spinning sidle

▶ **9. c.** $pmm2$ or spinning jump **e.** $p1g1$ or step   **h.** $p111$ or hop

▶ **10. a.** $pm11$ or sidle **e.** $p1g1$ or step

▶ **13.** $pm11$ or sidle

▶ **16.** $p1g1$ or step

◆ 5.4. WALLPAPER PATTERNS: PLANE SYMMETRY

▶ **4.** Draw the perpendicular bisectors of the lines connecting corresponding points on the images. The point where these perpendicular bisectors intersect is the rotation center.

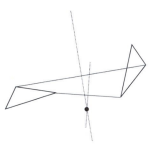

▶ **6.** Suppose that we had a 7-fold rotation point at point A. Let B be another 7-fold rotation point so that A and B are the closest two such points of all the 7-fold rotation centers. If B is rotated by

$\frac{360}{7} = 51.4°$ about A, then we have another 7-fold rotation point C with $AB = BC$, since the wallpaper pattern is symmetric by rotation about A. Similarly, if A is rotated 51.4° around B, we get another 7-fold rotation point at D with $DA = AB$. But then C and D are closer to each other than A and B were.

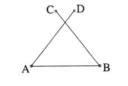

▶ **8.**

 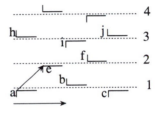

▶ **10. p1g1**

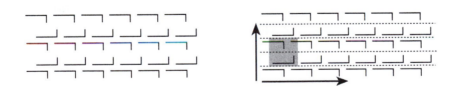

▶ **13. p2gg**

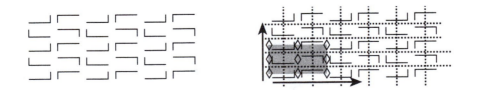

► **17. c1m1**

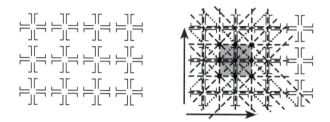

► **21. p4mm**

► **23. p3m1** is first, then **p31m**. Draw the lines of reflection and glide reflection and mark the centers of rotation.

► **26. a. p111   e. c1m1   i. p2mg   m. p4gm   q. p6**

◆ 5.5. ISLAMIC LATTICE PATTERNS

► **1.** The design has pattern **p4mm**, but the interlaced pattern is **p4**.

► **5.** The design has pattern **p6mm**, but the interlaced pattern is **p6**.

► **12.** Circles are arranged like the squares and octagons in the tiling 4.8.8.

► **13.**

◆ 6.1. FLATLANDS

▶ **4.** Hint: Nails will not work, since driving a nail into a line segment would break it in half.

▶ **7.** A 3 by 3 square moved 3 inches in a third direction will form a cube with volume 3^3.

▶ **11.** The square has 4 vertices representing the initial and terminal positions of the endpoints of the line segment, 4 edges, two of which represent the initial and terminal positions of the line segment and two of which are formed by the movement of the two endpoints of the line segment, and one two-dimensional face, formed by the movement of the line segment.

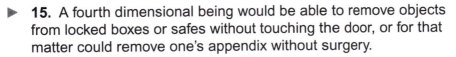

▶ **15.** A fourth dimensional being would be able to remove objects from locked boxes or safes without touching the door, or for that matter could remove one's appendix without surgery.

▶ **21.** Yendred spelled backwards is Derdney, or Dewdney (remember that Alice has a slight speech impediment).

▶ **22.** An Ardean would have to pinch the string.

▶ **26.** Drums and xylophones are quite easy to imagine. Try to think of others.

◆ 6.2. THE FOURTH DIMENSION

▶ **2. a.** The *xy*-plane cuts three-space in half.
  b. The three-dimensional space given by *x*, *y*, and *z* cuts four-dimensional space in half.

▶ **4. a.** A point.   **b.** The empty set.   **f.** A line.

▶ **8.** Unbolt the table top from the base, slip the rings on and replace the top. Alternatively, the rings might be previously cut in two, and then glued carefully together around the table base.

▶ **10.**

◆ **7.1. PYRAMIDS, PRISMS, AND ANTIPRISMS**

▶ **3.** Squares.

▶ **7.**

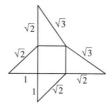

▶ **9.** A right regular hexagonal pyramid:

▶ **13.** $f = n + 1$, $e = 2n$, and $v = n + 1$

▶ **16.** A skew irregular hexagonal prism:

▶ **19.** Two congruent triangular prisms can be joined to form a right rectangular box, so $V = \frac{1}{2} lwh$.

◆ 7.2. THE PLATONIC SOLIDS

▶ **1. a.** tetrahedron: 4 faces   **b.** octahedron: 8 faces
c. icosahedron: 20 faces

▶ **5.** The tetrahedron is a triangular pyramid.

▶ **9.** The dual of the right square pyramid with equilateral sides is a smaller upside-down right square pyramid:

▶ **11.** This is the octahedron and so its dual is the cube.

▶ **15.** The Schlegel diagram of the tetrahedron:

▶ **17.** Square.

▶ **20.** Triangle.

▶ **23.** Euler's formula should give $\chi = 2$ for each of the Platonic solids.

◆ 7.3. ARCHIMEDEAN SOLIDS

▶ **1.** Every semiregular polyhedron must have at least three polygons meeting at each vertex so that it doesn't make a two-dimensional figure, but no more than five so that the angles will add up to less than 360°.

▶ **6.** At any vertex of the top *n*-gon, one has two triangles which share edges with the adjacent edges of the *n*-gon. A third triangle, with one edge shared with the bottom *n*-gon, meets these two triangles, as shown below:

▶ **8.** Icosidodecahedron, 3.5.3.5:

▶ **11.** The configurations 3.5.5 or 5.3.5 violate Rule 4. Since 3.n.n for n odd can be rewritten as n.3.n, it also violates Rule 4.

▶ **13.** Truncated cube, 3.8.8:

▶ **18.** The patterns 3.n.4.n for $n > 5$ will give an angle sum of at least $60° + 120° + 90° + 120° > 360°$, and 3.4.$n$.4 for $n > 5$ gives an angle sum of at least $60° + 90° + 120° + 90° \geq 360°$. Thus, neither can form a semiregular solid.

▶ **24.** Great rhombicuboctahedron, 4.6.8:

◆ 7.4. POLYHEDRAL TRANSFORMATIONS

▶ **3.** Halfway through the truncation process a truncated octahedron is formed. When the slices meet, a cuboctahedron is formed.

▶ **6.** For the truncated tetrahedron, each of the 4 vertices of the tetrahedron is replaced by three new vertices, so $v = 12$. Each vertex of the original tetrahedron is replaced by three new edges in addition to the 6 original ones, so $e = 18$. We have the 4 original faces plus 4 new ones replacing the original vertices, so $f = 8$. For the octahedron, the vertices of the truncated tetrahedron merge in pairs, so $v = 6$. The original edges of the tetrahedron disappear, so $e = 12$, but the octahedron has the same number of faces as the truncated tetrahedron so $f = 8$.

▶ **11.** The cuboctahedron is halfway between the octahedron and the cube, and the icosidodecahedron is halfway between the icosahedron and the dodecahedron.

▶ **13.** The great rhombicuboctahedron has $v = 48$, $e = 72$, and $f = 26$.

◆ 7.6. INFINITE POLYHEDRA

▶ **1.**

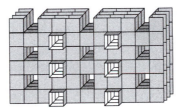

▶ **2.** This polyhedron can be denoted as 6.6.6.6. The spaces enclosed by the polyhedron and outside the polyhedron are identical.

▶ **14.**

▶ **16.** Build a multiple-layered infinite polyhedron from the tiling 4.8.8 by deleting half of the octagons and building tunnels alternately to the layer above and the layer below:

▶ **19.** To close-pack tetrahedra and octahedra, you will need two tetrahedra for every octahedron.

◆ 8.1. SYMMETRIES OF POLYHEDRA

▶ **1.** There are 3 more 3-fold rotation axes, for a total of 4: one for each vertex.

▶ **4.** This is a 3-fold axis, with a rotation angle of 120°. There are 3 other such axes, for a total of 4: one for each pair of vertices.

▶ **8.** There are 9 mirror planes: 3 of the first type parallel to each pair of faces, and 6 of the second type: one for each pair of edges.

▶ **11.** Truncated forms have the same symmetries as the parent polyhedra.

▶ **13.** The snub cube has the same rotational symmetry as the cube but does not have any reflectional symmetry. The snub dodecahedron has the same rotational symmetry as the dodecahedron but does not have any reflectional symmetry.

▶ **20.** The symmetries with at least one fixed point are the identity (2 reflections), reflection, rotation (2 reflections), and rotary reflection (3 reflections).

◆ 8.2. THREE-DIMENSIONAL KALEIDOSCOPES

▶ **4.**

▶ **8.** You should see a three-dimensional cube.

▶ **12.**

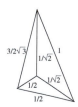

▶ **15.** An octahedron.

▶ **17.**

◆ 9.1. SPIRALS AND HELICES

▶ **4.** Note that answers are not unique. Here are two possibilities:

▶ **8.** $x = \frac{20}{3}$ and $y = \frac{16}{3}$

▶ **12.**   ▶ **15.**

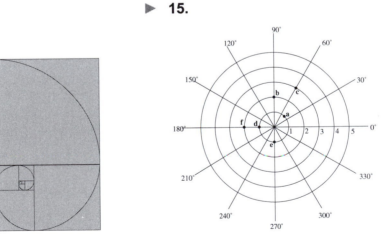

◆ 9.2. FIBONACCI NUMBERS AND PHYLLOTAXIS

▶ **2.** 1, 1, 2, 3, 5, 8, 13, 21, 34, 55, 89, 144, 233

▶ **4.** **a.** $1 + \frac{1}{1} = 2$ **b.** $1 + \dfrac{1}{1 + \frac{1}{1}} = \frac{3}{2} = 1.5$

c. $1 + \cfrac{1}{1 + \cfrac{1}{1 + \cfrac{1}{1 + \frac{1}{1}}}} = \frac{8}{5} = 1.6$

d. $1 + \cfrac{1}{1 + \cfrac{1}{1 + \cfrac{1}{1 + \cfrac{1}{1 + \frac{1}{1}}}}} = \frac{13}{8} = 1.625$

e. The numbers formed are ratios of Fibonacci numbers, with limit ϕ, the golden ratio.

▶ **16.** $.6(360°) = 216° = -144°$:

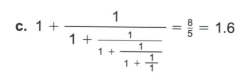

◆ 10.1. PERSPECTIVE

▶ **4.**

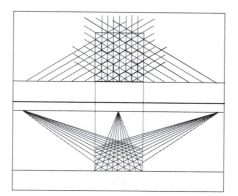

▶ **9.**

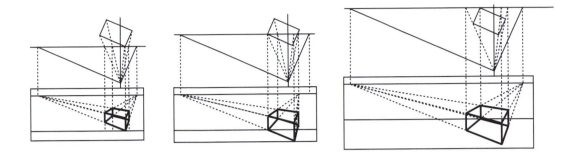

▶ **10.** Moving the picture plane only reduces or enlarges the picture.

◆ 10.2. OPTICAL ILLUSIONS

▶ **2.** Smaller angles seem to exaggerate the effect for most people.

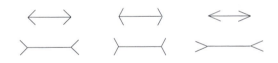

▶ **6.**

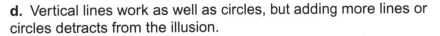

d. Vertical lines work as well as circles, but adding more lines or circles detracts from the illusion.

▶ **9.** First construct a hexagon as in Chapter 3.1. Find the midpoints of all of the sides and draw circles with centers at the vertices of the hexagon and radius one-half of a side, thus tangent to each other at the midpoints. Draw another circle with center at the center of the hexagon and radius one half the distance from this center to any vertex of the hexagon as shown on the next page.

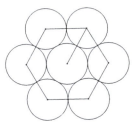

▶ **11.** The lines in Zöllner's illusion are actually parallel.

▶ **16.** You should get six cubes counting one way and seven counting the other. When only six cubes are counted, note there are enough bits left over to make a seventh cube.

◆ 11.1. NONEUCLIDEAN GEOMETRY

▶ **2.** Bump the center up by drawing the circumference together, while keeping the diameter the same length.

▶ **9.** On a sphere, lines are great circles: circles passing around the fattest part of the globe, for example, the equator.

▶ **10.** There are infinitely many lines (great circles) passing through any pair of antipodal points (points directly opposite one another, like the North and South Poles).

▶ **15.** There is a unique line through any two points on the floppy surface.

▶ **17.** Yes, given a line and a point not on the line, you find a line through the point that never intersects the original line.

▶ **20.** One example is the triangle with one vertex at the North Pole and the other two on the equator, chosen so that the angle formed at the North Pole is 90°.

▶ **24.** The angle sum of a hyperbolic triangle will be less than 180°.

◆ 11.2. MAP PROJECTIONS

▶ **5.** At the instant when the sun is directly over the point where the prime meridian and the equator intersect, it is 1 o'clock at longitude 15 E. It is midnight at the date line. It is 7 am in Washington, DC.

▶ **12.** *Cylindrical projections*: The projection of the latitudinal circles are shown.

A. Equal-area projection:

B. Equirectangular projection:

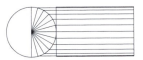

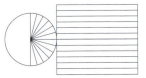

C. Central projection:

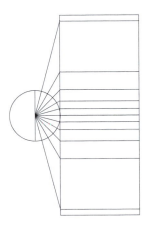

◆ **11.3. CURVATURE OF CURVES**

▶ **3.** A catenary:

▶ **7.** A limaçon:

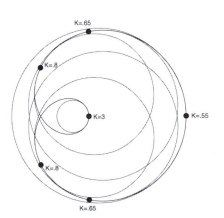

◆ 11.4. CURVATURE OF SURFACES

▶ **2.** The apex of the pyramid has $\delta = 120°$ and the four bottom vertices $\delta = 150°$, so the total angular deficit is $\triangle = 720°$.

▶ **6.** The sum of the angles is $270° = \frac{3\pi}{2}$ and the area of this spherical triangle is one-eighth of the sphere or $\frac{\pi}{2}r^2$.

▶ **10.** $\frac{\pi}{12}r^2$

▶ **14.** $r = \sqrt{\frac{720}{\pi}}$

▶ **21. a.** potato   **b.** celery   **c.** kale

◆ 11.5. SOAP BUBBLES

▶ **2.** It should be about 120°.

▶ **4.** The first form looks like a disc twisted to form a figure eight: there are two sheets of film. The third form pictured has a sheet of soap forming a twisted cylindrical wall between the rings. The second transitional form has this twisted wall with a central disc spanning the smaller circle of the figure 8.

▶ **11.**

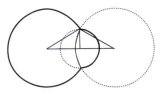

▶ **15. a.** $r < h$ **b.** $2h < r$

▶ **17.** The helicoid is approximately flat everywhere so both principle curvatures are zero and the mean curvature $H = 0$.

◆ 12.1. GRAPHS

▶ **4.**

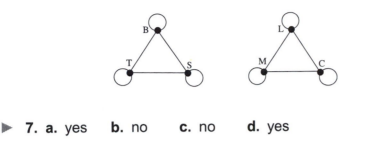

▶ **7. a.** yes   **b.** no   **c.** no   **d.** yes

▶ **9. a** yes   **b.** yes   **c.** no   **d.** yes   **e.** no   **f.** no

▶ **15.** The sum of the degrees must equal twice the number of edges, and so must be an even integer. Therefore, there must be an even number of vertices with odd degree.

▶ **17.**

▶ **19.** No. Explain why, using an appropriate diagram.

▶ **24.** The circuit is $A \to C \to B \to A$, so A moves to C's house, C moves to B's house, and B moves to A's house. D stays put (his third choice).

◆ 12.2. TREES

▶ **1. a.** $\chi = 0$ **b.** $\chi = -1$ **c.** $\chi = -2$

▶ **3.**

▶ **7.** Hint: Look at angles as well as lengths.

▶ **10.** If the square is 1 unit on each side, the first scheme has length 3. The second has length $2\sqrt{2} \approx 2.828$. The road plan that minimizes length is shown below and has length $1 + \sqrt{3} \approx 2.732$:

◆ **12.3. MAZES**

▶ **4.**   ▶ **8.**

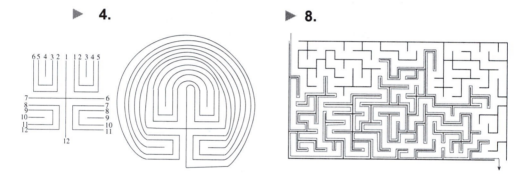

▶ **11.**

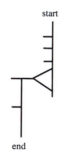

▶ **14.**

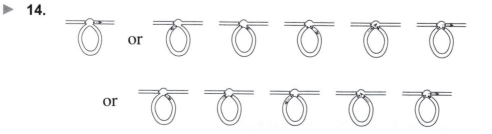

◆ 13.1. DIMENSION

▶ **1.** You get a line segment, which is one-dimensional.

▶ **4.** If your collection of points are spaced far enough apart (or if your pennies are small enough), the pennies will not overlap, so the thickest intersection is one penny.

▶ **6.** The thickest intersection of pennies should be three pennies thick.

▶ **11.** A circle with a knot in it cannot be fit back into one- or two-dimensional space, so that it will have an embedding dimension of 3.

◆ 13.2. SURFACES

▶ **2.** $\mathbb{S}\#\mathbb{S} = \mathbb{S}$

▶ **5.** This is a five-holed torus, $\mathbb{T}\#\mathbb{T}\#\mathbb{T}\#\mathbb{T}\#\mathbb{T}$.

▶ **9.** The boundary of the Möbius band is a circle.

▶ **13.** You get a thin Möbius band (the same length around as the original but one third as thick), linked with a band that is one third as thick as the original and twice as long. This band has two twists, and so is orientable.

◆ 13.3. MORE ABOUT SURFACES

▶ **1.** After removal of the two indicated points, the first figure falls into three disconnected components, while the second figure, no matter which two points are removed, will always fall into two pieces.

▶ **5.** When you remove a circle from a sphere, the result is disconnected, no matter which circle is removed.

▶ **6.** Removing a circle from a torus may give a connected or disconnected result, depending on which circle is removed, as shown following. Remember that we are looking for the maximum number of circles that can be removed without disconnecting the surface.

▶ **8.** $C(\mathbb{T}\#\mathbb{T}) = 5$, $C(\mathbb{T}\#\mathbb{T}\#\mathbb{T}) = 7$

▶ **11.** $C = 2$

▶ **15.** $C = 4$

▶ **17.** $\chi = 0$

▶ **20.** $\chi(\mathbb{T}\#\mathbb{T}) = -2$.

◆ 13.4. MAP COLORING PROBLEMS

▶ **3.**

▶ **8.** The disassembled puzzle has 10 faces, 34 interior edges, and 27 interior vertices. In the assembled map, $f = 10$, $e = 17$, and $v = 8$. Thus, the number of faces is the same, the number of edges in the disassembled puzzle is $2e$, and the number of vertices in the disassembled puzzle is greater than $3v$.

▶ **12.** Note that each country borders each of the other countries, so each must be colored differently. Thus, all seven colors are necessary.

▶ **15.** Five colors are necessary, but this does not contradict the Four-color Theorem which requires that countries be connected.

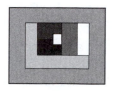

Bibliography

Syed Jan Abas and Amer Shaker Salman, *Symmetries of Islamic Geometric Patterns*, World Scientific Press, Singapore, 1995.

Edwin A. Abbott, *Flatland*, Princeton University Press, Princeton, 1991 (reprint of original of 1884, with an introduction by Thomas Banchoff).

Colin Adams, *The Knot Book*, W.H. Freeman, New York, 1994.

Frederick J. Almgren, *Plateau's Problem*, W.A. Benjamin, New York, 1966.

Frederick J. Almgren and Jean E. Taylor, "The Geometry of Soap Films and Soap Bubbles," *Scientific American* 235, July 1976.

I.I. Artobolevski, *Mechanisms for the Generation of Plane Curves*, Macmillan, New York, 1964.

Clifford W. Ashley, *The Ashley Book of Knots*, Doubleday, New York, 1944.

David Auckly and John Cleveland, "Totally Real Origami and Impossible Paper Folding," *American Mathematical Monthly* 102, March 1995.

George Bain, *Celtic Art: The Methods of Construction*, Dover, New York, 1973.

Iain Bain, *Celtic Knotwork*, Sterling Publ., New York, 1992.

W.W. Rouse Ball and H.S.M. Coxeter, *Mathematical Recreations and Essays*, Dover, New York, 1987.

Thomas Banchoff, *Beyond the Third Dimension*, W.H. Freeman, New York, 1990.

Thomas Banchoff, *From Flatland to Hypergraphics: Interacting with Higher Dimensions*, http://www.geom.umn.edu/~banchoff/ISR/ISR.html

Anatole Beck, Michael N. Bleicher, and Donald W. Crowe, *Excursions into Mathematics*, Worth Publishers, New York, 1969.

Alexander G. Bell, "The Tetrahedral Principle in Kite Structure," *National Geographic Magazine* XIV(6), June 1903.

N.L. Biggs, E.K. Lloyd, and R.J. Wilson, *Graph Theory: 1736–1936*, Clarendon Press, Oxford, 1986.

Donald W. Blackett, *Elementary Topology*, Academic Press, New York, 1982.

Ethan Bloch, *A First Course in Geometric Topology and Differential Geometry*, Birkhäuser, Boston, 1997.

Benjamin Bold, *Famous Problems of Geometry and How to Solve Them*, Dover, New York, 1969.

Brian Bolt, *Mathematics Meets Technology*, Cambridge University Press, New York, 1991.

F.H. Bool, J.R. Kist, J.L. Locher, and F. Wierda, *M.C. Escher: His Life and Complete Graphic Work*, Harry N. Abrams, New York, 1992.

Franco and Stefano Borsi, *Paolo Uccello*, Harry N. Abrams, New York, 1994.

Thom Boswell, *The Kaleidoscope Book*, Sterling, New York, 1992.

J. Bourgoin, *Arabic Geometrical Pattern and Design*, Dover, New York, 1973.

C.V. Boys, *Soap Bubbles: Their colors and the forces which mold them*, reprint of 1905 edition, Dover, New York, 1959.

Claude Bragdon, *A Primer of Higher Space*, Manas Press, Rochester, 1913.

Claude Bragdon, *Projective Ornament*, reprint of 1915 edition, Dover, New York, 1992.

Sir David Brewster, *The Kaleidoscope*, Van Cort Publishers, Holyoke, MA, 1987 (reprint of original text of 1819, Constable & Sons).

Titus Burckhardt, *Art of Islam: Language and Meaning*, World of Islam Festival Publishing Co., 1976.

Dionys Burger, *Sphereland*, translated by Cornelie Rheinbolt, HarperCollins, New York, 1983.

Pierre Cabanne, *Duchamp & Co.*, Terrail, Paris, 1997.

Italo Calvino, *All at One Point*, in *Cosmicomics*, translated by William Weaver, Harcourt Brace, New York, 1968.

John Canady, *Mainstreams of Modern Art*, Holt Rinehart and Winston, New York, 1959.

Hadley Cantril (ed.), *The Morning Notes of Adelbert Ames, Jr.*, Rutgers University Press, New Brunswick, 1960.

Lewis Carroll, *The Complete Sylvie and Bruno*, Mercury House, San Francisco, 1991.

James Casey, *Exploring Curvature*, Friedrich Vieweg & Sohn, Braunschweig, 1996.

Wellman Chamberlin, *The Round Earth on Flat Paper*, National Geographic Society, Washington, DC, 1950.

Gary Chartrand, *Introductory Graph Theory*, Dover, New York, 1977.

Ian Chilvers (ed.), *The Concise Oxford Dictionary of Art and Artists*, Oxford University Press, Oxford, 1996.

Helen Christensen, *Mathematical Modeling for the Marketplace: Applying Graph Theory in Liberal Arts and Social and Management Sciences*, Kendall/Hunt, Dubuque, 1988.

COMAP, *For All Practical Purposes: Introduction to Contemporary Mathematics*, W.H. Freeman, New York, 1994.

John H. Conway, "The Orbifold Notation for Surface Groups," in *Groups, Combinatorics, and Geometry*, ed. Liebeck and Saxl, Cambridge University Press, New York, 1992.

Theodore A. Cook, *The Curves of Life*, Dover, New York, 1978.

Sir Thomas Cook, *Curves of Life*, reprint of 1914 edition, Dover, New York, 1979.

Richard Courant and Herbert Robbins, *What Is Mathematics?*, Oxford University Press, London, 1941.

H.S.M. Coxeter, *Introduction to Geometry*, Wiley & Sons, New York, 1969.

H.S.M. Coxeter, *Regular Complex Polytopes*, Cambridge University Press, New York, 1991.

H.S.M. Coxeter, *Regular Polytopes*, Dover, New York, 1973.

H.S.M. Coxeter, "Regular skew polyhedra in three and four dimensions, and their topological analogues," *Proceedings of the London Mathematical Society* 43, 1937.

Keith Critchlow, *Islamic Patterns*, Thames and Hudson, New York, 1976.

Peter R. Cromwell, "Celtic Knotwork: Mathematical Art," *Mathematical Intelligencer* 15(1), 1993.

Peter R. Cromwell, *Polyhedra*, Cambridge University Press, New York, 1997.

Donald W. Crowe and Dorothy K. Washburn, *Symmetries of Culture*, University of Washington Press, Seattle, 1988.

H.M. Cundy and A.P. Rollett, *Mathematical Models*, Oxford University Press, New York, 1961.

Philip J. Davis, *Spirals: From Theodorus to Chaos*, A.K. Peters, Wellesley, 1993.

Charles H. Deetz and Oscar S. Adams, *Elements of Map Projection*, U.S. Government Printing Office, Washington, DC, 1938.

A.K. Dewdney, "Computer Recreations: Imagination meets geometry in the crystalline realm of latticeworks," *Scientific American* 258, June 1988.

A.K. Dewdney, *The Planiverse*, Poseidon Press, New York, 1984.

Stéphane Douady and Yves Couder, "Phyllotaxis as a Self-organized Growth Process," in *Growth Patterns in Physical Sciences and Biology*, ed. J.M. Garcia-Ruiz et al., Plenum Press, New York, 1993.

Peter Doyle, Jane Gilman, and William Thurston, *Geometry and the Imagination*, reprint, Geometry Center, University of Minnesota, 1991.

A.N. Dranishnikov, "On a Problem of P.S. Alexandroff," *Matematicheskii Sbornik* 135, 1988.

A.N. Dranishnikov, "On a Problem of P.S. Alexandroff," *Mathematics of the U.S.S.R. – Sbornik* 63, 1989.

H.E. Dudeney, *Amusements in Mathematics*, Dover, New York, 1970.

Issam El-Said and Ayşe Parman, *Geometric Concepts in Islamic Art*, Dale Seymour Publ., Palo Alto, 1976.

Bruno Ernst, *The Magic Mirror of M.C. Escher*, Taschen, New York, 1978.

Bruno Ernst, *Optical Illusions*, Taschen, New York, 1992.

M.C. Escher, *The Regular Division of the Plane*, from F.H. Bool, J.R. Kist, J.L. Locher, and F. Wierda, *M.C. Escher: His Life and Complete Graphic Work*, translated by T. Langham and P. Peters, Harry N. Abrams, New York, 1992.

Euclid, *The Elements*, translated by Sir Thomas Heath, Dover, New York, 1956.

David W. Farmer, *Groups and Symmetry*, AMS, Providence, 1991.

David W. Farmer and Theodore B. Stanford, *Knots and Surfaces*, AMS, Providence, 1996.

Robert Fathauer, *Kites and Darts Instruction Booklet*, Tessellations, Tempe, 1996.

J.V. Field, *The Invention of Infinity*, Oxford University Press, New York, 1997.

P.A. Firby and C.F. Gardiner, *Surface Topology*, Ellis Horwood Ltd., Chichester, 1982.

Edward F. Fry, *Cubism*, McGraw-Hill, New York, 1966.

Tomoko Fusé, *Unit Origami*, Japan Publ., Tokyo, 1990.

David Gale, "Egyptian Rope, Japanese Paper, and High School Math," *Math Horizons VI*, September 1998.

Martin Gardner, "Flatlands" in *The Unexpected Hanging and Other Mathematical Diversions*, Simon & Schuster, New York, 1969.

Martin Gardner, "The Island of Five Colors," in C. Fadiman, *Fantasia Mathematica*, Springer-Verlag, New York, 1958.

Martin Gardner, *The Last Recreations*, Springer-Verlag, New York, 1997.

Martin Gardner, "Mathematical Games: Casting a net on a checkerboard and other puzzles of the forest," *Scientific American* 254, June 1986.

Martin Gardner, "Mathematical Games: Extraordinary nonperiodic tiling that enriches the theory of tiles," *Scientific American* 236, January 1977.

Martin Gardner, "Mathematical Games: On tesselating the plane with convex polygon tiles," *Scientific American* 233, July 1975.

Martin Gardner, "Mathematical Games: Six sensational discoveries that somehow or another have escaped public attention," *Scientific American* 232, April 1975.

Martin Gardner, "On tessellating the plane with convex polygon titles", *Scientific American,* July 1975.

Martin Gardner, *Penrose Tiles to Trapdoor Ciphers*, W.H. Freeman, New York, 1989.

Martin Gardner, *Time Travel and Other Mathematical Bewilderments*, W.H. Freeman, New York, 1988.

Martin Gardner, *The Unexpected Hanging and Other Mathematical Diversions*, Simon & Schuster, New York, 1969.

Martin Gardner, "The Wonders of a Planiverse" in *The Last Recreations,* Springer-Verlag, New York, 1997.

David Gay, *Geometry by Discovery*, Wiley & Sons, New York, 1998.

Robert Geretschläger, "Euclidean Constructions and the Geometry of Origami," *Mathematics Magazine* 68(5), 1995.

Robert W. Gill, *Basic Perspective*, Thames and Hudson, New York, 1974.

Alfred Gray, *Modern Differential Geometry of Curves and Surfaces*, CRC Press, Boca Raton, 1993.

Christopher Gray, *Cubist Aesthetic Theories*, Johns Hopkins Press, Baltimore, 1953.

David Greenhood, *Mapping*, University of Chicago Press, Chicago, 1964.

Judith Gries, "Creating an Allée," *Fine Gardening* 54, March–April 1997.

Branko Grünbaum and G.C. Shephard, "Interlace Patterns in Islamic and Moorish Art," *Leonardo* 25, 1992.

Branko Grünbaum and G.C. Shephard, *Tilings and Patterns*, W.H. Freeman, New York, 1987.

Rona Gurkewitz and Bennett Arnstein, *3-D Geometric Origami*, Dover, New York, 1995.

Vagn Lundsgaard Hansen, *Geometry in Nature*, A.K. Peters, Wellesley, 1993.

David W. Henderson, *Experiencing Geometry on Plane and Sphere*, Prentice Hall, Upper Saddle River, 1996.

Linda Dalrymple Henderson, *The Fourth Dimension and Non-Euclidean Geometry in Modern Art*, Princeton University Press, Princeton, 1983.

D. Hilbert and S. Cohn-Vossen, *Geometry and the Imagination*, Chelsea Publishing Co., New York, 1990.

Stefan Hildebrandt and Anthony Tromba, *The Parsimonious Universe*, Springer-Verlag, New York, 1996.

Peter Hilton and Jean Pedersen, *Build Your Own Polyhedra*, Addison Wesley, Menlo Park, 1994.

Charles Howard Hinton, *An Episode in Flatland*, Swan Sonnenschein & Co., London, 1907.

David Hoffman, "The Computer-aided Discovery of New Embedded Minimal Surfaces," *Mathematical Intelligencer* 9(3), 1987.

Douglas Hofstadter, *Metamagical Themas: Questing for the Essence of Mind and Pattern*, Basic Books, New York, 1985.

Alan Holden, *Shapes, Space, and Symmetry*, Dover, New York, 1971.

Thomas Hull, "A Note on 'Impossible' Paper Folding," *American Mathematical Monthly* 103, March 1996.

H.E. Huntley, *The Divine Proportion*, Dover, New York, 1970.

Humiaki Huzita, "Drawing the Regular Heptagon and the Regular Nonagon by Origami," *Symmetry: Culture and Science* 5, 1994.

Cyril Isenberg, *The Science of Soap Films and Soap Bubbles*, Dover, New York, 1992.

William H. Ittelson, *The Ames Demonstrations in Perception*, Hafner Publishing Co., New York, 1968.

Harold R. Jacobs, *Mathematics: A Human Endeavor*, W.H. Freeman, New York, 1994.

Konrad Jacobs, *Invitation to Mathematics*, Princeton University Press, Princeton, 1992.

Roger V. Jean, *Phyllotaxis*, Cambridge University Press, Cambridge, England, 1994.

Scott Johnson and Hans Walser, "Pop-up Polyhedra," *Mathematics Magazine* 81, 1997.

Michio Kaku, *Hyperspace*, Oxford University Press, New York, 1994.

Jay Kappraff, *Connections: The Geometrical Bridge Between Art and Science*, McGraw-Hill, New York, 1991.

Kunihiko Kasahara, *Origami Omnibus*, Japan Publ., Tokyo, 1988.

Kunihiko Kasahara and Toshie Takahama, *Origami for the Connoisseur*, Japan Publ., Tokyo, 1987.

Martin Kemp (ed.), *Leonardo on Painting*, Yale University Press, New Haven, 1989.

Martin Kemp, *The Science of Art*, Yale University Press, New Haven, 1990.

A.B. Kempe, *How to Draw a Straight Line*, National Council of Teachers of Mathematics, Washington, D.C., 1977 (reprint of the Pentagon edition, which was a reprint of the 1877 edition of Macmillan, London).

Joe Kennedy and Diane Thomas, *Kaleidoscope Math*, Creative Publ., Sunnyvale, 1989.

L. Christine Kinsey, *Topology of Surfaces*, Springer-Verlag, New York, 1993.

Rosalind Krauss, *The Optical Unconscious*, MIT Press, Cambridge, 1993.

Fred Leeman, *Hidden Images*, Harry N. Abrams, New York, 1975.

Silvio Levy, "Automatic Generation of Hyperbolic Tilings," *Leonardo* 25(3), 1992.

W. Lietzmann, *Visual Topology*, American Elsevier, New York, 1965.

Elisha S. Loomis, *The Pythagorean Proposition*, NCTM, Washington, DC, 1968.

Sam Loyd, *The Eighth Book of Tan*, reprint of 1903 edition, Dover, New York, 1968.

M. Luckeish, *Visual Illusions*, Dover, New York, 1965.

Joseph Malkevitch (ed.), *Geometry's Future*, COMAP, Lexington, 1991.

Henry P. Manning (ed.), *The Fourth Dimension Simply Explained*, reprint of 1910 edition, Dover, New York, 1960.

Henry P. Manning, *Geometry of Four Dimensions*, reprint of 1914 edition, Dover, New York, 1956.

George Markowsky, "Misconceptions about the Golden Ratio," *College Mathematics Journal* 23(1), 1992.

George E. Martin, *Transformation Geometry: An Introduction to Symmetry*, Springer-Verlag, New York, 1982.

W.H. Matthews, *Mazes and Labyrinths*, Dover, New York, 1970.

Pamela K. McCracken and William S. Huff, "Wallpapers precisely 17: an eye-opening confirmation," *Symmetry: Culture and Science* 3, 1992.

Aidan Meehan, *Celtic Design: Knotwork*, Thames and Hudson, New York, 1991.

David Mitchell, *Mathematical Origami*, Tarquin Publ., Stradbroke, England, 1997.

William J. Mitchell, *The Reconfigured Eye*, MIT Press, Cambridge, 1992.

Mark Monmonier, *Drawing the Line*, Henry Holt, New York, 1995.

James R. Munkres, *Topology: A First Course*, Prentice Hall, Englewood Cliffs, 1975.

Dénes Nagy, "Symmetro-Graphy: Bibliography: Origami, paper-folding, and related topics in mathematics and science education," *Symmetry: Culture and Science* 5, 1994.

Dénes Nagy, "Symmet-Origami (Symmetry and Origami) in Art, Science, and Technology," *Symmetry: Culture and Science* 5, 1994.

National Research Council, *Everybody Counts: A Report to the Nation on the Future of Mathematics Education*, NRC, Washington, DC, 1989.

Michael Naylor, "Nonperiodic Tilings: The Irrational Numbers of the Tiling World," *The Mathematics Teacher* 92(1), January 1999.

Gülru Necipoğlu, *The Topkapi Scroll—Geometry and Ornament in Islamic Architecture*, The Getty Center for the History of Art and the Humanities, Santa Monica, 1995.

Gary Newlin, *Simple Kaleidoscopes*, Sterling, New York, 1996.

John M. Novak, "WorldPlot," *The Mathematica Journal* 3, 1993.

Dominic Olivastro, *Ancient Puzzles*, Bantam Books, New York, 1993.

John A.L. Osborn, "Amphography: The Art of Figurative Tiling," *Leonardo* 26(4), 1993.

Robert Osserman, *A Survey of Minimal Surfaces*, Dover, New York, 1986.

A.R. Pargeter, "Plaited polyhedra," *Mathematical Gazette* 43, 1959.

Frederick Pearson, II, *Map Projections: Theory and Applications*, CRC Press, Boca Raton, 1990.

Jean Pedersen, "Some Isonemal Fabrics on Polyhedral Surfaces," in *The Geometric Vein: The Coxeter Festschrift*, ed. C. Davis, B. Grünbaum, and F. Scherk, Springer-Verlag, New York, 1981.

Dan Pedoe, *Geometry and the Visual Arts*, Dover, New York, 1983 (a republication of *Geometry and the Liberal Arts,* St. Martin's Press, 1976).

Roger Penrose, "Escher and the Visual Representation of Mathematical Ideas," in *M.C. Escher: Art and Science,* ed. H.S.M. Coxeter et al., North Holland, Amsterdam, 1986.

Roger Penrose, "Pentaplexity: A Class of Non-Periodic Tilings of the Plane," *Mathematical Intelligencer* 1, 1974.

Ivars Peterson, "Clusters and Decagons: New rules for constructing a quasicrystal," *Science News* 150, October 12, 1996.

M.H. Pirenne, *Optics, Painting and Photography*, Cambridge University Press, New York, 1970.

Henri Poincaré, *Science and Hypothesis*, Dover, New York, 1952.

G. Polya, *How to Solve It*, Princeton University Press, Princeton, 1945.

V.V. Prasolov, *Intuitive Topology*, AMS, Providence, 1995.

Przemyslaw Prusinkiewicz and Aristid Lindenmayer, *The Algorithmic Beauty of Plants*, Springer-Verlag, New York, 1990.

Anthony Pugh, *Polyhedra: A Visual Approach*, University Of California Press, Berkeley, 1976.

Charles Radin, "Symmetry and Tilings," *Notices of the American Mathematical Society* 42(1), 1995.

Tony Robbin, *Fourfield: Computers, Art, and the Fourth Dimension*, Little, Brown and Company, Boston, 1992.

F.S. Roberts, *Discrete Mathematical Models, with Applications to Social, Biological, and Environmental Problems*, Prentice Hall, Upper Saddle River, 1976.

J.O. Robinson, *The Psychology of Visual Illusion*, Hitchinson & Co., London, 1972.

Nigel Rodgers, *Incredible Optical Illusions*, Barnes & Noble, New York, 1998.

C.P. Rourke and B.J. Sanderson, *Introduction to Piece-wise Linear Topology*, Springer-Verlag, New York, 1982.

Sundara Row, *Geometric Exercises in Paper Folding*, Open Court Publishing Co., La Salle, 1958.

Rudy Rucker, *The Fourth Dimension: Toward a Geometry of Higher Reality*, Houghton Mifflin, Boston, 1984.

Doris Schattschneider, "The Plane Symmetry Groups: Their Recognition and Notation," *American Mathematical Monthly* 85, 1978.

Doris Schattschneider, "Tiling the Plane with Congruent Pentagons," *Mathematics Magazine* 51(1), 1978.

Doris Schattschneider, *Visions of Symmetry: Notebooks, Periodic Drawings, and Related Work of M.C. Escher*, W.H. Freeman, New York, 1990.

Doris Schattschneider, "Will It Tile? Try the Conway Criterion!," *Mathematics Magazine* 53(4), 1980.

Michael Schneider, *A Beginner's Guide to Constructing the Universe*, HarperCollins, New York, 1994.

Marjorie Senechal, *Crystalline Symmetries*, Adam Hilger, Bristol, 1990.

Dale Seymour and Jill Britton, *Introduction to Tesselations*, Dale Seymour Publ., Palo Alto, 1989.

Chuck Shepherd, *News of the Weird*, The Tennessean, Nashville, July 6, 1997.

A.V. Shubnikov and V.A. Koptsik, *Symmetry in Science and Art*, translated by G.D. Archard, Plenum Press, New York, 1974.

John P. Snyder, "Delighting in Distortions," *Mercator's World* 1, 1996.

John P. Snyder, *Flattening the Earth*, University of Chicago Press, Chicago, 1993.

John P. Snyder and Philip M. Voxland, *An Album of Map Projections*, U.S. Geological Survey, Washington, DC, 1994.

Peter S. Stevens, *Handbook of Regular Patterns*, MIT Press, Cambridge, 1980.

Peter S. Stevens, *Patterns in Nature*, Little, Brown and Co., Boston, 1974.

Ian Stewart, "Daisy, Daisy, Give Me Your Answer, Do," *Scientific American* 272, January 1995.

Ian Stewart, *The Magical Maze: Seeing the World Through Mathematical Eyes*, Wiley & Sons, New York, 1997.

C.L. Strong, "The Amateur Scientist: How to blow soap bubbles that last for months and even years," *Scientific American* 220, May 1969.

Duncan Stuart, *Polyhedral and Mosaic Transformations*, North Carolina State University School of Design Publ., Raleigh, 1963.

Serge Tabachnikov, *Billiards*, Société Mathématique de France, Marseille, 1995.

Peter Tannenbaum and Robert Arnold, *Excursions in Modern Mathematics*, Prentice Hall, Upper Saddle River, 1998.

D'Arcy Thompson, *On Growth and Form*, abridged edition, Cambridge University Press, Cambridge, 1961.

Anthony Thyssen, *Tetrahedral Kite Using Straws*, http://www.sct.gu.edu.au/~ anthony/kites/straw_plan

Marvin Trachtenberg, *Architecture: From Prehistory to Post-Modern*, Harry N. Abrams, New York, 1986.

Richard J. Trudeau, *Introduction to Graph Theory*, Dover, New York, 1993.

Richard Trudeau, *The Noneuclidean Revolution*, Birkhauser, Boston, 1987.

E.B. Vinberg, "On Kaleidoscopes," *Quantum* 7, 1997.

A. Wachman, M. Burt, and M. Kleinmann, *Infinite Polyhedra*, Technion, Haifa, Israel, 1974.

Jearl Walker, "The Amateur Scientist: Methods for going through a maze without becoming lost or confused," *Scientific American* 259, December 1988.

Jearl Walker, "The Amateur Scientist: Mirrors make a maze so bewildering that the explorer must rely on map," *Scientific American* 254, June 1986.

Edward C. Wallace and Stephen F. West, *Roads to Geometry*, Prentice Hall, Englewood Cliffs, 1992.

Jeffrey R. Weeks, *The Shape of Space*, Marcel Dekker, New York, 1985.

Magnus Wenninger, *Polyhedron Models*, Cambridge University Press, New York, 1971.

Hermann Weyl, *Symmetry*, Princeton University Press, Princeton, 1952.

Robert Williams, *The Geometric Foundation of Natural Structure*, Dover, New York, 1979.

Denis Wood, *The Power of Maps*, Guilford Press, New York, 1992.

C.R. Wylie, *Introduction to Projective Geometry*, McGraw-Hill, New York, 1970.

Robert C. Yates, *Geometrical Tools*, Educational Publishers, St. Louis, 1949.

W. Zimmerman and S. Cunningham, *Visualization in Teaching and Learning Mathematics*, MAA, Washington, DC, 1991.

Index

A

Abbott, Edwin, 180
Ames, Adelbert, 324
Angle, 8, 13, 374
    constructible, 47, 65, 69
    interior, 18
    vertex, 16, 75, 86
Angular deficit, 360
Antiprism, 214
Archimedean solid, 224
Area, 6, 330, 363

B

Bell, Alexander G. , 248
Billiards, 20, 172, 301
Bragdon, Claude, 199

C

Calvino, Italo, 193
Carroll, Lewis, 428
Celtic knots, 30, 174
Circumference, 5
Congruence, 13
Connectedness, 386, 431
Connected sum, 422
Connectivity, 431
Constructions
    origami, 61
    ruler & compass, 40
Continued fraction, 292
Convexity, 14, 211
Conway, John H., 106, 117, 151
Conway criterion, 106
Coxeter, H.S.M., 135, 253, 293

Crystallographic notation, 151, 162
Cubism, 204
Curvature, 351, 359
    Gaussian, 354, 357, 367
    mean, 354
    principal, 355

D

Dali, Salvador, 202
Della Francesca, Piero, 307
Deltahedra, 222
Descartes' formula, 362
Dewdney, A.K., 171, 189
Digraph, 390
Dimension, 180, 411
    covering, 415
    embedding, 417
    fourth, 183, 192, 194
Dodgson, Charles, 428
Duchamp, Marcel, 206
Duplication of the cube, 47

E

Escher, M.C., 97, 105
    Circle Limit IV, 334
    Metamorphosis I, 110
    Metamorphosis III, 111
    Regular Division of the Plane, 109
    Reptiles, 105
    Sky and Water II, 108
    Sphere Spirals, 345
    Waterfall, 328

Euclid, 1, 40, 300, 330
Euler, Leonhard, 201, 383
    Euler characteristic, 201, 220, 234, 362, 365, 394, 436
    Euler circuit, 385
    Euler path, 385
Extrinsic property, 357, 417

F

Fermat, Pierre de, 4, 49, 287
    Fermat primes, 49
    Fermat's Last Theorem, 4
Fibonacci sequence, 285, 290
Flatland, 180
Four-color conjecture, 439
Fourth dimension, 183, 192, 194
Fox-Artin curve, 418
Frieze, 147
Fuller, Buckminster, 268, 347

G

Gardner, Martin, vii, 320, 444, 446
Gauss-Bonnet formula, 365, 370
Gauss' Theorem, 49, 70
Geometry
    elliptic, 330
    hyperbolic, 330
    noneuclidean, 329
Gnomon, 280
Golden angle, 298
Golden ratio, 51, 58, 66, 117, 291
Golden rectangle, 55
Golden triangle, 51, 57, 118, 281

Graph, 383
  complete, 388
  complete bipartite, 388
  planar, 389
Group
  cyclic, 143
  dihedral, 141
  rosette, 138, 141

H
Haga, Kazuo, 67, 240
Helix, 289
Hilbert, David, vii
Hinton, C.H., 188, 199
Homeomorphism, 411
Huff, William, 113
Hypercube, 184, 195, 200
Hyperspace. *See* Fourth dimension

I
Illusion, optical, 315
Intrinsic property, 357, 416
Isometry, 147, 266

J
Jerome, Jerome K., 411

K
Kaleidoscope, 127, 135, 270
Kepler, Johann, 224
Kite, 248
Klein bottle, 427, 433, 445
Knot, 30, 72, 197, 414
  closed, 33
Königsberg bridge problem, 383

L
Labyrinth, 401
Latitude, 338
Lattice, 156, 171
Lee, Kevin, xi, 100, 104
Leonardo da Vinci, 128, 142, 301, 302
Linkages, 78
Longitude, 337

M
Map
  coloring, 439
  projection, 336

Marvell, Andrew, 339
Maze, 401
Mercator, Gerardus, 344
Minimal surface, 378
Möbius band, 154, 373, 425, 435, 446
Mondrian, Piet, 321

N
Net, 202, 211, 218

O
Optical illusion, 315
Origami, 61, 239

P
Pantograph, 80
Parallelogram, 14
Parquet deformation, 108, 113
Parthenon, 58, 320
Peaucellier's linkage, 84
Penrose, Sir Roger, 117, 327
Perimeter, 5
Perspective, 300, 323
  one-point, 308
  two-point, 311
Phyllotaxis, 290
Picasso, Pablo, 204
Planar diagram, 426
Plateau, Joseph, 374
Platonic solids, 216, 265
Poincaré, Henri, 205, 333, 438
Polya, George, xiv
Polygon, 14, 17, 332
  constructible, 48, 51, 65, 72
  convex, 14
  Coxeter, 135
  nomenclature of, 17
  regular, 14, 51, 74, 181
  star, 73, 171
Polyhedron, convex, 211
  dual, 218, 266
  infinite, 253
  models of, 235
  origami, 239
  plaited, 246
  regular, 216, 223, 253
  semiregular, 224
  symmetry of, 263
  transformation of, 230

Prism, 211
Projective plane, 427, 446
Pyramid, 208
Pythagorean spiral, 46
Pythagorean Theorem, 2

Q
Quasicrystral, 125

R
Rabattement, 305
Regularity, 14, 74, 181, 223, 253
Reptile, 95
Rhombus, 14
Ruler & compass constructions, 40

S
Schlegel diagram, 219, 264
Slope, 10, 21
Snells' Law, 20
Soap bubbles, 372, 397
Spiral, 279
  Archimedean, 279
  golden, 284
  hyperbolic, 287
  logarithmic, 283, 288
  parabolic, 287
  Pythagorean, 46
Star polygons, 73, 171
Steiner, Jakob, 398
Surface, 378, 421
  minimal, 378
  nonorientable, 425
  orientable, 425
  with boundary, 424, 434
Symmetry
  bilateral, 127
  glide reflection, 99, 148, 267
  reflection, 24, 121, 129, 139, 147, 265
  rotary reflection, 267
  rotation, 24, 100, 121, 138, 147, 223, 263
  screw rotation, 268
  translation, 97, 148, 267

T
Tangent circle, 351
Tangent line, 10, 349

Tangram, 15
Tennessee, 439
Tesselation. *See* Tiling
Tesseract. *See* Hypercube
Texas, 337
Thompson, D'Arcy, 283
Tiling, 18, 85, 134, 334
    aperiodic, 115
    Archimedean, 85
    dual, 91
    irregular, 94

    Penrose, 115
    periodic, 115
    regular, 85
    semiregular, 85
Topology, 339, 411
Torus, 422, 433, 445
Trapezoid, 14
Tree, 393
Tremaux's algorithm, 406
Trisection of angle, 48, 69
Truncation, 230

U
Uccello, Paolo, 308

V
Vegetables, 349, 359
Volume, 212

W
Wagon, Stan, xi, 68, 381
Wallpaper, 156
Weeks, Jeffrey, 93, 192, 207

Permissions